Stories from the Skeleton

Interpreting the Remains of the Past

A series edited by *Mary K. Sandford*, University of North Carolina–Greensboro

Volume 1
STORIES FROM THE SKELETON
Behavioral Reconstruction in Human Osteology
Robert Jurmain

This book is part of a series. The publisher will accept continuation orders which may be cancelled at any time and which provide for automatic billing and shipping of each title in the series upon publication. Please write for details.

Stories from the Skeleton

Behavioral Reconstruction in Human Osteology

Robert Jurmain

San Jose State University
California, USA

Gordon and Breach Publishers

Australia Canada China France Germany India
Japan Luxembourg Malaysia The Netherlands
Russia Singapore Switzerland

Amsteldijk 166
1st Floor
1079 LH Amsterdam
The Netherlands

Front cover photo by Lynn Kilgore.

British Library Cataloguing in Publication Data

Jurmain, Robert
Stories from the skeleton : behavioral reconstruction in human osteology. – (Interpreting the remains of the past ; v. 1 – ISSN 1027-9334)
1. Bones 2. Skeleton 3. Anthropometry
I. Title
611.7

ISBN 90-5700-541-7

For

W. W. Howells and *T. Dale Stewart*

CONTENTS

INTRODUCTION TO THE SERIES

This series provides a forum for presenting innovative ideas and methods relative to our understanding of the human past. The concept developed during preparation of my edited work, *Investigations of Ancient Human Tissue: Chemical Analyses in Anthropology* (Gordon and Breach, 1993). While researching and writing that volume, I became acutely aware of the need for comprehensive and timely works focused on topics within the intersection of archaeology and physical anthropology.

That book examined the promise and pitfalls of using elemental and isotopic analyses in understanding past diets, nutritional patterns and disorders. Such topics, and the manner in which we elected to address them, influenced the scope and goals of the present series in several fundamental ways. The inauguration of these analytical techniques in anthropology signaled intensification of multi-disciplinary approaches. These techniques made their debut in anthropology during the 1970s and the decade itself was one of fervor and optimism, as students seized upon such new technologies in hopes of gaining a better and more accurate understanding of the human past.

The enthusiasm that marked the introduction of trace-element analysis in anthropology was tempered by recognition of the vast complications surrounding its use. Moreover, as with any method adopted from another discipline, most anthropologists simply lacked the training necessary for using the techniques or interpreting data. Remembering this as we prepared *Investigations of Ancient Human Tissue* some two decades later, we endeavored to contextualize our case studies with basic information on both theory and method, striving to make these techniques more accessible and understandable to a larger number of our colleagues.

The need for work with the requisite breadth to explore the reaches of a multi-disciplinary perspective, or the depth to probe the intricacies of a specialized technique, is even more compelling now. Indeed, what seemed to be quite extraordinary a mere twenty years ago has been far outpaced by innovations and discoveries of today. Scientific visualization and digital technology have revolutionized our ability to visually assess and quantify the objects of our investigations, while providing us with the means, through virtual technology, to share our latest findings with colleagues around the world. Advances in biotechnology have expanded the bounds of our imagination; the ability of extract DNA from tissue may help us resolve issues emanating from such concerns as the history of disease to the origins of humankind. As we approach the new millennium, it is staggering to contemplate the matters we will be discussing, debating and endeavoring to understand in another twenty years. In providing a forum for cutting-edge ideas and techniques, it is my hope that this series will serve to both chronicle and propel our understanding of the human past.

Mary K. Sandford

PREFACE

While youth fosters enthusiasm as well as occasional intemperance, age—and with it, one hopes, experience—promotes reflection and reassessment. So it can be also with scientific inquiry. Certainly during my twenty-five years of research in paleopathology, I have often found it necessary to reflect upon much of my earlier work. Beginning at the Smithsonian Institution, my studies of osteoarthritis, especially, formed the primary thrust of my research in the 1970s.

At the Smithsonian, following the lead and encouragement of Donald Ortner, I began investigating patterns of degenerative disease in human skeletal populations. From his own provisional and innovative analysis of elbow osteoarthritis, Ortner suggested I further pursue a similar line of inquiry, most especially focusing on patterns of arthritic involvement in Inuit elbows. Ortner's direction proved to be quite inspired, as I did find what appeared to be clear patterning of elbow involvement, most notably as evidenced among the Inuit.

The primary hypothesis guiding most of my early investigations centered on establishing a link between *activity* and specific skeletal involvement of osteoarthritis. It seemed to me (and others) that the Inuit elbow evidence *proved* what has been termed the "stress hypothesis." Extrapolating from this empirical foundation, along with others, I postulated that much of OA patterning reflected activity; this assumption was rather easily extended to a variety of joints in a variety of populations.

None of this is to deny that activity cannot influence the onset of osteoarthritis; considerable skeletal and clinical evidence argues that in certain circumstances it can. Such observations have merit and are potentially informative, both to osteologists and clinicians. What was not generally critically assessed by osteologists like myself was the complexity concerning etiology of the condition. Further, this complex etiology limits considerably the capacity for osteologists to *predict* activity from the skeletal patterning of osteoarthritis. For me, however, it would take some time and further experience to recognize these limitations.

Sources of this experience were varied. First, from ongoing research on skeletal remains of Central California Indians I came to realize the patterns of degenerative involvement apparently did not conform simply to expectations of the stress hypothesis. Second, as other investigators, including students, friends and my spouse, attempted to apply methodologies and interpretations similar to my own, I also grew more skeptical. It is perhaps easier to see equivocal components of certain research designs when they are attempted by others. Last, I had the good fortune to discuss such mutual concerns with all these colleagues, and their insights helped me immensely.

Beyond osteoarthritis, in more recent years I have also directed research on other types of skeletal conditions. In particular, the interpretation of traumatic injuries as seen in both human and great ape skeletons has become a focus of interest. Here, as well as with osteoarthritis, interpretations by me and colleagues have centered around potential behavioral influences. Most notably, the interpretations of "parry" fractures and craniofacial injuries have stimulated a critical evaluation, including reassessment of my own prior views.

In addition, a variety of other skeletal manifestations have fairly recently been proposed as "markers" of activity (e.g., stress fractures, enthesopathies and aspects of bone diaphyseal geometry). In order to provide a consistent and critical assessment of the value of these varied techniques and approaches, each is reviewed.

A basic perspective followed throughout these discussions is incorporation of *both* clinical and anthropological data. Both sources of information are necessary, since the bones alone do not (and cannot) provide an independent test of verification. We can never be sure solely from analysis of skeletal material what activities produce what osteological changes. If there is any confident means by which activity can ever be predicted from skeletal markers, such verification must first be demonstrated from controlled data sources.

Clinical data obviously are not perfect sources of information. Limitations arise not only because of sampling difficulties and imperfect research designs; most fundamentally, interpretations of osteoarthritis, enthesopathies and so forth are clouded because physiological (and environmental) processes leading to these conditions are complex. Indeed, this lesson is probably the most basic one to be learned from the rich literature produced by biomedical research. We all approach our investigations with an array of preconceptions, indeed, biases. Yet, if osteological interpretations are to achieve any real rigor, we must recognize these biases and seek to evaluate our data more critically. This is certainly a more difficult path to follow than one assuming simple relationships between activity and specific bone response. It is also a different path from the one hoped for in youth.

ACKNOWLEDGMENTS

In reality, this work reflects a considerable portion of my professional career in human osteology. Thus, there are a number of people to whom I am greatly indebted for assistance with research, access to collections, discussion, collegiality and friendship. The listing here is not comprehensive and I regret any omissions. It is both a rewarding and humbling experience to document the retinue of individuals who have contributed so fundamentally to one's life and work. And it is more than being able to stand on the shoulders of giants, although I certainly had that opportunity through the training and guidance of W. W. Howells, T. Dale Stewart and J. Lawrence Angel. Further support of students, friends, family and colleagues has made my professional life possible.

For assistance with my initial research on osteoarthritis at the Smithsonian Institution and Harvard I am indebted to Donald Ortner, Dale Stewart, Larry Angel, W. W. Howells, Jonathan Friedlaender, Henry McHenry, Jeff Froelich and Claudia Jurmain. My continued work on skeletal remains at San Jose State University has been possible because of the help of Lorna Pierce, Alan Leventhal, Tony Musladen, Viviana Bellifemine, Sandra Weldon, Rhonda Gillett, Muriel Maverick, Pat Rafter, Charlane Gross, David Calleri, Cynthia Arrington, Peggy Binns, and the staff of the photography laboratory, Instructional Resources Center. Bert Gerow at Stanford provided both assistance with research as well as access to skeletal material. The cooperation and collaborative support given by Rosemary Cambra, chairwoman of the Muwekma Ohlone Tribal Council, as well as other members of the Muwekma Ohlone Tribe, have made the osteological studies at San Jose State both possible and rewarding.

Analysis of the Nubian skeletal remains housed at the University of Colorado was made possible by the assistance and collaboration of Dennis van Gerven. Research on African ape materials was aided by M. E. Morbeck, Jane Goodall, Adrienne Zihlman, Betty and Harry Early, Derek Howlett (Powell-Cotton Museum), and Wim van Neer and Guy Teugels (Museé Royal de l'Afrique Centrale).

I am also grateful to numerous colleagues for sharing ideas and publications, including: Charlotte Roberts, Patricia Bridges, Nancy Lovell, Juliet Rogers, Allison Galloway, Clark Larsen, Bruce Rothschild, David Weaver and Robert Woods. Finally, my colleague and partner Lynn Kilgore has contributed intellectually and emotionally to most of what is contained in this work. For what is good here, she can justly share in the credit. For what is not, the responsibility is mine alone.

In addition, this work has been further enhanced by the thoughtful comments of reviewers Donald Ortner and Charlotte Roberts. I am grateful to them for their efforts and adherence to high professional standards. I am indebted to the editor of this book series, Mary K. Sandford, who has provided her time as well as motivation and encouragement. Without her, this book would not have been initiated, to say nothing of completed. Editors Carol Hollander and Lauren Shulsky Orenstein have also been of tremendous help and encouragement in bringing the book to fruition. K.B.S.S. Sundaram shepherded the volume through production.

Financial support for various aspects of research reported here was provided by Harvard University, the Smithsonian Institution, the College of Social Sciences, San Jose State University and the L. S. B. Leakey Foundation.

Chapter 1

Introduction: Of Science and Stories

"So What?"
Anonymous

BACKGROUND

Every researcher labors with some expectation, one could even say illusion, that their results will mean something. Thus, it came as somewhat of a rude awakening when, a few years ago, I received the query of "so what" from a reviewer of an article on osteoarthritis I had submitted for publication. The article was eventually published, but this inquiry did prompt an immediate rejoinder and considerable further reflection. In fact, in many ways, this reflection has led directly to this book.

My own human skeletal research has focused on behavioral reconstructions, especially relating to my earlier work on osteoarthritis as well as more recent investigations of traumatic involvement. It is, however, my experience with osteoarthritis research which has, in recent years, most stimulated a reappraisal of the nature of behavioral reconstruction as it is practiced by human osteologists.

Why behavioral inferences are so commonly and so eagerly pursued in skeletal analysis is not difficult to fathom. Popular conceptions of what skeletal biologists ostensibly *can* learn from osteological remains usually center on such behavioral inferences. The wide popularity (and, also, misunderstanding) of forensic anthropological investigation has further led to similar unrealistic expectations of what inferences osteologists can reasonably make. Moreover, drawing behavioral inferences from "dry" skeletal materials, obviously, *is* fascinating, both to scientist and to the general public. Also shared with forensic colleagues, there is the challenge of detective-like deduction as well as the motivation to "flesh-out" the individuals recovered from archaeological contexts, somehow to make them "real people." Thus, the intellectual compensations of pursuing behavioral reconstructions are considerable, but so too are the perils.

HISTORICAL PERSPECTIVES IN SKELETAL/BEHAVIORAL RESEARCH

In addition, it could be argued that the foundation of anthropology in the United States has further fostered a behaviorally oriented approach in osteology, and more generally, in physical anthropology. As usually a minority partner in a three-field (or four-field) general anthropology, physical anthropologists have been much encouraged to link osteological evidence with that obtained by our colleagues, be they archaeologists or cultural anthropologists. This is part of what has been termed the "biocultural perspective." Washburn succinctly stated the motivation for such an orientation: "If we would understand the process of human evolution, we need a modern dynamic biology and a deep appreciation of the history and functioning of culture. It is this necessity which gives all anthropology unity as a science" (1962: 13).

All this is well and good, and at a *general* level makes perfectly good sense. Ultimately, as anthropologists we wish to understand the human experience, as it has been shaped by human adaptation. Clearly, one cannot understand the biology of any organism without also understanding the environments to which it has adapted. Humans, and earlier hominids as well, obviously have adapted to a largely cultural environment. Yet, if one is to accept evidence of stone tools as a signature of early cultural behavior, then such cultural processes have been at work for at least 2.5 million years. Thus, to say that *Homo sapiens* shows biocultural adaptations is analogous to saying fish reflect aquatic adaptations. The statements are, of course, accurate, just not very informative.

What clearly is needed is *specificity*. That is, we wish to know what environmental factors influence human biology and how. These questions are, however, not as easy to address as are the broad glittering generalities. Over the last few decades, much of the paleoanthropological research of earlier hominids has attempted to wrestle with these issues, and so too with human osteological investigation of more recent populations.

A significant portion of human osteological research has focused on the reconstruction of certain behavioral attributes of earlier populations.

Some of this research has concerned dental variation, especially as it relates to dietary reconstruction. Similarly, much effort has been devoted to the investigation of chemical constituents of human bone, also aimed ultimately at dietary inference. This research, however, is not the focus here. Instead, this book reviews those techniques aimed at inferring *activity* from skeletal remains.

Assessing Activity

There is a fairly long history of such attempts by skeletal biologists, particularly by those who have come to be identified as paleopathologists. For example, in an early analysis of Nubian remains, Elliot Smith and Wood Jones (1910) drew behavioral inferences both from the patterning of osteoarthritic lesions as well as those resulting from trauma. In fact, the quality of documentation, especially as relating to traumatic involvement (Wood Jones, 1910), is still an excellent illustration of the necessary paleoepidemiological methodologies required in this type of research. Moreover, because these data remain so potentially informative, they have recently been reworked by Berger and Trinkaus (1995), and thus this research continues to provide useful data and to stimulate further investigation (see Chapter 6).

Probably the most central figure in the development of an explicit behaviorally oriented osteological perspective in the U.S. was J. Lawrence Angel. Angel's innovative perspectives have fundamentally influenced anthropological interpretations of osteoarthritis (1966; 1971); trauma (1976); enthesial reactions (1964; Angel et al., 1987; Kelley and Angel, 1987); parturition scars (1969; 1971; Angel et al., 1987), and Schmorl's nodes (1971; Angel et al., 1987). In fact, all of these topics are discussed in this book. Moreover, Angel's perspectives and enthusiasm provided a stimulus to many other researchers. Working with his colleague at the Smithsonian Institution, T. Dale Stewart, they initiated a remarkable series of skeletal studies and, in the process, launched an array of innovative techniques. This research was to have immediate derivative effects, for example, D.J. Ortner's investigation of osteoarthritis (1968). In fact, through the suggestion and direction of Don Ortner, and also influenced by Angel and Stewart at the Smithsonian, I began my own research on osteoarthritis.

As the balance of this book details, many others have followed in the footsteps of Angel, Stewart, and Ortner. Moreover, in a corresponding fashion, the influence of Calvin Wells (1963; 1964; 1965) in the U.K. also initiated a great interest in behaviorally oriented research, especially as concerning osteoarthritis.

CURRENT PERSPECTIVES

Both in the U.S. and the U.K. this interest has yet to wane, although it has recently been somewhat redirected. For example, at the 1997 American Association of Physical Anthropologists annual meeting numerous papers addressed a variety of skeletal "markers" of behavior/activity. Four of these contributions concerned trauma, all dealing with craniofacial injury and interpretations of interpersonal aggression. A fifth paper made a similar conclusion regarding cannibalism. Three papers dealt with activity inferences from bone geometric properties, one addressing nonhuman primates and another concerning behavioral reconstruction from late Paleolithic/Mesolithic contexts. Likewise, one contribution concerned Schmorl's nodes, two addressed skeletal asymmetries/handedness, one investigated squatting facets, and two others attempted to use osteoarthritic involvement to infer behavior. Indeed, with just two papers dealing specifically with osteoarthritis, there is an apparent trend away from using this once-popular approach as a grounds for behavioral inference (although a few other papers also incorporated degenerative lesions as part of a more "holistic" approach to such reconstructions).

The most obvious new emphasis for inferring activity clearly involves the use of "musculoskeletal markers," or, more specifically, enthesial reactions. In this recent national meeting, 11 papers focused on such evidence and did so specifically in an attempt to draw behavioral inferences, although one of these (Bridges, 1997), did report *negative* results. Much of this recent interest in enthesial reactions at the physical anthropology meeting was stimulated by a special symposium on this topic which included nine of these contributions.

Beyond the confines of professional conferences, it is quite apparent that such endeavors still attract much interest among osteologists. Given both the general public's eagerness to hear such stories as well as the biocultural emphasis of American anthropology (noted above), professional journals also commonly encourage such interpretations. A clear impression one gets from such an orientation is that evidence of osteoarthritis (and many other skeletal manifestations) is interesting only in so far as it helps reconstruct activity patterns in past groups. Unquestionably, my own earlier research helped fuel this unrealistic expectation. Moreover, it is an overly-limiting view, failing as it does to recognize the manifold other ways that skeletal data can contribute to understanding the biology of osteoarthritis and other significant conditions.

Despite the current fascination with enthesopathies, over the last two decades osteoarthritis has been the condition most frequently used by osteologists to infer behavior. For example, in Kennedy's eclectic review of "skeletal markers of occupational stress" (1989), the majority of characteristics relate to some form of joint or peri-articular involvement. Likewise, in Merbs' landmark publication on "activity-induced pathology" (1983) most of the emphasis also was placed on osteoarthritic patterning and its utility in reconstructing activity from skeletal remains.

Following considerable enthusiasm during the 1970s and 1980s, some criticisms have been voiced over the past few years (e.g., Bridges, 1992; 1993; Jurmain, 1990a; 1991b). Perhaps, these critiques have had a inhibiting effect on certain aspects of research relating to osteoarthritis as a skeletal marker of activity. Other researchers are, understandably, not as convinced and continue to evaluate osteoarthritis for this purpose, but they usually do so with more caution than many of us exercised in past years! Certainly, these differing perspectives stimulated some lively and informative discourse in 1996 at a special symposium (as part of the Paleopathology Association annual meeting).

In fact, it would be a detriment for osteologists to turn away from investigation of osteoarthritis *simply* because its skeletal manifestations do not usually relate directly to an unambiguous behavioral etiology. There is an amazingly deep and rich clinical literature on osteoarthritis, which is the primary reason we are now more aware of its inherently complex etiopathogenesis.

This present review focuses somewhat more on osteoarthritis, both in terms of its clinical manifestations and osteological interpretations, than on those other skeletal conditions recently utilized to infer activity patterns. The rationale for this emphasis relates partly to my own experience, but is also linked to other considerations as well:

(1) Osteoarthritis is among the most common pathological condition found in human skeletal remains.
(2) As mentioned above, osteologists, to date, have used osteoarthritis more than other behavioral markers.
(3) The clinical literature is more detailed than for any other similar skeletally manifested condition, and a host of studies have dealt specifically with issues pertaining directly to osteological/behavioral interpretations.
(4) Osteoarthritis can then serve as a useful model for other emerging, and less well-tested, perspectives.

I would argue that osteological studies have contributed to a greater understanding of osteoarthritis, providing a complementary perspective to that of clinical investigations. Indeed, as will be emphasized, the greatest potential for advances in understanding derive from collaborative research efforts, in which osteological data supplement those obtained from living subjects. This is a point well discussed by others (e.g., Buikstra and Cook, 1980) and has also been the central focus of the Paleopathology Association since its founding in 1973.

ORGANIZATION OF BOOK

Recognizing that osteologists have more consistently utilized osteoarthritis than other skeletal conditions to draw behavioral inferences, the next three chapters focus on osteoarthritis. Chapter 2 reviews clinical and osteological perspectives particularly relating to various hypotheses concerning the etiopathogenesis of the disease. In a sense, this chapter attempts to view these complex clinical data through the eyes of a skeletal biologist; the resulting approach thus emphasizes pathological processes affecting bone.

Chapter 3 reviews the occupational and sports literature on osteoarthritis. These data are clearly central to anthropologists, since they represent a direct clinical analogy to the types of activity-related inferences proposed by osteologists. Chapter 4 completes the discussion of osteoarthritis with a review specifically of anthropological perspectives. The focus here is on data derived from skeletal research, and the more explicit emphasis concerns how osteologists have utilized osteoarthritis patterning to reconstruct behavioral aspects of ancient societies.

In Chapter 5 the orientation shifts to a consideration of a range of other proposed osteological "markers" of activity. While a variety of skeletal manifestations including Schmorl's nodes, parturition scars, stress fractures (including spondylolysis), and auditory exostoses are discussed, the primary focus is on enthesopathies. As noted above, contemporary osteological research is emphasizing such enthesial reactions, and a review of this approach and its theoretical foundations thus seems prudent.

Chapter 6 addresses yet another commonly found skeletal condition, namely trauma. Behavioral interpretations of traumatic lesions have enjoyed a long history in anthropological research, and the osteological literature (as well as supporting clinical data) are reviewed. In particular, various proposals concerning the influence of interpersonal aggression affecting craniofacial involvement as well as "parry" fractures of the forearm are critically evaluated.

Probably the newest and potentially most innovative approach relating to behavioral reconstruction based upon skeletal material concerns analysis of bone geometry. Thus, Chapter 7 focuses on analysis of bone geometry, again emphasizing both theoretical bases and practical utility in osteological investigations.

Chapter 8 concludes the book with a synthesis of the varied perspectives presented in prior chapters. Moreover, this final chapter suggests how future methodological approaches can be refocused to avoid various problems inherent in current research.

OSTEOLOGY AS SCIENCE

Without some direct link between osteological indicators and well-documented morbid changes, skeletal biologists can only hypothesize in a theoretical vacuum. Relating bone changes only to *presumed* behavioral causation, and then assuming these very same bone "markers" corroborate said behavior, is obviously a circular piece of reasoning. Since osteologists usually lack any direct (and independent) evidence of behavioral attributes of earlier societies, they have occasionally sought other sources for such details. Most typically, this type of information has been extracted from ethnohistorical records. However, it should readily be apparent that this often-haphazard form of documentation is almost always inadequate to the task, as it lacks both the necessary precision and specificity. What osteologists then do (as exemplified by my own earlier assertions regarding Inuit osteoarthritis) is to selectively glean those bits from the ethnographic record that serve to "fit" a particular hypothesis. Such reasoning, however, is only slightly less circular than imaginative scenarios created solely from the skeletal evidence.

Another parallel line of evidence sometimes utilized for more recent archaeological contexts comes from historical accounts; unfortunately, these also are usually too imprecise to offer much in the way of rigorous insight. What such information clearly *cannot* usually do is provide an adequate means of independently testing the influence of behavior on skeletal response. For some recent historic contexts, there are, however, a few notable exceptions in which the quality of documentation has allowed a means of testing certain skeletal-behavioral relationships. In particular, the Spitalfields (Waldron and Cox, 1989) and the *Mary Rose* (Stirland, 1991; 1997) samples meet this standard.

Whatever their source of potential corroboratory evidence used to substantiate behavioral reconstructions, osteologists must always ask how accurate are these sources? Do they provide an independent test (from that of the skeletal material) to verify either general or specific relationships?

Without answering these questions satisfactorily, there is no rigorous methodology of verification. Ultimately, the best documented sources derive from clinical contexts. As noted, osteologists investigating osteoarthritis have been well aware of the manifestly complex clinical evidence relating to this condition. Similarly, substantial clinical support has been provided for a few other behaviorally based interpretations of skeletal lesions. The most notable of these better-substantiated explanations is that by Merbs relating to spondylolysis (Merbs, 1989; 1996b).

It is important to remember, of course, that clinical perspectives are frequently limited in their applicability as well. For example, the same types of constraints related to inadequate sampling that handicap osteologists are also inherent to many clinical research designs. Moreover, there is oftentimes a similar tendency to rush to judgment, usually resulting in overly simplistic explanations of complex phenomena. This impulse is commonly fostered by attempts to explain medical conditions that are perceived as of major contemporary significance (e.g., AIDS therapies, environmental carcinogens, dietary influences on heart disease). Certainly, the way the popular press disseminates and exploits both current research findings and normal scientific debate further distorts the public's views and expectations of scientific investigation.

Given such obvious shared restrictions, there are, nevertheless, a variety of lessons skeletal biologists can learn from our clinical colleagues. Most notably, numerous methodologies frequently utilized in medical epidemiology are instructive. For example, the rigorous control of samples, standard statistical testing, and attention to potential confounding variables are all methodological approaches not as widely applied in paleopathology as they might be.

Beyond the success of providing a good behavioral explanation for spondylolysis, what of other contemporary assertions of presumptive behavioral correlates inferred from skeletal data? At present, none of these other areas of investigation has provided adequate independent confirmation. In some cases, as with enthesial reactions and bone geometric variation, clinical evidence is still quite limited (although some useful data do exist). Nevertheless, there has, to date, not been enough relevant clinical data presented by osteologists from which even to begin validating most specific behavioral hypotheses. Moreover, if the lessons provided by the last few decades of osteoarthritis research are any guide, those seeking such validation have a long way yet to go.

Thus, this is a book about human osteology, and, more specifically, it concerns the subfield of paleopathology. More to the point, this work raises issues about the *science* of paleopathology. Given, of course, that paleopathology is a scientific pursuit, it thus becomes a question of how paleopathologists can go about testing their assertions in a rational and rigorous manner.

Naturally, scientific endeavors cannot be practiced as if in an ideal world. Not all data sources are perfect or, necessarily, even particularly good. Frequently, we simply evaluate those skeletons that are available. Furthermore, it is obvious in paleopathology that not all hypotheses are testable, at least not in the classical scientific sense. This limitation, of course, should not inhibit us from making the *best* interpretation we can offer from the data available.

In many other applications within the natural sciences, wherein hypotheses are not amenable to complete testing, great insight has nevertheless been gained. Paleontology and astronomy, for instance, both have produced admirable bodies of data and theory, frequently without fully explicit testing criteria. What has proven successful in such scientific pursuits requires that: data are rigorously compared between studies; based on current knowledge, predictions are made; and subsequent studies are used to help verify (or discard) earlier hypotheses. Furthermore, an essential component in this, and in *any* other, type of scientific approach is that all possible alternative explanations must be considered. In this way, as future studies are performed to expand the database, various views can be fairly evaluated. Ultimately, then, the "best" explanation is attained.

Accordingly, it is not here argued that all paleopathology must be judged by idealized and generally unattainable standards. What is suggested, however, is that methodological limitations be more explicitly recognized. Equally important, alternative explanations need to be more honestly entertained.

In practice these basic but somewhat abstract scientific goals are not always easy to reach. The opportunities for and commensurate rewards of appealing interpretations are usually obvious; the inherent constraints frequently are not so easily, or willingly, recognized. Certainly, before the limitations can be overcome, they must first be confronted.

Galloway has pointedly emphasized some of these constraints, particularly the tendency to over-interpret certain aspects of skeletal-behavioral relationships:

> Our own biases and desires to focus on certain features and processes predispose us to accept those studies or portions of studies which support our

> viewpoint. The ability to read life histories in the bones is extremely attractive—it "gives voice" to the deceased and permits us to contribute a more rounded picture of the individual. We must always remember, though, that the voice which is spoken is through our mouths and there may be discrepancies between the life events and what we think we see (Galloway, 1995: 95).

If this book appears concerned with likely mistakes in interpretation, it is, no doubt because I have committed them myself. Moreover, if this work seems overly concerned with the temptation to over-interpret data, it is because it is a constant struggle not to do so in my own research.

This book, thus, is not offered as a judgmental treatise, identifying the failures of others. All those who do paleopathology are aware of the inherent obstacles. Here are offered some signposts to highlight many of the pitfalls, and, with some luck, a chance of avoiding them.

Chapter 2

Osteoarthritis: Clinical and Osteological Approaches

> OA (osteoarthritis, degenerative joint disease) is a degenerative disease of the cartilage of joints. It is of diverse etiology and obscure pathogenesis (Mankin et al., 1986: 1132).

DEFINITION AND CLASSIFICATION

Because osteoarthritis is such a widespread and oftentimes debilitating condition, it has attracted intense scrutiny from a wide variety of clinicians. According to the most recently reported national health survey for the U.S., close to 40 million Americans are estimated to suffer from rheumatic complaints (the majority of which is at least loosely referable to osteoarthritis) (CDC, 1996a). As Hamerman (1989: 1322) emphasizes, "This prevalence and its costs—billions of dollars in medications, surgery, and days lost from work—account for the growing interest in uncovering the basic mechanisms by which this disease affects human joints."

While research interest has been intensive for almost a century, and the resulting accumulated data impressive, there is little in the way of consensus regarding the biology of osteoarthritis among clinical researchers. Indeed, there is even dispute regarding the proper name for the condition. Traditionally, in the United States it has been called "osteoarthritis" (OA), but this term was thought misleading by many (e.g., Bennett and Bauer, 1937), as it implied an inherently inflammatory condition. Among some contemporary researchers (Hough, 1993) the same criticism is still voiced, although others (Jasin, 1989) clearly recognize a significant inflammatory component.

As an alternative term, "degenerative joint disease" was suggested (Bennett et al., 1942), and this referent was widely used for a number of years. However, more recently, it too has been critiqued, since the term, "degenerative" is thought misleading and ill-defined (Mankin, 1982b), "implying a manifest, passive process associated with old age" (Dieppe, 1987: 16.76). Consequently, in the U.S. and the U.K., the preferred current terminology is once again "osteoarthritis," with the alternative term, "osteoarthrosis" used frequently on the continent (in the latter usage, modifying the "itis" suffix to de-emphasize inflammatory aspects). A variety of other terms have over the years also found their way into the literature, including: arthritis deformans, hypertrophic arthritis, traumatic arthritis, and for specific joints, gonarthrosis (knee) and coxarthrosis (hip). In this work, the term osteoarthritis (OA) will be used.

Among clinical workers, there is more general agreement concerning to which joints the term OA (or its synonyms) should be applied. Most researchers reserve the term for fully movable, diarthrodial, joints. Also called, "synovial joints," the articular surfaces are covered with hyaline cartilage and are contained within a joint cavity enclosed by a synovial membrane. Such joints include all the major articulations of the appendages (including those of the hands and feet) as well as the small interfacetal (also known as apophyseal or zygoapophyseal) joints of the spine.

However, OA is generally *not* the term preferred for the arthropathy associated with joints of the vertebral bodies. These slightly movable joints (amphiarthroses, also called syndesmoses) are separated by fibrocartilage as well as hyaline cartilage, the latter directly covering joint surfaces, and similar joint anatomy is also seen in the pubic symphysis and distal tibio-fibular joint.

The anatomy of the intervertebral disk joints is quite different from diarthrodial ones (Mankin and Radin, 1985). The fibrocartilage of the intervertebral disks differs from hyaline cartilage, as the former contains primarily Type I collagen (Mankin et al., 1986), and although these syndesmoses do show degenerative changes with aging, "Little is known of the process, however, except perhaps that facet joint OA and disk degeneration may be linked in some way" (Mankin et al., 1986: 1146). This association of apophyseal OA with disk disease has, in fact, also been well substantiated by osteological research (Merbs, 1983; Bridges, 1994; Kilgore, 1990). Regarding differences in synovial joints compared to intervertebral disk joints, Dieppe (1987) states that the clinical features are "quite different at the two sites," and he thus prefers the term spondylosis for the latter (to differentiate it from OA).

Commonly, in osteological studies, the conditions are kept distinct (both terminologically and in interpretation), and pathology of the intervertebral syndesmoses is thus usually termed "osteophytosis" or "vertebral osteophytosis (VOP)." This latter term will be used here.

Among clinicians there is, again, little consensus regarding an appropriate and general definition for OA. Hough (1993: 1699) describes OA as "an inherently non-inflammatory disorder," but Howell and Pelletier (1993: 1723) also emphasize a more dynamic series of secondary phenomena and define it "as a complex of interactive degradative and repair processes in cartilage, bone, and synovia, with secondary components of inflammation."

In efforts in the mid-1980s to reach greater consensus, two broad collaborative endeavors sought to define and characterize OA. In one attempt, a "workshop on the etiopathogenesis of osteoarthritis" (Mankin et al., 1986) with contributions by approximately 60 leading researchers, the following was offered by three of the participants (p. 1132):

> OA (osteoarthritis, degenerative joint disease) is a degenerative disease of the cartilage of joints. It is of diverse etiology and obscure pathogenesis ... *Pathologically*, the disease is characterized by irregularly distributed loss of cartilage more frequent in areas of increased load, sclerosis of subchondral bone, subchondral cysts, marginal osteophytes, increased metaphyseal blood flow, and variable synovial inflammation." [Note: OA was, additionally, characterized according to clinical, histological, biomechanical, biochemical and therapeutic criteria.]

The reaction of the workshop participants as reported by Mankin and colleagues (1986: 1132) is informative, "The response to this definition was overwhelming and unanimous: no one cared much for it (some felt stronger)." On a more positive note, they, however, add, "On further reflection, this challenge to the definition of OA is a symptom of the need for further research that may clarify conceptions of the nature of the disorder" (Mankin et al., 1986: 1132).

The other attempt at a consensus definition was developed and published as part of a collaborative effort of 25 clinicians participating as part of the Subcommittee on Classification Criteria of Osteoarthritis (of the American Rheumatism Association):

> OA is defined as a heterogeneous group of conditions that lead to joint symptoms and signs which are associated with defective integrity of articular cartilage, in addition to related changes in the underlying bone and at the joint margins (Altman et al., 1986: 1039).

As reflected clearly in this last effort, many of the problems of definition, no doubt, reflect the variable causes and pathologic consequences of what is subsumed under the clinical umbrella of "osteoarthritis." This depiction of the heterogeneity of the disease is also emphasized by Hough (1993) and Panush and Lane (1994). Paul Dieppe has perhaps stated the concept best in describing OA not as a single entity, but more of a "joint failure" "analogous to heart failure—the state of an organ which can result from a number of different diseases or physiological changes" (quoted in Mankin et al., 1986: 1132).

Dieppe in 1990 further elaborated this position,

> The term OA describes an abnormal state of a synovial joint. It is not a disease. Each of several different 'diseases' can trigger a reaction pattern leading to the characteristic features of the OA joint: focal loss of articular cartilage and hypertrophy of subchondral bone. Not surprisingly, attempts to define and classify OA as a single disease entity have not been helpful. (Dieppe, 1990: 262.)

From the above, it is clear from the outset that any discussion of osteoarthritis, other than the most superficial, requires consideration of a complex (and poorly understood) etiopathogenesis. In this and the following chapter the discussion will attempt to grapple with the complexities concerning the causation and patterning of OA. Osteologists must obviously not forget or gloss over these issues and resulting fundamental uncertainties which swirl around and obscure an understanding of OA. Osteological approaches that fail to do so are destined for failure, if not also some embarrassment in the eyes of our clinical colleagues.

Classification of OA

As with definitions, a variety of classifications has been proposed, although here perhaps with some wider agreement. Most commonly, the condition is divided most basically into primary and secondary forms. Primary OA (also referred to as "idiopathic OA") is recognized, "when it occurs in the absence of any underlying predisposing factor" (Moskowitz, 1993: 1735).

It is somewhat less than reassuring that this most basic form of OA (and the form by far most commonly studied by osteologists) is only defined by negative criteria. This point is further reinforced by the Subcommittee of the American Rheumatism Association, which characterizes primary OA as "of unknown origin" (Altman et al., 1986).

By contrast, secondary disease is "related to a known medical condition or event" (Altman et al., 1986). Such predisposing conditions could include acute trauma (Figure 2-1); inflammatory disease;

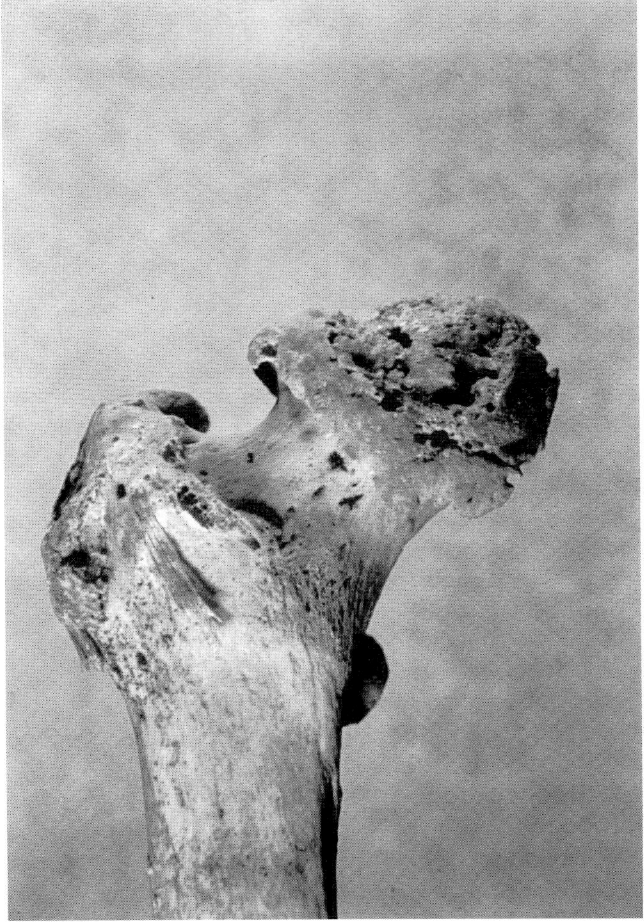

Figure 2-1 Secondary osteoarthritis. Right proximal femur. Kulubnarti, Sudanese Nubia; female, 50+years. The severe degeneration of the head of the femur is most likely secondary to trauma and aseptic necrosis. Photograph by the author.

metabolic, endocrine, or neuropathic problems; congenital malformations; or other arthropathies of still-unknown etiology, such as diffuse idiopathic skeletal hyperostosis (DISH) or calcium pyrophosphate deposition disease (CPPD). An often-cited classification (as developed by the American Rheumatism Association Subcommittee) is shown in Table 2-1.

Primary OA, as shown, is further divided into localized entities as contrasted with a manifestation, generalized osteoarthritis (GOA),

Table 2-1. A classification of OA

I. Idiopathic
  A. Localized
    1. Hands: e.g., Heberden's and Bouchard's nodes (nodal), erosive interphalangeal arthritis (non-nodal), schaphometacarpal, scapho-trapezoidal
    2. Feet: e.g., hallux valgus, hallux rigidus, contracted toes (hammer toes), talonavicular
    3. Knee
      a. Medial compartment
      b. Lateral compartment
      c. Patellofemoral compartment (e.g., chondromalacia)
    4. Hip
      a. Eccentric (superior)
      b. Concentric (axial, medial)
      c. Diffuse (coxae senilis)
    5. Spine (particularly cervical and lumbar)
      a. Apophyseal
      b. Intervertebral (disk)
      c. Spondylosis (osteophytes)
      d. Ligamentous (hyperostosis—or DISH)
  B. Generalized: includes 3 or more areas listed above
    1. Small (peripheral) and spine
    2. Large (central) and spine
    3. Mixed (peripheral and central) and spine

II. Secondary
  A. Post-traumatic
  B. Congenital or developmental diseases
    1. Localized
      a. Hip diseases: e.g., Legg-Calve-Perthes, congenital hip dislocation, slipped capital femoral epiphysis, shallow acetabulum
      b. Mechanical and local factors: e.g., obesity (?), unequal lower extremity length, extreme valgus/varus deformity, hypermobility syndromes, scoliosis
    2. Generalized
      a. Bone dysplasias: e.g., epiphyseal dysplasia, spondylo-apophyseal dysplasia
      b. Metabolic diseases: e.g., hemochomotosis, ochronosis, Gaucher's disease, hemoglobinopathy, Ehlers-Danlos disease
  C. Calcium deposition disease
    1. calcium pyrophosphate deposition disease
    2. apatite arthropathy
    3. Destructive arthropathy (shoulder, knee)

Table 2-1. *(Continued)*

D. Other bone and joint disorders: e.g., avascular necrosis, rheumatoid arthritis, septic arthritis, Paget's disease, osteopetrosis, osteochondritis

E. Other diseases
   1. Endocrine diseases: e.g., diabetes mellitus, acromegaly, hypothyroidism, hypoparathyroidism
   2. Neuropathic arthropathy (Charcot joints)
   3. Miscellaneous: e.g., frostbite, Kashin-Beck Disease, Caisson disease

(Altman et al., 1986: 1040).

which includes more widespread, polyarticular, involvement. GOA was argued forcefully as a distinct entity by Stecher (1940; 1959), emphasizing the polyarticular involvement of Heberden's nodes in distal interphalangeal joints of the hands (DIPs). The development of Heberden's nodes, that is, small bony and cartilaginous exostoses of the finger DIPs, was central in Stecher's description of GOA as well as later characterizations of the syndrome. Following intensive investigation by Kellgren and Lawrence (Kellgren et al., 1964), the definition of GOA was further refined to include a polyarthritis involving three or more joints (encompassing DIPs of hands, but also extended to other joints including 1st carpo-metacarpal, 1st tarso-metatarsal, vertebral apophyseal, knee, and hip joints).

Further, GOA was subdivided into nodal (i.e., with Heberden's nodes) and non-nodal variants. The generalized nodal expression of OA, particularly, is thought to stand out as a distinct entity with pronounced hand involvement (and especially as seen in white females) (Dieppe and Watt, 1985). Early on (Stecher, 1959) pointed out the strong familial association of GOA and even suggested a possible Mendelian mode of inheritance may underlie it. Several epidemiological studies have further confirmed this polyarticular manifestation, and some very tentative evidence suggests it may be tied to mutations (ultimately producing polymorphism) in the locus coding for Type II collagen (Silman and Hochberg, 1993). Moskowitz (1993) concurred that the nodal form of GOA appears to be inherited as a Mendelian trait, but argued that the non-nodal forms appear more polygenic in pattern. While seemingly a distinct entity, "which implies some systemic factor of cartilage vulnerability" (Peyron, 1986: 17), the differences between GOA and more localized manifestations may not be as obvious as once believed. For example, in one study of severe hip and knee disease, requiring surgical replacement, 62% of patients showed polyarticular involvement indicative of GOA (Cooke, 1983).

Nevertheless, recent studies have further confirmed a clear polyarticular pattern in various populations as well as suggesting strong genetic influences. Waldron (1997) found in a skeletal population from 18th/19th Century London a marked correlation of multiple site hand involvement with OA of the knee. A similar result was reported in a contemporary group from Australia (Cicuttini et al., 1997), here correlating hand OA with involvement of the tiobiofemoral compartment of the knee (but *not* the patellofemoral joint). In fact, these authors concluded that there may well be different etiological agents influencing the two compartments of the knee joint.

Biochemical and endocrine factors may also play a differential role in polyarticular OA, as compared to more localized manifestations of the disease. Dequeker et al. (1997) found increased levels of growth factors and hypermineralization in individuals with generalized OA.

Further emphasizing the blurred lines in differentiating subsets of OA, another well-known classification (Mankin et al., 1986) placed *all* traumatic variants (acute and microtraumata from "occupation" and "sports") as secondary conditions. Osteologists who assume they are clearly focused on primary disease may thus be accurate, but perhaps only to the point of which classification is employed, and frequently ignoring many of the unavoidable ambiguities. Indeed, Moscowitz (1993) pointed out that the distinction (primary/secondary) may be artificial, and many causes of so-called primary disease may well result from predisposing conditions (especially as seen in the hip; see below).

Another interesting wrinkle for osteologists is apparent in the subsets listed in Table 1 and as further refined by Altman (1991). As seen, the subsets of primary/idiopathic (localized) OA include hands, feet, knee, hip, spine (particularly cervical and lumbar segments), as well as "other sites" (shoulder, TMJ, sacroiliac, ankle, wrist, acromioclavicular). The elbow, however, does not receive even a mention. As will be discussed in Chapter 4, the elbow has probably figured more prominently in behavioral reconstructions by osteologists than any other joint.

The relative silence concerning elbow disease by most clinicians is, no doubt, explained by the very low prevalence of elbow involvement in contemporary populations. Dieppe (1990) lists the most commonly affected joints as those of the hand (particularly DIPs), base of the thumb, knee, hip, and apophyseal joints of the cervical and lumbar spine. Panush and Lane (1994) provide much the same order of joints most frequently affected, but they also specifically mention the proximal interphalangeal joints (PIPs) of the hand (second only to DIPs) and list hip involvement prior to that of the knee. Further epidemiological patterns of OA involvement will be discussed below, but the elbow is rarely mentioned and almost never emphasized in studies of contemporary populations.

NORMAL JOINT ANATOMY AND FUNCTION

While OA is a very common condition, with a moderate to high prevalence in all contemporary populations thus far sampled, it must be remembered that it is not an inevitable condition (even in older individuals). In fact, most people, even those who work hard most of their lives, never experience symptoms or show signs of OA. In a normally functioning joint the surfaces move against each other with remarkably little friction, with friction resistance measured in experimental studies of animals as low as 0.002 (indicating friction resistance in cartilage twice as low as rubber on steel and 10 times lower than an ice skate on ice) (Mankin and Radin, 1985).

The articular cartilage achieves this nearly friction-free state by releasing water under load (under normal conditions the cartilage is hydrated, freely exchanging water with the surrounding synovial fluid). The cartilage surfaces maintain this superior accommodation even under markedly varying stress conditions by employing two different types of lubrication. When exposed to high loads, a hydrostatic mode is used, creating a fluid film separating the surfaces; under lower loads, lubricating glycoproteins keep the surfaces from touching in a form of boundary lubrication (Mankin and Radin, 1985). This dynamic interaction is facilitated by action of the synovial fluid, which is well adapted for both secretory and resorptive function. Thus, joint physiology should be viewed as dynamic, and, as discussed below, capable of considerable repair. Nevertheless, metabolic activity is usually quite restrained, as indicated by a joint's hypothermic environment (temperatures vary from 90.3 to 94.3 degrees) (McKeag, 1992).

Joint integrity is further enhanced by the anatomical structures of the joint providing stability through:

(1) close-fitting joint congruity (e.g., the ankle; alternatively, gliding joints like the shoulder provide much less stability)
(2) ligaments, some of which are contained within or merge with the joint capsule (others are more discrete)
(3) fibrous joint capsule consisting of dense connective tissue, covering the entire joint area and firmly attached to the bone
(4) muscles

(5) fibrocartilaginous partitions, e.g., menisci (particularly seen in the knee, TMJ, and sternoclavicular joints, but less consistently in acromioclavicular and distal radio-ulnar joints).

The tissues most centrally involved in osteoarthritic changes are the articular cartilage and subchondral bone. Articular cartilage has been the primary focus of most clinicians, as this tissue shows the most consistent and obvious gross osteoarthritic pathological changes. Because clinical approaches mostly emphasize articular cartilage changes in the pathogenesis of OA, such changes will be discussed first. However, the subchondral bone is obviously of major import to osteological researchers, and, in fact, several clinicians have also focused attention on degenerative aspects of this tissue. Specific components of bone involvement in OA will be discussed below.

Articular Cartilage

In its normally functioning state, articular cartilage must retain its dimensional stability; that is, it needs to be able to deform and elastically respond under potentially high loads (Mankin et al., 1986). As noted above, it is able to accomplish this by releasing water (when loaded), and it can compress up to 40% of its depth (Mankin and Radin, 1985). In fact, the deformation of cartilage is what permits under normal circumstances such excellent joint congruity, being thinner in better fitting joints and thicker in less well-fitting ones (Mankin and Radin, 1985). At no point in its life cycle, however, is the cartilage very thick, ranging in thickness from 2 to 5 mm (Mankin and Radin, 1985; Dieppe, 1987), including the articular cartilage of the patellofemoral joint, which is the thickest in the body (Mankin et al., 1986).

Articular cartilage is a highly unusual connective tissue, being avascular, alymphatic, aneural, and quite hypocellular (Mankin and Radin, 1985; McKeag, 1992). By volume, the tissue is composed of:

water	68–78%
collagen fibers	15%
proteoglycan ground substance	10%
cells (chrondrocytes)	<5%

(Mankin et al., 1986; Dieppe, 1987; McKeag, 1992)

More than 90% of the collagen is Type II (Hamerman, 1989), which is distinct from the collagen found in bone or skin (Mankin and Radin, 1985). At a fairly gross level (under low magnification), the cartilage can be viewed in four zones characterized primarily by differing shapes and orientations of the cells: (1) the tangential (gliding) zone; (2) the transitional zone; (3) the radial zone; and (4) the calcified zone (attached to the subchondral bone, and cells in this zone are contained within a heavy calcification of hydroxyapatite crystals) (Mankin and Radin, 1985). Most notably, between the transitional zone and the calcified zone there is a distinct histological line (usually staining blue) called the *tidemark*. This tidemark is similar to cement lines in bone and its location in the cartilage suggests it functions possibly as a limit to calcification (occurring deeper) and/or to demarcate the organization of the more superficial collagen fibers, which assist in resistance to shearing forces (Mankin and Radin, 1985; Hamerman, 1989). The tidemark is also a boundary separating vascularized *vs.* non-vascularized portions of the cartilage, and thus may be central in understanding the invasion of vascularization ("neo-vascularization") that has been implicated in the pathogenesis of OA (Brown and Weiss, 1988; Hamerman, 1989).

In addition, the tidemark is developmentally a crucial feature of articular cartilage, as it seems to act as a growth proliferation area (i.e., like a metaphysis). Moreover, as with a metaphysis, it may be liable to injury, especially at specific ages (i.e., during childhood and adolescence).

It is here, at least in certain of the more dramatic manifestations of OA, perhaps that the osteoarthritis process *begins*. Some data from clinical studies and, significantly, *also* from osteological analyses, appear to support this hypothesis. More on this presently.

While the tidemark, between zones 3 and 4, is perhaps most crucial to normal cartilage growth and maintenance, and perhaps explains failure as well, it is just superficial to the osteochondral junction (i.e., the surface of the subchondral bone). Rather than functioning like a metaphysis, the osteochondral junction acts more like an epiphysis, permitting "actual enlargement of the bone through sequential ossification of the calcified cartilage" (Hough, 1993: 1705). This region may thus also be at risk to injury during growth, but the effects of such damage might differ from those involving the tidemark.

At the osteochondral junction, disruptive mechanical (and perhaps biochemical) forces during development might produce an alteration in joint congruity, a pathologic feature imputed as significant in the pathogenesis of hip OA especially (Murray and Duncan, 1975). By altering the substrate rather that the cartilage itself, the latency period, prior to symptom onset, might well be considerably longer for changes to the osteochondral junction as opposed to those affecting the tidemark; in the latter, the latency period may indeed be *quite* short.

Osteologists cannot, of course, see the tidemark or even the remnants of where it once was. But they can largely see the osteochondral junction and vestiges of physiological/mechanical activity once occurring there. Thus, the osseous responses labeled as "porosity" and "eburnation" are thought by many skeletal biologists to relate directly to significant aspects of joint function, in both normal and pathological joints. While eburnation almost certainly fulfills this expectation, lesions recognized as "porosity" probably do not (see below).

The intercellular matrix (comprising the vast majority of the volume of articular cartilage) is produced and maintained by the chondrocytes, which potentially can rapidly process the proteoglycans as well as synthesize collagen. Rapid production of the proteoglycans, particularly, is suggestive of an internal remodeling system, and one capable of considerable repair; additionally, this process is probably regulated by lysosomal enzymes, also secreted by the chondrocytes (Mankin and Radin, 1985).

Because of its integral role in maintaining cartilage *functional* capacity, the proteoglycan component has in the last two decades received a great share of the attention from researchers. From a structural point of view, the intercellular matrix of articular cartilage can be viewed as a network of collagen fibers embedded in a proteoglycan gel. The proteoglycans contain the water, and the collagen retains the proteoglycans. Under compressive load, the hydrated proteoglycan gel attempts to move from regions under high pressure to those under lower pressure, but is restricted by the collagen fiber mesh (Weightman, 1976).

Biochemically, the proteoglycans are macromolecules organized structurally at hierarchical levels of size and complexity. Polysaccharide chains (called glycosaminoglycans) are most basic, two forms of which are significant: chondroitin sulfate and keratin sulfate. These chains attach to a protein core to form proteoglycan subunits (monomers), and, in turn, these subunits combine to form enormous proteoglycan aggregates. The size and integrity of these aggregate proteoglycan units are tied directly to the ability of the gel to retain and release water under load (like a sponge), and these aggregates are most probably crucial in normal cartilage function and health. Numerous researchers (e.g., Hamerman and Klagsbrun, 1985) suggest that the hydrolytic degradation of the proteoglycan aggregates is a major feature in the pathogenesis of OA.

The biochemical and structural integrity of the collagen fibers is, no doubt, likewise important. For the proteoglycans to be able to maintain their high water content, especially under high loads, the collagen mesh must provide stiffness, particularly compressive stiffness.

If the stiffness decreases, the matrix deforms excessively and thereby makes rapid fatigue more likely (Mankin et al., 1986).

PATHOLOGICAL CHANGES IN ARTICULAR CARTILAGE

Mankin et al. (1986), in describing the essential biochemical alterations of OA, emphasize clearly the role of articular cartilage and the proteoglycans within it:

> *Biochemically*, the disease is characterized by reduction in the proteoglycan concentration, possible alterations in the size and aggregation of proteoglycans, alteration in collagen fibril size and weave, and increases in synthesis and degradation of matrix macromolecules (Mankin et al., 1986: 1132).

Howell and Pelletier (1993) also emphasized the combined pathological effects of biochemical changes (i.e., decrease in proteoglycan content) in association with architectural features (damaged collagen structure) leading to "functional loss of normal matrix physiologic properties."

Another well-documented physical change in the cartilage is the marked increase in water content (i.e., becomes "hyperhydrated"), which is thought to be an early and key event in the pathogenesis of OA, perhaps resulting from an altered network of collagen fibers (Hamerman, 1989). As the network loses its tight weave, increased water absorption might occur. Further, this change might help explain the loss of proteoglycans, as functional deterioration of the matrix might promote loss of proteoglycans, at least partially by diffusion (Hamerman, 1989). Mankin et al. (1986) agree that hyperhydration is a crucial feature in the pathogenesis of OA, but state its causes are not totally understood. They concur with Hamerman's scenario as a possible pathway, but also suggest an alternative: that initial changes in the proteoglycans may produce the increased water content. Whatever the causes of hyperhydration, its effects are most likely quite serious. As a result of hyperhydration the cartilage apparently loses its ability to respond dynamically, and with further mechanical stress, more deformation and less elastic return ensues. Moreover, as the vicious cycle progresses, there would be increased pressure on the subchondral bone and possible fissures (i.e., fibrillation) of the cartilage surface (Hamerman, 1989).

Another significant component contributing to the disruption of articular cartilage formation has also been emphasized in the last decade, most notably the role of the chondrocytes, especially as it is related to increased synthesis of degradative enzymes. "Much recent research implicates enzymatic breakdown of the cartilage as a key feature of disease progression" (Howell and Pelletier, 1993: 1725). The most commonly implicated enzymes, which are known to degrade cartilage, are the proteases, but collegenase also appears to play a role. Moreover protease inhibitors have also been identified, and OA progression might be influenced through an imbalance in various enzymes, "which is due to a relative deficiency of inhibitor" (Howell and Pelletier, 1993: 1726). Such a lack of inhibitor might, for example, lead to activity of what had been latent (inactive) collegenase.

The effects of these degradative enzymes are probably crucial in most, if not all, manifestations of OA. Indeed, they appear to be secreted as part of cartilage physiological response as the tissue attempts to repair damage. But, by doing so, it appears that in many cases the pathological condition is only compounded, condemning the physiology of the joint, and ultimately the affected individual into a "Faustian compromise" (Hamerman and Klagsbrun, 1985). The crucial dynamics of joint repair and their contribution to OA will be discussed further below.

A possible role played by biochemical mediators, especially the interlukins, has also been suggested (Hamerman and Klagsbrun, 1985), probably secreted in the synovium by macrophages or macrophage-like cells. Lastly, growth hormone has been suggested as playing a role in mediating OA pathogenesis (Hamerman and Klagsbrun, 1985). The mediating effect of the interlukins, and perhaps growth hormone, is thought to contribute to the degradation of collagen and proteoglycans in the matrix, perhaps by influencing the chondrocytes or the interface between the chondrocytes and the matrix.

The chondrocytes, and their role in OA, have become increasingly a focus of attention by researchers in recent years (Hamerman, 1989; Howell and Pellitier, 1993). Any disruption of the chondrocytes or in their interaction with the matrix would likely be a crucial element in the pathogenesis of OA, if not in initiation, at least in further progression. As noted earlier for the tidemark and osteochondral junction, the chondrocytes also may be a key to understanding *age-related* components of OA. Indeed, they are most likely central to OA pathogenesis, because probably what most differentiates the tidemark and the osteochondral junction are the *cells* within them.

Thus, several changes in the pathogenesis of OA cartilage have been recognized, and many are now well documented (summarized in Table 2-2). But what are the *earliest* changes? Hough (1993) noted that any correlation of microscopic and biochemical alterations with cell biological information is very difficult to ascertain because of regional

Table 2-2. Summary of pathological changes in articular cartilage associated with OA

Decrease in proteoglycan content
Changes in (disruption of) proteoglycan aggregates
Physical changes in collagen fiber structure (alteration in weave and integrity)
Hyperhydration
Increase in degradative enzyme activity (associated with possible decrease in activity of inhibitors)
Possible damage to chondrocytes
Possible role of mediators (interlukins and/or growth factor)

variations within the cartilage as well as continuous attempts at repair, some of which probably foster further degeneration.

At a more macroscopic level, the factor of regional variation, both within and between joints, has been further supported by analysis of human articular cartilage analyzed in cadaver specimens. Swann and Seedhom (1993) measured the stiffness of articular cartilage derived at necropsy from the knee and ankle. Their findings indicate considerable variation in the degree of "softening" within joints as well as between joints. They were able to plot these changes using an innovative topographical mapping methodology similar to that also employed by some osteologists (Merbs, 1983; Woods, 1995; Nagy, 1996).

The difficulties of untangling a precise sequence in the pathogenesis of OA articular cartilage is due to a variety of factors, most notably the complexity of the tissue and the multi-faceted nature of the disease. Indeed, as discussed above regarding classification of OA, Dieppe's comments concerning the variable nature of OA again come to mind. There may not be a *single* initiating factor in all joints in all circumstances, for individuals of all ages. Almost certainly, there is not such a simple, uniform process to be discovered and documented, because one does not exist.

BONE CHANGES

While clinicians have understandably paid most attention to the pathological changes within articular cartilage, osteologists are, of necessity, more concerned with those joint changes that occur in the adjacent bone. The bone, articular cartilage, synovium, or the capsule do not, of course, act independently of each other. It would be incorrect to view subchondral bone as an isolated structure, because it should be understood as functionally and anatomically integrated with the calcified cartilage and tidemark (on the superficial side) and with such phenomena as cysts, increased vascularity, microfractures and altered trabecular structure on the deep side (Mankin et al., 1986; Sokoloff, 1987).

Osteophytes

In OA joints new bone can develop at two primary (but distinct) locations: (1) at the joint surface; and (2) at the joint margins (i.e., osteophytes). The most significant ossification at the joint surface results from focal endochondral processes, but can also take place below in the subjacent cancellous bone (Hough, 1993). Such remodeling has been noted as a consistent feature of OA. Indeed, perhaps the most distinctive feature of OA pathogenesis is its "mixed" pattern, wherein bone is added in some portions of the joint, while in others resorption of bone (and cartilage) occurs.

Marginal osteophytes, likewise, can develop variably, either as protuberances within the joint space, or, periarticularly, within the capsule and ligamentous attachments to the joint capsule. The periarticular foci of ossification have been occasionally evaluated as part of the skeletal manifestation of OA (e.g., in my early work, Jurmain, 1977a,b), but it now seems they, and other osteophytes as well, are more peripheral to the OA process, both anatomically and pathogenically. Periarticular osteophytes develop *outside* the joint capsule, for example, within the trochanteric fossa of the proximal femur (insertion of obturator externis) or at the quadriceps insertion on the ventral patella.

A further application of periarticular hypertrophic lesions is also intimately linked to understanding the etiology of "enthesopathies." However, the same difficulties in interpreting the periarticular changes as a consistent component of OA also apply to any understanding of the etiology of enthesopathies (namely, the confounding influence of systemic factors such as age and genetic influences). Appropriately so, periarticular manifestations have not been a major focus in osteological studies of OA.

While periarticular lesions have not been emphasized, a major macroscopic osteological alteration ostensibly linked to OA, and used commonly by osteologists, is the presence of marginal osteophytes (Figure 2-2(a) and (b)). These bony outgrowths develop, as their name implies, around margins of joints, frequently adjacent to the attachment of the joint capsule. *In vivo,* they consist mostly of bone that merges imperceptibly with the tissue of the subchondral bone. Moreover, the osteophyte is usually capped by both hyaline and fibrocartilage

Figure 2-2(a) Marginal osteophyte (moderate). Distal femur, margin of condyle.

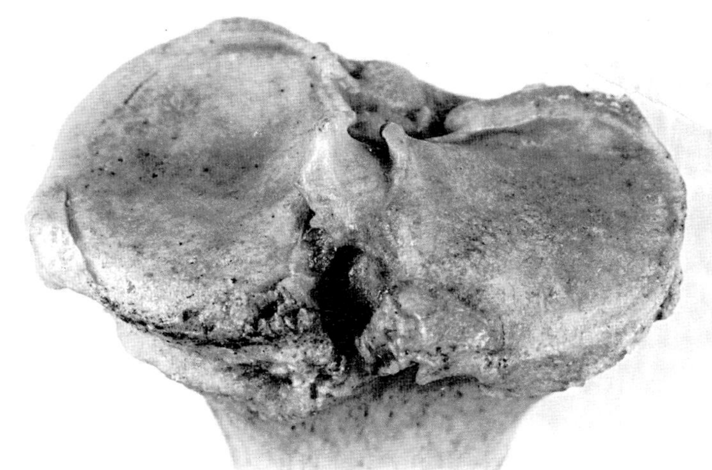

Figure 2-2(b) Severe marginal lipping, proximal tibia. Note the marginal changes are present in the absence of surface changes. Photographs by the author.

(the latter indicative of joint repair response), and this cartilage cap is continuous with adjacent synovial lining (Hough, 1993). Many of these histological features may, however, become obliterated in advanced OA, as the osteophytes themselves are subjected to further degenerative modifications.

The initial stimulation for osteophyte development may arise as capillaries penetrate into the subchondral plate and the deep (calcified) zone of articular cartilage. According to one view, emphasizing the central role of neovascularization (Brown and Weiss, 1988), this process could only be initiated when the normal vascular barrier of articular cartilage was somehow breached.

For most of this century clinicians, mostly employing radiographic imaging, as well as osteologists have recognized marginal osteophytes as a significant component (and indicator) of OA. However, in recent years this view has been seriously questioned. First, several detailed clinical/radiological studies assessing the correlation of osteophytes with other presumably diagnostic indicators of OA have found osteophytes to be largely uncorrelated with other changes (including narrowing of joint space and eburnation) (Hernborg and Nilsson, 1973; 1977; Moskowitz, 1993). This lack of correlation has been documented in radiographic studies of the hip (Danielsson, 1964; Danielsson and Hernborg, 1970) as well as the knee (Hernborg and Nilsson, 1977); however, a smaller study ($N=30$) of patients with hip replacements did find a stronger correlation (Cuccurollo and Croce, 1980). Nevertheless, the consensus is that osteophytes in the knee joint "are probably most of all related to age and are not necessarily an early sign of osteoarthritis" (Hernborg and Nilsson, 1973: 73).

Likewise, controlled osteological studies found poor correspondence of osteophyte development with other signs of OA in the knee (Rogers et al., 1990). Further, in my research, osteophytes were found to occur largely independently of changes on the joint surfaces. These patterns were most clearly demonstrated in multivariate (principle components) analysis of the individual indicators of OA involvement for each of the four large peripheral joints (i.e., shoulder, elbow, hip, and knee) (Jurmain, 1977b; 1978; 1991b). The sample used was White and African Americans from the Terry Collection (National Museum of Natural History; $N=444$), and, as age was quite well documented, the contribution of age was also factored into many of the analyses. Results show that, for all four joints, measures of osteophytes are the variables with the highest loadings on the first principle component (see Table 2-3). This pattern is especially obvious for the shoulder, hip, and knee, where osteophytes are separated clearly from indicators of

Table 2-3. Variables with highest loadings in principle components analysis (1st principle component) of major peripheral joints (Terry collection; N = 444)

Shoulder
Marginal lipping of glenoid
Periarticular osteophyte-Gtr/Lsr tubercles
Marginal lipping of humeral head
Age
Elbow
Lipping of inferior margin of radial head
Lipping of margin of capitulum
Marginal lipping of olecranon process
Superior articular surface of radial head
Hip
Marginal lipping of acetabulum
Age
Marginal lipping of femoral head
Lipping of fovea for ligamentum teres
Knee
Marginal condylar lipping of femur
Marginal lipping of lateral condyle of tibia
Marginal lipping of patella
Marginal lipping of medial condyle of tibia
Lipping of superior rim, patellar surface of femur

surface OA (in these joints, these latter variables have the lowest loadings). The elbow is a partial exception, as clustering with the osteophyte indicators are also the variables indicating surface degeneration of the radio-humeral joint. The role of age is also interesting, as it is highly loaded with osteophyte development in the hip and shoulder, to a lesser degree in the knee, and least of all in the elbow.

Given these various lines of evidence showing poor correspondence of osteophyte development with other indicators of OA (and, conversely, apparent strong correlation with age), should it continue to be used as an osteological criterion of OA? Duncan (1979) states the situation quite unambiguously,

> Although osteophyte production in many instances parallels the degree of osteoarthritis, the size, number and distribution of osteophytes should *not* be used as a major criterion for deciding on the degree, type, and magnitude of osteoarthritis (Duncan, 1979: 8).

This is a view that I and several other osteologists (Bourke, 1969; Cassidy, 1979; Rogers et al., 1990; Waldron, 1992; Rogers and Waldron, 1995; Nagy, 1996) have come to accept in recent years. In this view, *only* severe disease (especially as evidenced by eburnation) should be used as the major criterion for identifying the skeletal presence of OA, and this methodology would most certainly also assist in inter-populational comparisons.

Subchondral Bone

The subchondral bone, for obvious reasons, has been a major focus of osteological investigation. Histologically, this subchondral bone is organized generally in the same manner as other bone, although the cortical portion is both thinner and more deformable (up to 10 times so) than diaphyseal cortical bone (Mankin and Radin, 1985). In the underlying cancellous area the most notable feature is the organization of the trabecular plates, whose orientation appears to vary from joint to joint. As the plates are consistently aligned either along the end plate or perpendicular to the line of major forces acting on the joint, it thus seems probable that trabecular orientation is governed by biomechanical factors (Mankin and Radin, 1985).

The pathogenesis of bone changes within the joint produces initially quite subtle manifestations, which may be (but are probably not) informative to the osteologist. It is, however, the more severe osseous changes that are more readily identifiable and relate most directly to clinical documentation of morbid involvement. Indeed, Sokoloff emphasizes that, "it seems sensible to bypass the initial event and recognize disintegration of the osteoarticular surface as the distinctive feature of OA" (Sokoloff, 1987: 10). Once the cartilage has been denuded (i.e., repair mechanisms have failed), greater proliferation of subchondral bone is stimulated (and is, in fact, most marked in areas denuded of cartilaginous covering) (Hough, 1993).

As the ossification process continues, in the presence of further biomechanical stresses, particularly compressive ones, the bone becomes compacted and (on X-ray) sclerotic. With continued use the surface will become polished, that is, producing the typical and osteologically pathognomonic lesion called *eburnation* (Figure 2-3).

In the presence of yet further and persistent loading, the joint surface can eventually become grooved, with indentations developing parallel to the line of motion. Interestingly, while osteologists are familiar with such lesions, seen particularly on the femoral condyles, but occasionally seen in other articulations, such as the 1st carpo-metacarpal joint, they

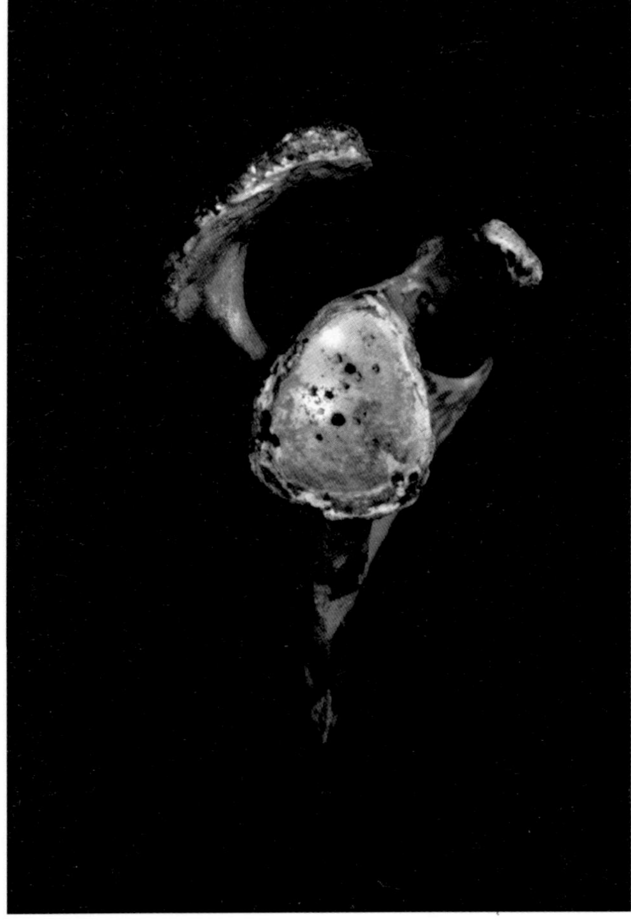

Figure 2-3 Eburnation, shoulder joint. Articular (glenoid) surface of scapula. Kulubnarti, Sudanese Nubia. Photograph courtesy, Lynn Kilgore.

have not been documented by clinical researchers (a point well made by Rogers and Dieppe, 1993).

Further severe osseous remodeling can occur, changing the entire contour of the articular surface; this deformity most likely results from osteonecrosis and collapse of the joint surface (Hough, 1993). Milgram (1983) suggests the massive deformation occurs (but rarely) in those cases where large confluent subchondral cysts weaken the structure sufficiently to cause collapse. Such a major anatomical and biomechanical alteration is obviously clinically important, especially in weight-bearing joints, and is a significant feature of advanced hip OA (Hough, 1993). Changes in bony contour have also been used as a classificatory criterion of skeletal OA by some researchers (notably, Rogers et al., 1987). Such changes are occasionally seen skeletally in severe cases involving small joints (e.g., spinal apophyseal articulations; 1st carpo-metacarpal; 1st tarso-metatarsal), but they are rarely seen in larger joints. The reason Rogers and colleagues were so aware of this particular pathological manifestation is perhaps explained by the source of some of their comparative material, derived from contemporary necropsy or surgical specimens (following arthroplasty). In individuals of advanced age, especially as seen in the hip, such lesions occur with some frequency. However, in archaeological samples this type of hip lesion is extremely rare (it is possible, of course, that in some historic samples a higher proportion of individuals could be senescent, and thus would greatly increase the likelihood that such lesions would be observed).

Another sequel of severe joint disease involving the bone develops as "nubbins of newly proliferated cartilage [which] usually protrude though minute gaps in the eburnated bone" (Hough, 1993: 1704). The manner in which these holes form is further supported by analysis of a large sample ($N=535$) of surgically replaced proximal femora (Milgram, 1983). Of those cases showing eburnation (which in the proximal femur is not common), most also displayed "tuffs" of white soft tissue studding the eburnated area. Milgram terms these lesions "chondrous tuffs," and histological analysis showed these to be continuous with underlying cysts containing myxomatous cells. The tuffs themselves may be reparative cartilage tissue, forming in the marrow of *exposed* articular bone (penetrating perhaps through existing holes, but causing them to enlarge) and representing abortive attempts at repair.

Osteologically, these lesions would appear as multiple, irregularly shaped holes with sharp edges. In other words, they represent what some researchers have called, "porosity." In fact, various osteolytic manifestations on joint surfaces occurring almost anywhere in the OA process have been labeled porosity. In the osteological literature, including my own publications, porosity is frequently utilized as a criterion for assessment of OA. It is, however, poorly defined and, as a result, inconsistently applied. There appear to be at least three different, and possibly etiologically separate, pathways by which "holes" can appear on bony articular surfaces. In reality, in osteological contexts, the articular surface, that is the cartilage, is usually gone; but *in vivo* the bone only becomes the "articular surface" in advanced cases. Various types of holes may develop due to:

(1) thinning of the articular plate (some of which may be post-mortem), thus exposing vascular channels. This process is passive and is probably not related to OA.
(2) active vascular invasion of the calcified cartilage (and tidemark) from the underlying subchondral bone. This neovascularization, as it violates the avascular cartilage, may well be a distinctive feature of OA (Brown and Weiss, 1988; Hamerman, 1989). Another aspect of an active invasion of the articular plate might occur as result of attempts at joint lubrication, that is, as a pathway for fat to lubricate the joint (Ragsdale, 1991).
(3) perforation through the articular plate subsequent to eburnation (Figure 2-4).

How is it possible to distinguish among these processes and what, if anything, does evaluation of porosity contribute to the skeletal recognition of OA? Firstly, it may be difficult to distinguish (in the absence of eburnation) the types of holes produced by thinning as compared to those resulting from vascular invasion. In some cases, best observed under magnification, the holes produced by active

Figure 2-4 Porosity secondary to eburnation, left patella, dorsal surface. Characteristic of severe knee involvement, grooving of articular surface is also present (see Rogers and Dieppe, 1993). Kulubnarti, Sudanese Nubia, male, 41–50. Photograph courtesy, Lynn Kilgore.

vascular invasion *may* appear more rounded. Also, in some cases a mixed pattern of hypertrophy/resorption may be evident which means that sometimes small patches of raised hypertrophic bone are clustered among holes, together appearing as isolated "islands" on the articular surface. It is possible these lesions represent focal areas of degeneration and healing. Such bony reactions are most easily recognized in the knee, especially on the femoral condyles, but may occasionally be seen in other areas (e.g., the capitulum and radial head). However, whether such lesions actually presage more severe ones is unknown, and thus their import is questionable, even if properly identified.

Another difficulty in interpreting porosity is pointed out by Woods' (1995) recent analysis of the distribution of such lesions in the knee. Using a highly detailed method of topographical mapping of the regional distribution of different types of lesions (eburnation *vs.* porosity, similar to that of Swann and Seedhom, 1993), Woods was able precisely to plot the location of lesions. Accordingly, the distribution of porosity compared to eburnation could be understood relative to the "geography" and associated biomechanical function of the joint. Woods' results clearly show that for the knee the distribution of porosity is *not* concordant with that of eburnation. In fact, the former is most often found in areas of the joint habitually in non-contact and presumably under the least load-bearing stress. Woods interprets these findings to indicate that porosity results from different pathogenic processes to those of eburnation, most likely from vascular invasion to supply undernourished cartilage. He conjectures, nevertheless, that the lesions still are significant signs of OA.

Another interpretation of the apparent independence of porosity from eburnation is that the former might result from processes separate from those producing pathological osteoarthritic change. Evaluation of porosity may, therefore, have little to contribute to understanding skeletal OA (Rothschild, 1997). At the very least, it seems porosity does not inform us very well regarding severe OA, and such severe disease is almost certainly what is most informative, both to osteologists and clinicians.

Woods further argues that porosity must be described more carefully, plotted by location much more precisely, and its scoring kept separate in the assessment of skeletal OA. All these points are well taken, and have also been applied to some degree by other osteologists (Merbs, 1983; Bridges, 1993; Nagy, 1996). In addition, Buikstra and Ubelaker (1994), in their recommended standards for data collection, also suggest collecting all these data independently. Following their suggestion, it would obviously be advisable to continue to collect and record *separate* data on eburnation, osteophytes, and porosity. For interpretive purposes, in the diagnosis of severe OA, it is argued here that eburnation be the sole criterion. Nevertheless, these other bone changes are probably informative of other skeletal processes. Further, and especially concerning materials which are to be reinterred, the collection of the maximum possible amount of information is obviously advisable. Indeed, if the view proposed here should prove insupportable, it would be inexcusable to lose these other potentially informative data.

Another skeletal manifestation frequently employed in clinical studies is the presence of bone cysts; however, such areas of rarefication are not classically "cysts," as they do not usually contain mucoid tissue (Hough, 1993). Since these lesions are recognizable only radiographically, they are rarely evaluated by osteologists. The approach advocated by Rogers et al. (1987) is an exception, exemplifying, at least methodologically, the advantages of a team-oriented effort in which skeletal biologists and rheumatologists work in close cooperation with radiologists. Still, whether analysis of such lesions can offer any benefit to osteologists is problematic because such bone changes are seen clinically occasionally in advanced cases of hip OA, but rarely in other joints (Hough, 1993).

Moreover, as noted above, the value of using osteophytes is also questionable; thus it is recommended here that for most analyses OA be assessed *solely* on the basis of eburnation. Nevertheless, this approach is still far from universally accepted. Clearly, the difficulties in developing greater consistency between studies is a major obstacle for osteologists (and a point discussed in Chapter 4). Agreeing on criteria (and applying those that are most applicable, and, at the same time, easiest to recognize) should be a major step in overcoming the challenge of methodological inconsistency and thus reaching some consensus.

Role of Bone Changes in the Pathogenesis of OA

Not suprisingly, since research efforts by clinicians have emphasized changes in articular cartilage, the majority of publications concentrate on this tissue rather than bone. Nevertheless, it is generally recognized that, "changes in the activity of subchondral bone are a very early feature of OA" (Dieppe, 1987: 16.78); indeed, most researchers acknowledge the varied expression of OA, involving *both* articular cartilage and bone (as the definitions quoted earlier in the chapter demonstrate). Still, in discussion of *primary* factors in the etiology of OA, most models continue to emphasize crucial changes in articular cartilage. One group of clinicians, led by Radin, is an exception in their focus on the subchondral bone in the initial pathogenesis of OA (Radin and Paul, 1971; Radin et al., 1972; Radin, 1976; Mankin and Radin, 1985; Radin and Rose, 1986).

The hypothesis that the subchondral bone, rather than the articular cartilage, may be the site of primary lesions is argued on three basic grounds (Hough, 1993):

(1) Articular cartilage has little impact-absorbing capacity.
(2) Microfractures of subchondral trabeculae and sclerosis may precede apparent changes in cartilage.
(3) Cartilage is more susceptible to impact loading (and thus to shearing forces), leading to deformation of subchondral bone.

According to this scenario, as the result of *impact loading*, with poorly accommodated forces being transmitted through the cartilage to the underlying bone, deformation of the bone occurs, producing multiple microfractures of the trabeculae. As a result of healing of these fractures, the trabeculae become thickened and the bone becomes stiffer. Further impulse loading and less shock absorption lead ultimately to cartilage failure and the full suite of OA pathological changes. Most of the substantiation for this hypothesis comes from experimental work, artificially exposing the cartilage and bone to various types of loading. However, considerable disagreement has arisen because of failure to demonstrate the sequence of pathogenic events *in vivo*. In more recent publications Radin is less dogmatic in asserting the primacy of bone changes, arguing, "*One* of the mechanisms of initiation *may* be a sharp stiffness gradient in the underlying subchondral bone" [emphasis added] (Radin and Rose, 1986: 34). Further, in the same year an authoritative review, of which Radin was a co-author, stated, "The scientific documentation supporting the concept that increased subchondral bone density leads to increased stress in articular cartilage is, at best, meager" (Mankin et al., 1986: 1145).

Thus, one would not necessarily *expect* much change in cartilage as a result of changes in bone stiffness, given the normal fluid dynamics of joints and elastic response of cartilage (*contra* point #1 above). Accordingly, there could be considerable alteration of subchondral bone stiffness without affecting cartilage. Mankin and co-authors (1986: 1145) conclude: "The methodology by which the hypothesis has been approached has been a major problem. Considerable refinement in technology will be required to establish a role for stiffening of the bone in the pathogenesis of the disease."

Specific epidemiological data also question the importance of impulse loading in influencing degenerative disease in humans, even in fairly stressful repetitive activities. Particularly, the data relating to pneumatic tool use (and upper appendage OA) as well as to running (and lower appendage OA), do not support the impulse loading hypothesis.

Indeed, especially the data regarding running and OA, the results actually suggest the opposite (see Chapter 3 for further discussion).

Some researchers have argued that perhaps responses of bone and cartilage may be largely separate, representing, "the independent response of each tissue to cyclic loading" (Weightman, 1976: 194). More likely the tissues do respond in a coordinated fashion, but rather than arguing primacy within the sequence, it might be more productive to isolate more accurately the timing of onset within the life cycle as well as the anatomic regions of responses, especially those within cartilage.

Radin and Rose (1986) suggest an intriguing possibility in focusing on alterations and forces (especially shear) near the cartilage tidemark and resulting cracks: "The propagation of such cracks leads to denudation of the articular cartilage and exposure of the bone" (Radin and Rose, 1986: 37). This opinion is shared by Hough (1993) and is further supported by Hamerman and Klagsbrun (1985), who note that superficial fissures of cartilage frequently heal or do not progress. Those changes that penetrate deepest and "that extend into the subchondral bone are associated with gross loss of cartilage and a more complex multi-tissue attempt at repair and remodeling of the articular end of the bone, producing deformities so characteristic of osteoarthritis" (Hamerman and Klagsbrun, 1985: 495).

Dequeker et al. (1997) did not find an increase in microfractures of subchondral trabecular bone in OA patients, but these individuals actually show fewer such lesions than seen in non-involved patients. Nevertheless these authors argue that subchondral bone changes occur very early in the pathogenesis of OA and thus "cartilage fibrillation could not be disassociated from bony changes even in the earliest stages of osteoarthritis" (p. 358). Perhaps of even wider etiological import, Dequeker and colleagues found widely distributed bone changes at all skeletal sites showing OA, most especially in generalized OA. Accordingly, they further conclude "a more generalized bone alteration may be the basis of the pathogenesis of osteoarthritis" (ibid.). In other words, bone remodeling due to impulse loading and consequent repair may itself be physiologically based in broader systemic phenomena.

Possibly, the key to understanding the initiation, and certainly, augmentation, of OA is not so much in looking separately at bone and cartilage, but at *where* they come together. Moreover, *when* stesses and failure take place may also be relevant. As alluded above, the tidemark and osteochondral junction may well be more prone to injury at certain phase of the life cycle than at others.

A final form of gross bone alteration that has attracted much clinical interest are Heberden's nodes found at the DIPs, particularly as associated with generalized OA. Systematic analysis of hand joints, especially for the large samples that would be required, obviously poses a tremendous challenge for osteologists, and no doubt explains why such lesions are only rarely addressed in skeletal research. Nevertheless, since the clinical and especially epidemiological identification of a polyarticular form of OA is now well documented in some population segments (with obvious genetic implications), further work in this area could be most informative. It should be noted that some osteological research has focused more attention on hand lesions (Waldron and Cox, 1989; Rogers et al., 1995; Waldron, 1997).

Moreover, there *may* be other intriguing osseous manifestations of primary generalized OA. Moskowitz (1993) notes that in this form of OA the spinal apophyseal joints become enlarged, along with the neural arches and spinous processes (the latter producing the radiographic appearance of "kissing spines"). He mentions further that in the knee of individuals affected with GOA osteophytes may take on more of a "molten wax" appearance, as opposed to the more typical sharply pointed form seen in localized idiopathic OA.

Further demonstration of the presence, and especially the distribution, of these lesions in skeletal populations, along with parallel data on hand involvement, would be highly interesting. Indeed, where skeletal samples permit, a more anatomically holistic approach to the patterning of OA would be of great value.

REPAIR MECHANISMS

From the preceding discussion, it is apparent that joints have capacity for repair. Remodeling of subchondral bone obviously takes place, but some repair is also possible in articular cartilage, although, given its avascularity, less so. The areas of cartilage most likely to show accommodation are those closest to the bone, that is, the tidemark and other components of the deep zone.

As noted above, neovascularization from the bone can invade the deep zone of cartilage, and, in some views (Brown and Weiss, 1988), this event is crucial in the later pathogenesis of OA. Such external (to the cartilage) features are what Radin and Burr (1984) label as "extrinsic" mechanisms, which they argue are more significant than intrinsic ones. Within the cartilage itself, any repair mechanisms must be rooted in the cells (chondrocytes), and the response of joints to injury (or surgery) suggests that "under circumstances of chronic injury such as seen in osteoarthritis, chondrocytes are capable of mounting a significant reparative response and can replicate their DNA to form new cells" (Mankin, 1982b: 461). Moreover, it appears that the chondrocytes' level of response may be partly age-dependent (Hamerman, 1989), although even in older individuals, there is some evidence that the cells can even "revert" to a chondroblastic state (Mankin, 1974).

Repair cartilage, especially that seen following trauma, is fibrocartilage, *not* hyaline cartilage and is composed mostly of Type I collagen (not Type II as in normal articular cartilage). That the cell has changed at a basic level, perhaps permanently, is further suggested by experimental models, "chondrocyte cultures derived from such cartilage appear to maintain their heightened metabolic activities, apparently even in subculture, which suggests a genetic change" (Hamerman, 1989: 1329). Moreover, these modified chondrocytes almost certainly modulate the rate of matrix production, most especially proteoglycan synthesis (Mankin, 1982).

As mentioned above, attempts at repair (which are normally initiated only *after* severe and deep defects develop in articular cartilage) may, ironically, initiate a series of further pathological changes which actually make the condition worse. In this view, the pathogenesis of OA is not viewed simply as resulting from cumulative external (i.e., mostly mechanical) stimuli: "Osteoarthritis is not a passive phenomenon. It most certainly is not merely the product of wear and tear. Rather, osteoarthritis (as anatomically and roentgengraphically defined) appears in large part to be a reparative mechanism" (Ehrlich, 1981: 123).

SYMPTOMS, SIGNS AND RADIOGRAPHIC FEATURES

Perhaps the biggest difficulty in documenting population patterns of OA is the *lack* of correlation between symptoms and radiographic features. Symptoms and signs of OA include: pain, tenderness, stiffness, limitation of movement, crepitus, swelling, bony enlargement, joint instability, and occasional effusion (Gresham and Rathey, 1975; Mankin et al., 1986; Dieppe, 1987; Moskowitz, 1993; Preidler and Resnick, 1996). Systematic diagnosis of OA is most commonly assessed by radiography and includes the observation of the formation of osteophytes (at joint margins and periarticularly at ligamentous/capsule insertions), periarticular ossicles (especially seen in interphalangeal joints), narrowing of the joint space, subchondral bone sclerosis, cystic areas in subchondral bone, and altered shapes of articular surfaces (particularly the femoral head) (Silman and Hochberg, 1993). Standard radiological scoring of these changes has been applied for several years using particularly the system developed by Kellgren and Lawrence (1957) and the *Atlas of Standard Radiographs* (University of Manchester Department of Rheumatology and Medical Illustration, 1973).

There has thus been for several years good consensus on both the symptoms and standard radiographic assessment of OA. Similarly, for several decades researchers have realized the disturbing lack of correspondence of one type of clinical information (patient reports/complaints) with the other (changes seen by radiologists). Cobb and colleagues as early as 1957 suggested a correlation as low as 30% of individuals with radiographic changes reporting pain at relevant sites. Peyron (1986) adds that consistently more than 15% of radiologically positive cases are clinically silent. In a classic, very large radiographic survey ($N = 2296$), Lawrence et al. (1966) found significant but variable correlation of symptoms with X-ray changes (but always $< 50\%$, even in the most severe cases). Further, Gresham and Rathey (1975) found correlations (with radiographic alterations) as high as 76% for crepitus, but as low as 28% for joint instability in a study of knee involvement in patients older than 60 years.

Moreover, little correlation has been found between joint symptoms and the *degree* of pathological or radiological change (Hadler, 1985; Moscowitz, 1993). One obvious complication is the presence of different musculoskeletal syndromes that could produce symptoms, and such difficulties of differential diagnosis almost certainly vary from joint to joint. In the shoulder, especially, where variable conditions affecting the rotator cuff may present clinically, correlation of symptoms with radiographic features has proven elusive (Petersson, 1983). Commenting further on the difficulties of making comparisons between different studies, Moskowitz states,

> Moreover, studies focused on different joints cannot be compared, even when similar case-finding methods are used. Variations that are due to interobserver error may be significant, despite the use of essentially identical protocols.

He does, however, conclude on a more positive note,

> Nevertheless, sufficient data permit a reasonable approximation of the natural course of the disease (Moskowitz, 1993: 1737).

Most of the detailed reporting on radiographic prevalence of OA has been done on the knee, which is understandable given the risks of radiation exposure. In a detailed statistical analysis Altman and colleagues, representing a subcommittee of the American Rheumatology Association, concluded, "No single set of classification criteria could satisfy all circumstances to which the criteria for OA of the knee would be applied" (Altman et al., 1986: 1045). In another survey of knee OA, Claessens et al. (1990) found somewhat higher overall correspondence of symptoms with radiographic features, but again found generally poor correlation for any *one* X-ray indicator of OA change. That is, the radiographic features (the best type of assessment currently available in clinical research) lack both sensitivity and specificity.

In a review of a variety of studies of knee OA Spector and Hart (1992) found a range of 40–80% correlation of symptoms with radiographic features and comment that somewhat better results are obtained when assessment of the patello-femoral compartment is also done, a point also made by McAlindon et al. (1992; 1994). These differences in correlation within various compartments of the knee is perhaps explained by the variable etiological factors operating in the two compartments (Cicuttini et al., 1997). In fact, while most radiographic evaluations of knee OA usually rely on antero-posterior radiographs, a lateral view allowing analysis of the patello-femoral compartment would provide considerable added information.

In my study of knee OA patterning in the Terry Collection, however, factor analysis failed to separate the patterns between tibio-femoral and patello-femoral compartments. But within the tibio-femoral component, variation did sort separately for medial and lateral sides (Jurmain, 1977b).

Given the obvious difficulties in contemporary epidemiological survey data in establishing what exactly constitutes OA at a population level, osteologists should be cautioned to make comparisons of skeletal prevalence with that found in modern samples only with great care. Various osteologists have voiced the opinion that skeletal markers might be more sensitive than clinical (especially radiographic) markers, since bone is more clearly observable. However, from the discussion above regarding macroscopic bone lesions, this seems an overly optimistic assessment, and in archaeological samples, one not independently verifiable. Perhaps, we might be on sound diagnostic footing, if we focus on severe disease only, that is, through presence of eburnation. Some results of controlled comparisons from osteological analysis of 20th century samples with matched epidemiological data will be discussed below.

EPIDEMIOLOGY

Recognizing from the above discussion concerning various problems in the way OA is assessed in individuals, those data used to establish population *prevalence* must be approached cautiously. Epidemiological information is usually collected using three different sources: (1) questionnaire (regarding self-reported symptoms; also background information, e.g., occupation); (2) clinical examination (frequently also including medical history); and (3) radiographic assessment (usually one or two joints, most often collected for the knee, hands, or hip). The most systematic epidemiological studies use a combination of all three types of information.

The largest studies that have been done on population prevalence of OA in the United States are part of the National Health and Nutrition Evaluation Survey (NHANES or HANES). Two surveys have thus far been completed, HANES I (data collected 1971–1975) (U.S. Department of H.E.W., 1979) and HANES II (preliminary results are contained within the National Health Interview Survey, data collected 1989–1991). For the latter survey, only preliminary results have thus far been published as part of CDC weekly reports (1995; 1996a,b). HANES I data include individuals ($N=6913$) between 25–74 years of age who were non-institutionalized at the time of the survey. Information was obtained from clinical exams, medical histories, and X-rays of the hips and knees, as well as some data on the cervical and lumbar spine. Results for the oldest age group for moderate and severe prevalence are shown in Table 2-4. This first national survey did not categorize completely as to ethnic group (although white and black prevalence were separately reported). However, HANES II protocols more consistently controlled for ethnic affiliation. To this point, HANES II reports (National Health Interview Survey) include self-reported prevalence with some highly intriguing patterns, but subject to considerable bias of reporting.

Another large earlier survey of the U.S. population (1962 U.S. Public Health Survey) included radiographic data for hand and foot involvement. Results from this survey have been extrapolated to estimate that 40.5 million Americans had (radiographic) OA in the early 1960s (including mild disease). Of these, 27 million were estimated to have had symptoms, and by the eighth decade, 85% of individuals were predicted to be affected (McKeag, 1992). For individuals in this older age group (age 75–79) it has been estimated that (including mild disease) 84% of hands, 51% of feet, 13.8% of knees, and 3.1% of hips showed *some* involvement (Lawrence et al., 1989). It must be remembered, however, that most of these cases were asymptomatic (Moskowitz, 1993).

Table 2-4. Prevalence data for OA of knee and hip, U.S. population (HANES I)—non-institutionalized, ages 55–64 and 65–74, moderate and severe involvement

	Males		Females	
	Knee %	Hip %	Knee %	Hip %
55–64 yrs.	1.0	0.7	0.9	1.6
65–74 yrs.	2.0	2.3	6.6	1.2

(U.S. Department, H.E.W., 1979).

As mentioned at the beginning of this chapter, the most recent (HANES II) estimates suggest about 40 million Americans are now symptomatic with arthritis (all forms), and this survey further projects that by 2020 60 million individuals will be affected (CDC, 1996a). Arthritic complaints, as of the early 1990s, already represented in the U.S. the most common self-reported chronic condition, ranking ahead of heart disease, hearing impairment, chronic bronchitis, asthma, and diabetes (CDC, 1996a).

In two large samples from Europe, one in England (Lawrence et al., 1966) and the other in the Netherlands (Van Saase et al., 1989), results were similar to those of the national U.S. surveys. In the English study 20% of all adults and 85% of those in the 55–64 age group showed some radiographic evidence of OA in one or more joints. The Dutch study, a comprehensive analysis of a suburban sample near the Hague ($N=6585$), surveyed radiographically a wide variety of joints (spine, shoulder, hand, foot, hip, knee, and sacroiliac). The results showed the highest incidence of degenerative disease in the cervical and lumbar spine and DIPs, followed by the knee (moderately high), but only rarely in the hip, shoulder, and sacroiliac. Van Saase and colleagues also made a useful comparative review of epidemiological samples, together comprising more than 22,000 individuals.

For comparative purposes data on knee OA from Van Saase et al. (1989) are shown in Figure 2-5(a) and (b). The population samples are from Holland (Zoetermeer), N. England (Leigh and Wensleydale), Bulgaria (Sofia), and the U.S. (HANES I white; HANES I non-white). For further comparison, osteological data derived from contemporary dissecting room samples are also included. These sample are from my study of the Terry Collection (white and black Americans) (Jurmain, 1977a,b; 1980) and from the Todd Collection (also white and black Americans) (Woods, 1995).

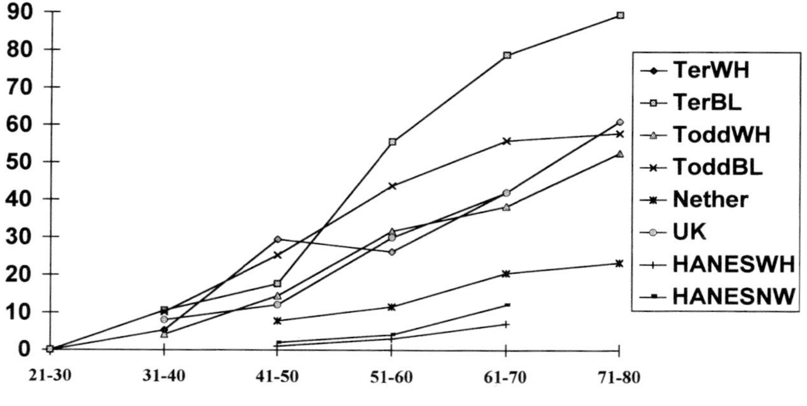

Figure 2-5a Males.

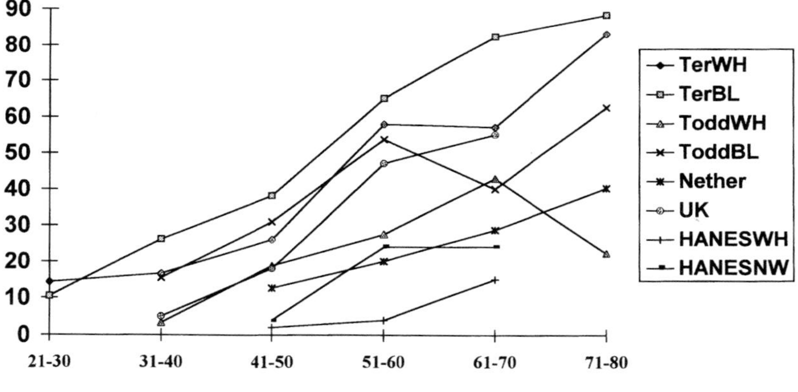

Figure 2-5b Females.

Figure 2-5 Age-specific prevalence of osteoarthritis in contemporary populations (epidemiological and skeletal samples). Knee joint, moderate and severe involvement (combined).
Notes: TerWH = Terry White (Jurmain, 1975; 1977a); TerBL = Terry Black (Jurmain, 1975; 1977a); ToddWH = Todd White (Woods, 1995); ToddBL = Todd Black (Woods, 1995); Nether = Netherlands (Zoetermeer) (Van Saase et al., 1989); UK = United Kingdom (Leigh/Wensleydale) (Van Saase et al., 1989); HanesWH = Hanes I Whites (Van Saase et al., 1989); HanesNW = Hanes I Non-White (Van Saase et al., 1989); Terry Samples and Netherlands sample, right knee only; other samples, combined left and right data. Terry samples, 71–80 = 71 + years.

Inspection of the comparative results shows that there are marked differences in prevalence in the clinical samples, with the Dutch and English samples showing higher frequencies than the other samples. Van Saase and colleagues are not overly concerned with these differences (similar data with much less marked inter-sample variation are also presented for the cervical spine and hand joints). They explain the variation as one mostly representing "differences in level." That is, "These differences of prevalence of radiological osteoarthritis may be attributed to interobserver differences—that is, different criteria used to establish radiological osteoarthritis, in addition to genetic or environmental factors, or both" (Van Saase et al., 1989: 271).

How well do the osteological results compare? At first impression, especially when combined frequencies of moderate and severe disease are tabulated, the prevalence rates for the osteological samples appear inflated. While Woods does not indicate to which level(s) of severity his data apply, they probably include aspects of moderate and severe involvement, and are thus included with the first set of comparisons. For both the Terry and Todd samples all results appear greatly inflated for early age groups, and these apparently very high prevalence rates are especially marked in comparison to HANES data, to which they *should* most directly compare (See Appendix A for tabulations of age-specific prevalence rates in osteological samples). An obvious conclusion is that the osteological criteria used to assess moderate involvement may be *too* sensitive and thus overestimate the likely prevalence of OA (The clinical data themselves, based on radiographic assessment, are already probably inflated to a considerable degree).

In order better to evaluate whether the osteological data might "fit" more closely with radiographically determined prevalence rates, the second set of comparisons (Figure 2-6(a,b)) displays only *severe* involvement for the Terry Samples of Black and White Americans (results from the Todd Collection were not reported in this fashion). In these comparisons the osteological samples compare much more closely with the clinical results derived from ostensibly similar samples of U.S. residents (i.e., HANES I data; but they and the clinical results are considerably lower than the results from Holland and England).

If epidemiologists recognize and accommodate such levels of difference *within* their own comparisons, perhaps osteologists should not overly despair in judging the comparability of their results to clinical data. Nevertheless, the shape of the prevalence rate curves for osteologically derived samples (especially ages 21–50) are too steep and thus indicate oversensitivity of some skeletal indicators, particularly those representing moderate involvement. Thus, in addition to the questions discussed above relating to pathogenic accuracy of certain osteological indicators (osteophytes, porosity) as well as better methodological reliability, broader issues of epidemiological comparability further argue that severe skeletal involvement (i.e., eburnation) is what osteologists should most consistently record and report.

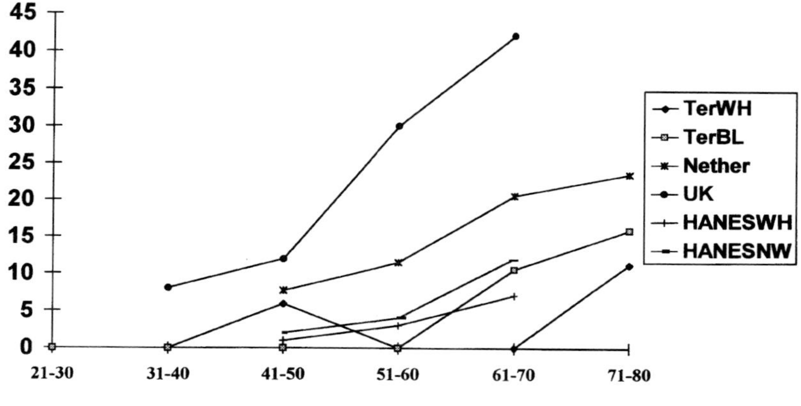

Figure 2-6a Males.

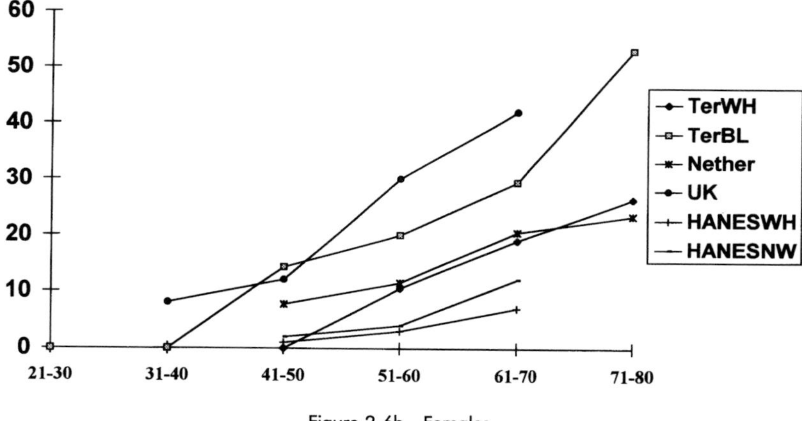

Figure 2-6b Females.

Figure 2-6 Age-specific prevalence of osteoarthritis in contemporary populations (epidemiological and skeletal samples). Knee joint, osteological samples (Terry Collection), severe involvement; other samples, moderate and severe involvement (combined).
Notes: TerWH = Terry White (Jurmain, 1975; 1977a); TerBL = Terry Black (Jurmain, 1975; 1977a); Nether = Netherlands (Zoetermeer) (Van Saase et al., 1989); UK = United Kingdom (Leigh/Wensleydale) (Van Saase et al., 1989); HanesWH = Hanes I Whites (Van Saase et al., 1989); HanesNW = Hanes I Non-White (Van Saase et al., 1989); Terry Samples and Netherlands sample, right knee only; other samples, combined left and right data. Terry samples, 71–80 = 71 + years.

Given the ambiguities in comparing clinically derived epidemiological prevalence rates, Van Saase and colleagues are understandably conservative in drawing conclusions from such data, especially as relating to the differential influence of long-standing mechanical and/or metabolic factors. They thus conclude: "The relative importance of these processes can, unfortunately, not be compared because they are rarely available and even when available lack methodological standardisation" (Van Saase et al., 1989: 279–280). Moscowitz (1993) also suggests differences in pattern between studies could in part be explained by how radiographs and clinical criteria were applied. These same issues of representativeness of samples, reliability of morbid indicators, and standardization of methods also apply equally to cross-populational comparisons utilizing skeletal samples. The constraints on analysis and necessary limits of interpretation seen in contemporary (clinical) research should stand as clear warnings to those postulating behavioral scenarios from skeletons!

There, too, is the larger issue of what such population data, especially using radiographic indicators, mean relative to actual morbidity of the population, given that only perhaps 30% of those with such changes show symptoms. Even a smaller proportion of individuals become significantly handicapped. Nevertheless, referring to English populational data, Dieppe reminds us that, "three-quarters of those disabled by arthritis are said to have OA and about 30 per cent of sickness incapacity from rheumatic diseases is attributed to this condition" (Dieppe, 1987: 16.76).

Specific Population Comparisons

Keeping in mind the caveats mentioned above, some specific epidemiological comparative data are available. In the HANES I survey ethnicity/race was not broadly controlled, but using these data comparisons between Black and White Americans has been possible. The most notable difference, after controlling for age and weight, was found in knee prevalence between Black and White females, with Black females showing increased risk (odds ratio = 2.12; Anderson and Felson, 1988). The HANES II survey is the first to systematically compile information nationally on a broad spectrum of what initial reports have termed, "racial, ethnic and hispanic groups." In the initial results (CDC, 1996b) of self-reported involvement (i.e., individuals responding to questionnaires regarding symptoms), whites, blacks, and American Indians/Alaskan Natives all show prevalence rates of 15–16%, Hispanics show a lower prevalence (11%), and Asians/Pacific Islanders markedly less yet (7%). These results must, however, be viewed critically, as self-reporting of disease affliction could easily be influenced by a variety of cultural factors.

Other studies reporting population prevalence in developing countries or in ethnic minorities (for example, in the U.S.) are not numerous and frequently suffer from small sample size. When compared to European samples, prevalence rates among African groups from Liberia (Valkenburg, 1983), Nigeria (Ebong, 1985; Ali-Gombe et al., 1996), and South Africa (Solomon et al., 1976) are consistently lower for hip OA. Among the Nigerian and Liberian samples, however, for some hand joints (metacarpo-phalangeal, PIPs) the rate of involvement was approximately the same as in European groups, and in South African males a higher prevalence for these joints was found. Moreover, in analysis of a Jamaican population (largely of African descent) the prevalence of knee involvement was also found to be higher than that in selected European samples (Bremner et al., 1968).

There are several confounding factors that could significantly influence the variability noted above, especially for hip involvement. Moskowitz (1993) suggests predisposing conditions (such as hip dysplasia) could vary between populations, thus influencing the prevalence rates of hip OA; a similar view was previously voiced by Solomon et al. (1976). Low prevalence rates of hip OA were also found in a Southern Chinese sample from Hong Kong (Hoaglund et al., 1973). Findings from this fairly large sample ($N=500$) also included data on finger joints and knees, but the authors emphasized the low prevalence found in the hip (DIPs were similar to that in European samples; all other joints showed a lower prevalence). To explain the low hip OA prevalence, a behavioral hypothesis was proposed: Due to the typical squatting posture among Chinese, increasing levels of flexibility *protect* against hip disease (low rates of hip dysplasia and generally low body weights with absence of obesity were not considered as alternative explanations). It is interesting in this instance to find such a behavioral explanation for contemporary patterning of OA. For skeletal involvement, osteologists are frequently tempted to make an opposite hypothesis—that repetitively employed, increased ranges of joint motion could lead to *increased* OA. A more recent study (Lau et al., 1995) has further confirmed these general findings for prevalence of OA in Chinese men from Hong Kong. Moreover, these latter data show low prevalence in both hands and hips, and in the latter, OA was not associated with hip dysplasia.

A recent study of patients requiring total hip replacements in San Francisco has also reported dramatic population differences among this sample (Hoaglund et al., 1995). Paralleling the results from the HANES I reports (noted above), Whites showed dramatically more frequent and severe hip disease than that found among Blacks, Hispanics, or Asians. In fact, Asian-Americans in San Francisco required this form of hip surgery only at about 10% the frequency of that among Whites.

Data on indigenous New World populations are limited, but a few studies have reported OA prevalence among the Blackfeet and Pima (Bennett and Burch, 1968) as well as for the Haida and Inuit (Blumberg et al., 1961; Anderson and Winckler, 1979). Much of these data are for hand involvement and show mixed results, but usually show equal or slightly greater involvement in indigenous New World groups compared with European samples. In one study of Alaskan Inuit, however (Blumberg et al., 1961), lower hand OA prevalence was found; moreover, in a more recent survey of knee OA in Greenland populations very low rates of severe OA were observed, especially in those samples presumably with the greatest amount of Inuit ancestry (Anderson and Winckler, 1979). Given the high *osteologically determined* prevalence of OA among Inuit (discussed in Chapter 4), these findings are surprising, especially since some of the groups sampled by Anderson and Winckler were said to have been still practicing traditional hunting and fishing.

Since there are inherent problems in comparing results from different studies compiled by different researchers (likely using somewhat different methodologies), systematic data recorded by a limited number of researchers is preferable. In a remarkable compilation of data from 17 different samples (9 European; 4 African; 4 native American), Lawrence and Sebo (1980) reread radiographs from more than 13,000 individuals (including samples cited here). Their results are suggestive of potentially important genetic and occupational factors. Consistently in the New World samples multiple-joint hand involvement among females was low, suggesting a low prevalence of GOA; however, among male Pima Indians, Lawrence and Sebo found a relatively high incidence of hand OA, including presence of Heberden's nodes. In this case, the authors suggest the increased incidence may be due to processing cotton, a contention certainly not easy to establish. They also found a low rate of OA in the African samples, here again leading them to suggest potentially important genetic contributions. Moreover, as with the individual studies discussed above, they found much lower rates of hip involvement in the African groups compared with European samples.

As noted, Lawrence and Sebo postulate hypotheses to account for population differences which include components of both mechanical (occupational) and genetic explanations. They are quite clear, however, that any such hypotheses must account for considerable complexity:

> The factors influencing prevalence of osteoarthrosis in populations are clearly very complex, depending on occupation, heredity and possibly diet and resultant body build (Lawrence and Sebo, 1980: 179).

How, then, is one to choose among these alternative factors, especially in less well-documented archaeological contexts? This question is of particular import, given the central place of similar population comparisons in osteological research (in my own publications as well as that of others; see Chapter 4 for further discussion).

ETIOLOGICAL FACTORS

Probably the most contentious issue relating to OA is that of trying to identify and assess the relative significance of a variety of possible etiological agents. Moreover, osteologists have sought to add to this discussion. Indeed, specifically in this realm, skeletal perspectives can make their greatest contribution.

Etiological agents can be subdivided as either localized, especially biomechanical, or systemic, including most importantly, age. A variety of other possible systemic factors are discussed, including sex, metabolic influence on bone turnover rates, other arthropathies, obesity, and diet. It is best, however, to begin with what are probably the two most central etiological influences, namely mechanical factors and age.

Mechanical Factors

There is no question that markedly altered joint biomechanics can initiate OA, as demonstrated by the onset of degenerative changes following severe trauma. The evidence indicating the impact of trauma comes from animal models, particularly from experiments on dogs and rabbits (Mankin et al., 1986); the effects of various repetitive types of stress (e.g., running) have also been studied in guinea pigs, horses, sheep, tigers, and lions (Mckeag, 1992).

In humans the best documented studies relating to trauma are those on knee OA following injury (especially to menisci). It, however, remains to be demonstrated, "whether a common pathway leading to tissue degeneration is associated with all forms of trauma" (Mankin et al., 1986: 1142). Not just *which* joints might be affected, but when in the life cycle trauma occurs is also an important consideration. For example, while knee OA has been shown to follow injury in adults, early (pre-adult) traumatic modifications may play a larger role in subsequent hip OA (Murray and Duncan, 1975).

However, in the absence of demonstrable *severe* trauma (i.e., leading to secondary OA), there is little consensus among researchers regarding the role of repetitive stress on the initiation of idiopathic OA. Firstly, *some* mechanical function is necessary to maintain healthy joint function, especially the articular cartilage. Studies of patients with paralyzed limbs or astronauts/cosmonauts subjected to prolonged weightlessness have demonstrated the deleterious effects of joint immobility (or lack of weight-bearing). Moreover, during growth joint tissues respond to applied stress and, "in fact, do not grow normally without it" (Mankin et al., 1986: 1141). Such results have also been confirmed from experimental work with animals (Barnett et al., 1961).

Indeed, as mentioned above, under normal loading, human joints function remarkably well and nearly friction-free. In a normal human lifetime a typical joint will be subjected to thousands of millions of load cycles (Weightman, 1976). "As all activities of daily living involve intermittent loading and oscillation of joints, in both upper and lower extremities—the mystery is why all articular cartilage in the load bearing areas does not wear out with age" (Radin, 1975: 862).

But, of course, joints do sometimes wear out, and many researchers argue that chronic overuse is a major cause. The etiological contribution of such long-term, repetitive wear-and-tear has been termed the "stress hypothesis" (Pommer, 1920; Cobb, 1971). This perspective is at the heart of osteological interpretations of OA, and certainly was a major focus of my own early work (Jurmain, 1977a,b; 1980).

It must be remembered, however, that, from a clinical perspective, the stress *hypothesis* is just that, a proposed explanation of the observed patterning of degenerative variation. Moreover, as indicated by some data already presented above, as well as the balance of this and the subsequent chapter will further show, the hypothesis is far from widely accepted by clinical researchers. In decades of attempts to establish its *general* validity, only very minimal success has been attained. In further attempts to demonstrate more specific *predictive* components relative to the influence of repetitive biomechanical stress, results have been highly contradictory and continue to frustrate even the best clinical research designs (see Chapter 3). Obviously, the message for osteologists is to view the stress hypothesis and its *assumed* applications with considerable caution. How well such behavioral models work in contemporary contexts and the various stories told by osteologists (derived from these models) are the central topics of this book.

The crucial deficiencies of the stress hypothesis are linked to its lack of precision. Firstly, for biomechanical stress to have a major, even a determinant, influence, it would follow that *other* etiological agents do not much affect the variable patterning of degenerative involvement. And, if stress/mechanical factors do contribute significantly *at least* in some cases, specifically what are those circumstances? As noted above, mechanical factors subsumed under "trauma" (including "microtrauma") almost certainly differ in their effects from joint to joint. In addition, mechanical factors variously influence different regions *within* joints. For example, an innovative method was devised to measure mechanical loading in a human hip *in vivo*. Following replacement surgery, a pressure sensor was fitted into a 73-year-old woman's prosthetic proximal femur (Lewin, 1986). Results confirmed that pressure was not evenly distributed across the joint; indeed, most of the extreme loading was *not* due to weight bearing, but resulted from muscle contraction, especially when rising from a sitting position, producing loads up to three times that during walking. Such mechanical effects might be especially deleterious in joints already compromised by pre-existing conditions, leading to compromised joint congruity. With a poorly fitting femoral head immensely high pressure (up to 1000 psi) can be generated (Lewin, 1986). Such huge (secondary) biomechanical forces might thus help explain the rapid development of OA following traumatic hip dislocation (or consequent to a poorly fitting prosthesis). Muscle contraction is also thought to produce quite high (intermittent) loads in the upper limb (Mankin and Radin, 1985), although not of the amplitude measured in the hip. Impact loading is also considerable in the lower limb, especially during such activities as running. Compared to normal walking, running is estimated to produce twice the impact load in the ankle and hip and up to six times the load in the knee (Note: Again, as will be shown, even in long-distance runners, usually without degenerative consequences).

Within the knee the various types of mechanical stresses have also been shown not to be equally distributed. In the patello-femoral compartment, particularly, stresses appear to differ from those in the tibiofemoral compartment (Seedhom et al., 1979; Froimson et al., 1990; Howell and Pelletier, 1993). The importance of recording the regional distribution of knee OA was noted above, following the research of McAlindon et al. (1992) and has also been approached through a more precise method of topographical mapping of joint surfaces in cadaver specimens (Swann and Seedhom, 1993) as well as in osteological remains (Wood, 1995).

Documentation of regional variation in the elbow has been a significant contribution of osteologists particularly (Ortner, 1968; Merbs, 1983), most likely, since elbow OA is usually so uncommon in contemporary populations (however, this situation is changing as more interest focuses on "little league" and "tennis" elbow; see Chapter 3). As mentioned above, multivariate patterning in my analysis of the Terry Collection also separated degenerative patterning within the functional compartments of the elbow (i.e., rotary *vs.* hinge; Jurmain, 1978). The same pattern was seen even more distinctively in a principle components analysis of degenerative variation in Inuit elbows (Jurmain, 1975; 1978). As in the knee, a productive and more precise (topographical) approach to recording this regional patterning within the elbow has been suggested by Merbs (1983) and more recently refined by Nagy (1996).

The stress hypothesis, as broadly defined, is probably too general to be amenable to testing, and is thus of limited scientific utility. In order more precisely to address the undoubtedly complex influence of mechanical stress on the development of OA, several factors should be considered:

(1) amplitude
(2) periodicity (i.e., how constant is the stress; are there adequate intervals for recovery?)
(3) duration
(4) age of onset
(5) predisposing biomechanical factors (especially affecting joint congruity in the hip)
(6) other systemic influences, such as obesity (important not just biomechanically, but also perhaps biochemically)
(7) regional variations in loading
    a. between different joints
    b. within joints (especially for the elbow and knee)

Age

That age greatly influences the onset of OA in all joints is not in dispute. Its *relative* contribution in different joints and its overall effect, independent of other possible etiological agents, are the less well understood issues. All epidemiological surveys in which age is controlled have consistently shown the strong correlation of both prevalence and severity of OA involvement with age. "This striking increase of OA with aging is consistent with other studies and when individuals with severe grades of OA only are considered, the increase with age is exponential" (Moscowitz, 1993: 1737).

Understandably, much of the clinical focus on specific components within aging joints has focused on the articular cartilage. Several changes in the cartilage concomitant with aging have been identified and include: decrease in cellularity, disorganization of collagen fibers, reduced fatigue life, reduction in water content, changes in chondroitin sulfate units, reduction in size of proteoglycan aggregates, loss of non-aggregated proteoglycans, and reduced capacity of the proteoglycans to form new aggregates (Peyron, 1986). In summarizing various lines of research, Hamerman (1989) drew distinctions between ("normal") age-related changes in articular cartilage and those which develop pathologically during OA. Hamerman suggested a useful distinction, stating: "Development entails aging, and since it is arbitrary to say when 'aging' occurs, maturation may be a more appropriate term to describe the changes that occur in cartilage over time" (Hamerman, 1989: 1325).

While articular cartilage shows fairly dramatic biochemical changes with age, it is not clear that these changes predispose the joint to the development of OA.

> Do these changes arise on a biological basis? Is there a 'running down' of the cellular machinery with age? Does the chondrocyte become senescent losing its capacity to maintain a normal extracellular matrix? Or do age-related changes in normal articular cartilage reflect the consequences of microtrauma to which the joint has been subjected during the lifetime of the individual? We do not know (Brandt and Fife, 1986: 117).

No doubt, there are multiple factors interacting in aging joints which, in some circumstances, ultimately produce degenerative changes. Hamerman emphasized the role played by altered cartilage biochemical properties (especially changes in the proteoglycans and collagen) which with increasing age lead to "impaired mechanical properties" (most notably, decrease in tensile strength).

Whatever its ultimate cause, the marked age-correlation of OA is strikingly illustrated in Figures 2 and 3, for all the clinical survey samples *as well* as for the osteological data (discussed further below). Other studies as well have yielded similar results. In the large sample from the New Haven survey, a multivariate analysis of potential systemic factors found age to be the single most important factor (Acheson and Collart, 1975). Likewise, in the long-term, well-controlled prospective study of knee OA in the Framingham sample

Felson et al. (1987) report that prevalence in individuals <70 years of age was 27%, but increased to 44% after 80 years of age. From more general data, including several joints, Peyron (1986) lists the following general age-stratified prevalence rates:

	Women	Men
<45 years	2%	3%
45–64	30%	24%
>65	68%	58%

Peyron (1986) also makes the important point that the onset and rate of progression *varies* between joints. The earliest onset is usually of the 1st metatarso-phalangeal joint, followed by the wrist and spinal apophyseal joints. After age 45 degenerative onset also is seen in the interphalangeal joints and carpo-metacarpal joints; later the knee (tibio-femoral component) becomes involved, and last, the hip shows OA changes.

That the progression of degenerative changes varies between joints is now well established. However, the specific rates of changes within individual joints still need greater elucidation. Given the capacities for repair discussed above, it seems obvious that in no joint is the progression of OA inevitable. However, for the joints best studied (hip and knee) it seems, that once OA begins, there is a difference in the prognosis (with the knee faring worse). Perhaps, this difference is due in part to the role of previous injury in knee OA. The best evidence showing lack of progression in the hip was presented by Danielsson (1964) and Nilsson et al. (1982), but for the knee results are more mixed. Some research shows the condition to be mostly progressive, with little or no repair or remission of symptoms (Hernborg and Nilsson, 1977; Massardo et al., 1989), but other studies of knee OA indicate at least a leveling off of symptoms (Lawrence et al., 1966), if not even some regression (Forman et al., 1983; Spector et al., 1992).

An interesting further wrinkle has been reported for age-related involvement within the knee joint. In a prospective radiographic study of knee involvement Nilsson et al. (1982) also tracked the *location* of degenerative changes within the tibiofemoral compartment. They found that at initial examination the vast majority of individuals (92%) displayed exclusively medial involvement; but at follow-up (10–18 years later) a significantly greater proportion showed a mixed pattern of both medial and lateral radiographic changes (increasing from 1% initially to 26% at follow-up).

Mankin et al. (1986) summarized the role of age in relationship to other factors emphasizing that, "age is the strongest predictor of risk for OA," but the influence of age as a factor likely combines with the "cumulative effects of exercise, trauma, metabolism, endocrine, and other factors acting over a lifespan." (Mankin et al., 1986: 1152)

The osteological data concerning age-related phenomena correspond extremely well with the clinical reports. As shown in Figures 2 and 3, contemporary osteological samples show marked increase in involvement with age; such dissecting room samples are the most appropriately applied here, since age is quite accurately documented, a situation almost never attained archaeologically. Moreover, further statistical analysis of the Terry Collection, wherein correlation coefficients for total joint scores with age were calculated (Jurmain, 1980), shows that the influence of age *varies* by joint:

	Right	Left
Shoulder	0.656	0.668
Hip	0.627	0.626
Knee	0.513	0.500
Elbow	0.397	0.403

As seen, OA in the more proximal ball-and-socket joints (shoulder and hip) is most age-correlated, but is less so in the more peripheral knee and elbow (and least so in the latter). Multivariate analysis (PCA) of these same materials, already mentioned above, further confirms this pattern. Age was relatively highest loaded (2nd) in the shoulder, next highest (4th) in the hip, less prominent (4th lowest) in the knee, and least loaded (2nd lowest) in the elbow (Jurmain, 1975; 1977b).

Age of Onset

An important point raised above is that age is not only a key to understanding the pattern of most contemporary idiopathic OA (especially marked in older age groups), but also an *early onset* subset of the condition may also be etiologically significant, at least in certain circumstances and in certain joints. The greater physiological plasticity of younger individuals is not a surprising aspect of development, and it is one that has been pointed out by various researchers dealing specifically with bone development. Garn (1970), for example, discusses how different effects of age influence normal bone growth in the hand. Stuart-Macadam (1985) has documented the differential development of porotic hyperostosis in young individuals. Lastly, Ruff et al. (1994) carefully document developmental components shaping bone diaphyseal geometry. Thus, it is a reasonable possibility that for OA, and for other conditions discussed in this book, as well, the differential influence of what can be called "age of onset" can produce significant effects on the developing musculoskeletal system.

Nevertheless, most of the available clinical evidence to support such an hypothesis are at present either mostly theoretical or anecdotal in nature. Commenting on the age-related contribution of various factors (biological *vs.* biomechanical; generalized *vs.* localized), Sokoloff emphasizes they, "might well vary with the time in life they come to play on the cartilage, i.e., before or after closure of the epiphysis" (Sokoloff, 1985: 199). It should also be added that maturational changes in the physiology, and perhaps, susceptibility, of the tidemark may also be significant. The stronger correlation of osteophyte development (see above) with age than is seen for other OA indicators may also suggest somewhat different etiological agents acting on joint surfaces as opposed to the marginal rim, and these different agents may vary substantially in their age of onset. For example, it is possible, although far from established, that osteophytes *might* be more correlated with a late-onset, and more typical, form of idiopathic OA.

Conversely, some of the more destructive processes affecting the articular plate, beginning perhaps at the tidemark and/or osteochondral junction, may develop mostly independently of osteophyte formation and also do so *earlier* in life. Moscowitz (1993), for example, noted that in the hip a more aggressive form of OA has been documented where degenerative pseudocysts develop in the absence of osteophytes. Duncan (1979) described a somewhat similar process in the knee, and here one occurring quite early. In young athletes (10–13 years of age), occasionally early cartilage disturbance develops particularly in the patellofemoral compartment. Duncan suggested this type of lesion, sometimes called "chondromalacia," may actually be misclassified and may rather represent an early form of OA.

Indeed, of the meager data currently available, the best supporting evidence for such an early onset form of degenerative disease come from the sports literature. For example, as noted above, Murray and Duncan (1971) found a higher prevalence of later hip OA in those males who had been more athletically active early in life. They suggested the etiological factor operating may be one that changes joint congruity, leading to OA later in life; this same view was also argued by Howell et al. (1979), especially as relating to weight-bearing joints. For the hip, particularly, early modifications may produce what appears as idiopathic OA, but only after a long latency period.

O'Neill and Micheli (1988) suggest that repetitive trauma to the growth plate may predispose to later OA in both the hip and knee.

Chantraine (1985) did, for example, find that those soccer players who began playing very early (in childhood) showed the greatest amount of knee involvement later on. In addition to the articular plate itself, which is perhaps most susceptible to injury of the developing epiphysis (i.e., the osteochondral junction), O'Neill and Micheli also discuss two other areas where growth cartilage can be disrupted: (1) in the articular cartilage itself (particularly vulnerable to shear forces in young athletes in the elbow, knee, and ankle), and (2) at apophyses (here constituting "stress fractures;" to be discussed in more detail in Chapter 5).

The elbow, in particular, may be highly vulnerable to early onset of degenerative changes. Increasing documentation of "little league elbow" has focused attention on elbow OA, a joint (as noted earlier) that previously was rarely emphasized by clinicians. Severe stresses early in life can lead to various elbow complications, both in the medial compartment (where changes are chronic but not serious) and in the lateral compartment (where changes can be quite severe). Adams (1965), for example, found a high prevalence of elbow OA in young baseball pitchers, especially those who began pitching early in life (A more detailed discussion of sports activity and its relationship to OA comprises a major portion of the next chapter).

In the elbow, early onset degenerative changes have sometimes been labeled as "ostechondritis dissecans," but it is possible (as with Duncan's suggestion regarding osteomalacia in the knee) that what these lesions represent is (more basically) an early manifestation of elbow OA. Further, O'Neill and Micheli (1988) note that similar lesions have also been described in the talus and proximal and distal femora of young athletes.

Unlike the pattern seen in adult-onset, typical idiopathic OA, the pattern of relatively greater elbow involvement found in archaeological samples appears to parallel more similarly the situation seen in certain athletes. That the elbow holds a central place in osteological descriptions of OA, and consequent behavioral scenarios, is now well known (Angel, 1966; Ortner, 1968; Jurmain, 1977a; 1978; Merbs, 1983). Moreover, and probably not strictly a coincidence, the description and interpretation of elbow disease among the Inuit has had the greatest influence. This osteological pattern of elbow disease, especially as documented in the Inuit, may, in fact, provide new insights into understanding the etiopathogenesis of certain forms of OA.

Of course, explanations based on archaeologically derived materials cannot, in themselves, conclusively *prove* any hypothesis. Ultimately, osteologists must validate their hypotheses using the accumulated data from clinical contexts. Nevertheless, the fruitful interchange of approaches to data, perspectives, and theoretical models between osteologists and clinical researchers is perhaps the most useful product of this type of research.

Other Etiological Agents

A variety of other etiological agents have also been postulated as potentially important in initiating OA. These include the role of genetics, other arthropathies, sex, bone turnover rates, obesity, and diet. Each of these will be discussed briefly.

Genetic Factors

The possible role of genetic influences is derived from five sources: (1) Comparisons between populations documenting different prevalence and/or a different pattern of involvement (see above). (2) Hypotheses relating to the indirect influence of inherited aspects of joint biomechanics; such features have been strongly suggested in the hip. Interestingly, also in animals, particularly in domestic dogs, the development of early onset hip OA is certainly related to congenital hip dysplasia (due to the effects of inbreeding). Additionally, in sub-clinical cases, disrupted hip biomechanics probably contribute to the common appearance of (later onset) hip OA in dogs. (3) The varied distribution of generalized OA (GOA) in different families and different populations. As noted above, there is good evidence of a familial pattern of GOA. Moreover, there is also considerable epidemiological evidence showing GOA varies between populations, with the highest prevalence among White females. (4) Evidence of genetic markers, particularly new evidence of mutations in the production of Type II collagen, producing some rare (familial) forms of OA (Cicuttini and Spector, 1996) as well as evidence of vitamin D receptor gene polymorphism (Keen et al., 1997; Uitterlinden et al., 1997). (5) Twin studies.

In particular, recent evidence of genetic influences of a polymorphism at a vitamin D receptor locus as it affects risk of knee OA is intriguing. Keen et al. (1997) observe an increased risk associated with one allelic variant, with an odds ratio of 2.82 over an alternative allele. Similar results are also recently reported by Uitterlinden and colleagues (1997), and these authors further note the genetic influence of the allele appears more correlated with osteophyte development as opposed to joint space narrowing. Such a *systemic* influence on bone apposition might have wide implications, not just for understanding

OA, but perhaps for other musculoskeletal manifestations as well (see Chapter 5).

Recent results of heritability studies of twins have also suggested quite strong genetic influences. In a study of monozygotic and dizygotic twins, Spector, Cicuttini, Baker et al. (1996) suggested heritability estimates for radiographically diagnosed OA varying (by site) between 0.39 and 0.65. These estimates argue for a considerable genotypic influences on OA, but also ones that vary by joint, with greater genetic influences manifested for hand OA as compared to knee involvement (and thus, in accord with prior data on generalized OA).

Similarly, recent controlled investigation of variable influences on the development of OA in different compartments of the knee joint have shown genetic factors appeared to be more important in the patellofemoral compartment, as compared to the medial tibiofemoral compartment, where previous injury/surgery played a differentially greater role (McAlindon et al., 1994).

Osteological data (Jurmain, 1980) also provide some further support to understanding patterns of polyarticular involvement. Data for population involvement for four skeletal samples are shown in Table 2-5. As the table shows, a higher prevalence of polyarticular involvement is seen in females of both contemporary samples (Terry Collection Black and White Americans). Only among the Inuit is there more polyarticular involvement in males. Moreover, for the most widespread forms of polyarticular involvement in this analysis (three or more joints severely involved), White females have the highest prevalence of any of the subgroups sampled (i.e., exactly paralleling clinical results).

Table 2-5. Polyarticular joint involvement in skeletal samples

	At least 2 joints involved %	At least 3 involved %	4 involved %	At least 2 severe %	3 severe %
White males	46.6	20.5	4.7	0.9	0.0
White females	49.5	32.0	8.7	5.9	3.9
Black males	50.8	31.9	14.7	2.6	0.9
Black females	58.5	39.0	13.5	5.9	1.7
Pecos males	17.1	9.9	0.9	0.0	0.0
Pecos females	9.2	4.1	1.0	0.0	0.0
Inuit males	38.7	18.7	6.2	3.7	1.2
Inuit females	16.7	9.1	1.5	1.5	0.0

(from Jurmain, 1980: 148).

Other Arthropathies

Clearly, the presence of other arthropathies can, not only complicate a diagnosis, but may also affect the relative prevalence of OA in certain joints in certain populations. Several arthropathies have been suggested as possibly influencing the recognition of OA, and potentially leading to its secondary development, including:

(1) Erosive arthritis (e.g., septic arthritis).
(2) Inflammatory arthritis (e.g., rheumatoid arthritis, RA); in contemporary populations this condition can introduce a major complication, but, as RA may be of fairly recent origin, in most archaeological populations RA probably has minimal influence on perceived patterning of OA.
(3) Endemic cartilage arthropathies (e.g., Kashin-Beck syndrome; Mseleni Disease) (Sokoloff, 1985).
(4) CPPD (Calcium pyrophosphate deposition disease).
(5) DISH (Diffuse idiopathic skeletal hyperostosis).

Because of possible osseous involvement, the last two conditions are of particular interest to osteologists. The characteristic features of CPPD are:

(1) Unusual articular distribution (including wrists, shoulders, and occasionally elbows).
(2) Unusual pattern within joints (e.g., patello-femoral involvement).
(3) Prominent subchondral cyst formation (lesions usually large and numerous).
(4) Severe (destructive) bone changes may be present.
(5) Variable osteophyte formation (may be quite large and irregular).
(6) Often develops subsequent to joint injury (or surgery).

(after Resnick and Niwayama, 1981; Dieppe and Watt, 1985)

There are, in fact, several different varieties of crystal deposition disease potentially leading to joint involvement. Some are acute, others more chronic, and some conditions (including gout) are known to be age related (Doherty and Dieppe, 1986). The influence of CPPD, particularly, has been postulated as influencing secondary onset of OA, but little data are currently available to validate this hypothesis. However, for knee OA, two studies (Hochberg, 1984; Felson et al., 1989a) have established an association of OA with CPPD, when articular chondrocalcinosis is also present. Nevertheless, in most cases CPPD is probably unrelated to OA (Dieppe and Watt, 1985), and, even in cases where some association may exist, it has recently been argued that crystal deposition is probably secondary to OA (Rogers and Waldron, 1995).

DISH is characterized by exuberant ("flowing") vertebral osteophyte development, especially involving the anterior and right lateral aspects of the longitudinal ligament, suggesting a role of the descending aorta in stimulating bone deposition (Rogers and Waldron, 1995). Pain and other symptoms are usually either mild or absent, and when present, usually involve spinal stiffness (Moscowitz, 1993). In addition to the pathognomonic spinal lesions, in DISH extraspinal manifestations (especially bony exostoses) are also usually seen. These bony growths are seen especially on the plantar aspect of the calcaneus and the olecranon process of the ulna. Severe ligamentous calcifications involving the sacrotuberous, iliofemoral and patellar ligaments may also be present; periarticular osteophytes often are conspicuous as well (Resnick, et al., 1978; Resnick and Niwayama, 1981). While growth hormone levels appear normal, several studies have indicated an increased level of insulin in DISH patients (Moscowitz, 1993). The etiology of the condition is unknown.

Clinically diagnostic cases of DISH are not particularly common, but subclinical manifestations may be closely linked to individual variability in rates of bone turnover. Indeed, systemic influences related to control of bone turnover may not only significantly influence OA and DISH, but several other conditions discussed in this book as well.

Rates of Bone Turnover

The best documented association relating to OA and bone turnover comes from ascertainment of bone density, especially as indicative of osteopenia/osteoporosis. Several studies have established an increased risk of OA with greater bone density (and an inverse association of OA with osteoporosis) (Dequeker et al., 1977; Solomon et al., 1982; Gevers et al., 1989; Howell and Pelletier, 1993; Moscowitz, 1993; Sowers et al., 1996). Osteological data have also helped confirm this relationship (Burr et al., 1983).

Perhaps, the most intriguing possibility of all concerning the rate of bone turnover is its potentially wide systemic influence. Continued positive turnover (indicating continued high osteoblast activity, in some individuals perhaps persisting late into life) leads to higher bone density. In such individuals the risk of (late onset) OA might be increased. Moreover, the higher rates of bone deposition in some individuals might produce bone involvement more dominated by osteophytes, but less so by other indicators of OA. What might be present, in fact, is a generalized bone syndrome in which some individuals not only produce more and larger osteophytes, but also display more enthesophytes as well (see Chapter 5). Lastly, such individuals may developmentally deposit bone differentially on bone surfaces, thus also complicating interpretation of bone geometry (see Chapter 7).

Osteological researchers, in particular, have recently focused on the identification of such a syndrome, and affected individuals have been termed, "bone formers" (Rogers et al., 1997). In fact, DISH may well be the end product of the same process, and more modest bone formers, who typically display marginal osteophytes and enthesophytes, may simply be sub-clinical cases of DISH. Osteologists, however, have routinely interpreted marginal joint osteophytes as indicative of OA and enthesophytes as indicative of increased muscle mass. In both cases scenarios have been constructed emphasizing the role of activity (see Chapter 4 for further discussion). The possible confounding influence of systemic factors would obviously throw such behavioral interpretations into serious doubt.

Moreover, it seems probable that basic rates of bone turnover are under the influence of relatively simple (i.e., Mendelian) genetic mechanisms, perhaps governed by only a few regulatory genes. As noted, some of these loci might well include those related to vitamin D receptors (Keen et al., 1997; Uitterlinden et al., 1997). The observation that individuals vary considerably in bone turnover would argue that these loci are probably polymorphic (already, in fact, confirmed for a vitamin D receptor locus). Thus, behavioral interpretations based on bone formation of osteophytes, enthesophytes, or, perhaps, bone geometry, may be largely incorrect. If individuals likely differ in these variable osseous expressions of gene action, so too might populations, who could easily differ in frequency of the alleles at the respective loci.

Sex

Epidemiological data have shown consistent patterns of gender differences in OA prevalence by age. Prior to age 55 males and females have similar prevalence rates, although many studies have shown males to be slightly more involved in these earlier decades (Peyron, 1986). After age 55, however, females show a higher prevalence, with more joints affected, and more severe disease, especially in hand joints (Moscowitz, 1993). These age-specific gender differences in prevalence are well confirmed in the HANES I survey (U.S. Department of H.E.W., 1979) and are shown for the knee joint for this and other large epidemiological surveys in Figures 2 and 3. Of note, the osteological prevalence data from the Terry Collection (Jurmain, 1977b) shows the same trend to a modest degree, but only when severe and moderate involvement are considered together. The Todd Collection results for the knee joint (Woods, 1995) fail to show this trend, and, in fact, both Blacks and Whites seem to show the reverse pattern. Since such osteological materials derived from 20th Century dissecting room contexts should be sampling essentially the same populations as those included in the HANES I survey, these differences in pattern are not easily interpretable.

Obesity

For a number of years obesity has been reported as an important systemic contributor to OA. Experimental work with mice (Silberberg et al., 1956) had earlier suggested a possible role of obesity, and more recent epidemiological research on humans has largely confirmed the relationship. HANES I data, while controlling for other possible confounding factors (age, sex, occupation), showed a very strong correlation, even resembling that of a "dosage effect" (Anderson and Felson, 1988). Recently reported HANES II data (CDC, 1996b) have further demonstrated a very strong relationship. Likewise, the well-controlled longitudinal Framingham study (with follow-ups after 18 years) also demonstrated a highly significant relationship (Felson et al., 1988). The authors of this report suggested that mechanical factors probably are the most important consideration as, "increased force per unit area in obese patients is probably the cause of the disease," but these researchers also emphasized the possibility of other factors affecting the way obesity interacts with OA. "The pathogenetic pathway by which obesity causes osteoarthritis of the knee needs further study"(Felson et al., 1988: 23). Indeed, Felson (1988) in a review of the epidemiology of OA noted that obesity may be a highly important systemic factor helping explain population patterning of *both* OA and osteoporosis. "Thus, the link between osteoarthritis and bone mass may be explained by the fact that both are associated with obesity. ... In this view, obesity causes osteoarthritis in postmenopausal women by preventing osteoporosis and maintaining high bone mass" (Felson, 1988: 10). Another large and well-controlled longitudinal study in Finland has also shown a strong (indeed, "linear") relationship of knee OA with obesity in women (Mannienen et al., 1996). Finally, a survey of a large population sample also found a strong correlation of obesity with knee OA, even after controlling for levels of serum cholesterol and uric acid as well as for diabetes and blood pressure (Davis et al., 1988).

While the above evidence clearly points to a strong relationship of obesity and knee OA, not all results from other joints have been this consistent (Moskowitz, 1993). In fact, the effects of obesity on hip OA are less obvious, and in are less clear yet in the hand (Silman and Hochberg, 1993). The inconsistencies in the findings are, no doubt, partly explained by the multitude of factors influencing the pathogenesis of OA in different joints. Moreover, obesity itself may have differential effects. As noted above, in weight-bearing joints biomechanical factors may predominate, but in other joints, "constitutional effects may be as important" (Hough, 1993: 1715). Additionally, there are also possible metabolic influences, perhaps tied to diet. Some experimental work, for instance, has shown that OA increased in mice fed saturated fat, but not in those fed unsaturated fat (Moscowitz, 1993). Finally, potentially linked to obesity, some other data suggest OA predilects in individuals of particular body conformations (Solomon et al., 1982).

Results obtained in skeletal populations could be influenced by some of these factors. Certainly, the high prevalence of OA among the Inuit might, at least in part, be influenced by diet, body conformation, and obesity. Without adequate controls, of course, we cannot be sure. However, just as obviously, we cannot be sure of mechanical explanations either.

Diet

For a number of years dietary influences on the development and patterning of OA in humans have been suggested, but have proven difficult to demonstrate. However, in two recent reports derived from the Framingham longitudinal investigations, intriguing dietary associations of micronutrients have been provisionally demonstrated. McAlindon and colleagues (McAlindon, Jacques, Zhang, et al., 1996) found that intake of Vitamin C appeared to slow the progression of knee OA and thus suggested that "high intake of antioxidant micronutrients, especially Vitamin C, may reduce the risk of cartilage loss and disease progression in people with OA" (p. 648).

Moreover, a second report (McAlindon, Felson, Zhang et al., 1996) indicated that reduced levels of Vitamin D were associated with increased likelihood of symptomatic OA progressing in the same population.

ETIOPATHOGENESIS OF OA: CONCLUDING COMMENTS

As the preceding pages make clear, OA is a condition influenced by a multitude of factors. Its etiology is strikingly multi-factorial and its pathogenic consequences variable. Both intrinsic joint components and extrinsic influences operate (Dieppe, 1990) although "their relative contribution may vary in different joints and in different forms of OA" (Hough, 1993: 1714).

The essential points relating to the complexity of understanding the etiopathogenesis of OA were nicely stated by Duncan:

> When one considers the number of joints which could be involved in different types of motion and trauma and then considers the duration of activity, the age, the sex, the genetic influence, the metabolic disturbances, etc., the complex etiology of osteoarthritis is evident (Duncan, 1979: 6).

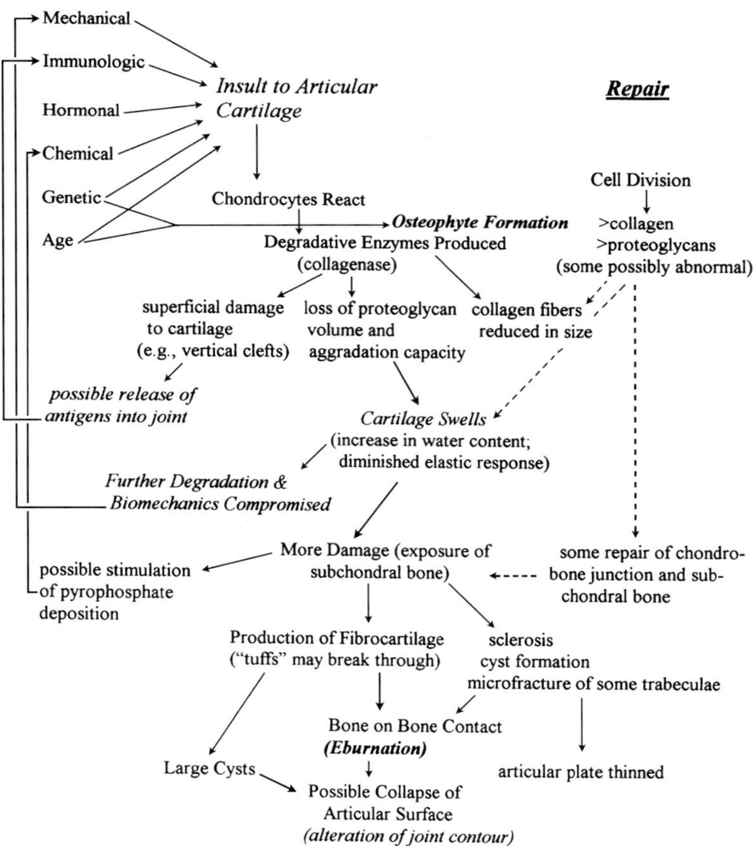

Figure 2-7 A model of the etiopathogenesis of osteoarthritis.

A schematic summary, reviewing many of the points raised earlier, and emphasizing several aspects of osseous involvement is shown in Figure 2-7. Osteologists, however, are not the only researchers who have ardently pursued evidence to substantiate the stress hypothesis. A considerable clinical literature concerning relative risks of OA from specific occupational and sports activities has developed. It is to this literature that we turn in the next chapter.

Chapter 3

Osteoarthritis and Activity: Occupational and Sports Studies

> The connection between sports and osteoarthritis is well known and it is assumed that what is well known is therefore true (Adams, 1976: 523).

INTRODUCTION

The relationship of osteoarthritis to activity has long been intensively researched in both industrial and sports settings. As the above quote indicates, and not just for sports activities, but for occupational ones as well, there has persisted a general *assumption* that a clear-cut relationship has been established. Certainly, osteologists routinely refer to the voluminous literature of this field, but, unfortunately, usually rather superficially. In my own early publications (Jurmain, 1977a,b), for example, only nine citations relating to epidemiological/occupational evidence of increased OA prevalence were included, and these were all studies finding positive results.

Nevertheless, there has for some time been anything but consensus among clinical researchers who have investigated links between activity and OA incidence. Moreover, even the most recent studies continue to produce mixed results, thus further emphasizing the complex nature of the etiopathogenesis of OA. It, therefore, appears somewhat overdue for a thorough review of the central literature, discussed in a format relevant to skeletal biologists.

Why have osteologists been particularly drawn to some of this literature, granted, however, from a somewhat biased perspective?

Waldron nicely summarizes both the underlying motivation and inherent bias:

> Were it really possible to be able to determine the occupation of past populations from the pathology in their skeletons this would be immensely useful and would be of great value to the paleoepidemiologist. And because the prospect is so alluring, the premises which underlie this assumption are not always carefully examined; if they are, then the proponents of the idea have not always wished to advertise their conclusions very widely (Waldron, 1994a: 92–93).

Recognizing what Waldron labels the "alluring prospect," it is all the more important for osteologists to be familiar with contemporary research, both that attempting to link degenerative changes with specific occupations/sports activities and, more generally, with increased levels of biomechanical stress.

An immediate issue raised repeatedly in Chapter 2, and recognized now for several decades among epidemiologists, is the multifactorial nature of OA. Adams commenting on simplistic interpretations of OA notes "wear and tear alone as a cause of osteoarthritis no longer appears acceptable" (1979: 187). He goes on to emphasize numerous other contributory mechanical factors such as altered joint biomechanics (resulting from ligamentous laxity), intra-articular loose bodies, and congenital/developmental abnormalities.

Writing at about the same time, an epidemiologist often cited by osteologists further cautioned:

> When one comes to consider degenerative joint disease itself, the role of usage becomes an issue of some debate .... Whether usage accelerates cartilage degeneration or is even a primary pathogenetic result is not clear. What is further confusing is the role, if any, of the pattern of usage on the pattern of DJD (Hadler, 1977: 1019).

The obviously complex etiopathogenesis of OA makes any simple conclusions regarding functional aspects all the more tenuous, as pointed out in a more recent review:

> A traditional understanding of DJD was that this was a 'wear and tear' phenomenon that accompanied aging or unusual circumstances. As noted, it is now appreciated that a variety of physical, biomechanical, occupational, and sports related conditions, environmental conditions, genetic factors, trauma, inflammation, congenital abnormalities, and nutritional and metabolic conditions are relevant (Panush and Inzinna, 1994: 1).

Several other recent reviews have further addressed these issues and have almost universally reached quite discouraging conclusions, as Mankin and associates stated succinctly in one collaborative effort: "Very little epidemiologic data exist to show that chronic occupational trauma or even acute injuries lead to OA" (Mankin et al., 1986: 1139). Moreover, as pointed out in the prior chapter, the uncertainties relate most directly to lack of understanding of joint physiology and OA pathogenesis, points further reinforced by Panush and Brown (1987: 58–59): "The effects of long term exercise on joint lubrication, local inflammation, microfractures, and ageing of cartilage are unknown."

It is no surprise, then, that the results of epidemiological studies are frequently contradictory, prompting two researchers to observe: "... when reviewing recent literature on the implied relation between exercise and degenerative joint disease, one is amazed in observing how contradictory the current opinions are" (Gremion and Chantraine, 1990).

Another leading epidemiologist further noted the frequent lack of good correlations and some of the inherent difficulties:

> Taken as a whole, currently available epidemiologic evidence is somewhat disappointing in that it has not fulfilled the hopes it raised. In spite of the considerable amount of work involved in this type of study and the vast quantity of data accumulated, it has mainly confirmed commonplace knowledge, i.e., that OA is influenced by age, sex, and, to some extent, mechanical factors. Even the *possibility* that OA may be due solely to mechanical overuse has not been established convincingly [emphasis added] (Peyron, 1986: 17).

In his overview of the literature, Tony Waldron also reached a similar conclusion:

> The inescapable impression which one gets from a review of the available literature, however, is that there is no consistent evidence of a *consistent* relationship between a particular occupation and a particular form of osteoarthritis and, given the multifactorial nature of the condition, that is probably what one ought to expect. Perhaps the most which can be said is that in some occupations in which repetitive and strenuous movements of a few joints occurs this movement will determine which joints are affected in those predisposed to the disease (Waldron, 1994a: 94–95).

In yet another recent review Oddis (1996) similarly concluded that OA is a degenerative process "acquired because of metabolism, mechanical, genetic, and other influences." He went on to emphasize that OA is apparently not increased by normal joint use, "but persons who participate in competitive sports, or who play with abnormal or injured joints are at increased risk." This latter point is in agreement with summary views expressed by Panush and Inzinna who argued that many studies "support the concept that certain physical activities over many years can accelerate osteoarthritis in selected joints of susceptible individuals" (Panush and Inzinna, 1994: 2).

Further, Lane (1995) also concluded that individuals with abnormal or injured joints, who participate in sports, particularly at an elite level "appear to be at increased risk of developing OA." Genti (1989) similarly suggested that in *some* cases forces common in certain occupations or sports might cause tissue injury, "thus increasing the development of OA." However, given the considerable disagreement within the field and conflicting results, he remained cautious, reminding us again that, as of yet, "Epidemiological investigations fail to provide sufficient proof as to whether mechanical overloading of occupational origin, or due to sports, will lead to a higher incidence of OA."

Indeed, even the association of presumably clear cases of repetitive trauma related to OA has not been particularly well confirmed and is still largely anecdotal. "In fact, although anecdotal evidence for the association of OA with trauma is quite strong, there are few reliable and definitive epidemiologic correlative studies" (Mankin et al., 1986: 1139).

Clearly, most people do not develop OA, even when apparent environmental conditions would suggest risk is high. The opposite hypothesis, i.e., that joints are *well-adapted*, even to unusually high stress loads, however, is also not yet fully established. "Although it appears that normal joints tolerate prolonged and vigorous exercise without adverse consequences, data to substantiate this assertion are not yet available" (Mankin et al., 1986: 1148).

Nevertheless, there are now considerable data to show that in certain instances involving long-term and relatively high levels of mechanical loading OA does *not* usually develop and certainly does not do so at an increased incidence relative to controls. These negative findings concerning the relationship of OA and activity are crucial in weakening central tenets of the "stress hypothesis," at least as conventionally stated and utilized by osteologists.

EXPERIMENTAL STUDIES

There has been a variety of experimental work performed on laboratory animals, usually involving some form of surgical intervention (as in the knees of dogs) or of artificial immobilization (as in rabbits). Some interesting, although anecdotal and poorly documented associations have been suggested relating purported increased OA in the hip and shoulder of sled dogs; the forelimbs of racehorses; hindlimbs of workhorses; and forelimbs of tigers and lions (Panush and Brown, 1987; Eichner, 1989).

One study of sheep locomoting on asphalt surfaces (particularly focusing on changes in subchondral trabecular bone) found a correlation (Radin et al., 1979), but these results were not confirmed in a later study of rabbits exercised on treadmills (Videman, 1982). Indeed, one review concluded from such contradictory results that "experimental observations have not found normal joint motion in exercising animals to be harmful to joints" (Panush and Brown, 1987: 54).

This same view was recently further confirmed in a comprehensive longitudinal study assessing the effect of weight-bearing exercise over a full ten-year span (Newton et al., 1997). Observations, even after such a long period of exposure (75 minutes per day, five days per week, for 527 weeks) showed that the exercised animals had no increase in OA compared to caged controls. The lack of any degenerative correlation was found on gross, microscopically, and in terms of functional (mechanical) capacity of the articular cartilage. The authors thus concluded: "These results show that a lifetime of regular weight-bearing exercise in dogs with normal joints did not cause alterations in the structure and mechanical properties of articular cartilage that might lead to joint degeneration" (Newton et al., 1997: 282).

PRIOR INJURY

Despite certain reservations by a few researchers relating to the *quality* of documentation (as noted above), most investigators agree that prior injury is an important contributor to the onset of OA, especially in individuals engaged in strenuous activities. J.A.D. Anderson (1984) recognized two different levels of trauma, either of which could initiate OA:

(1) Acute, i.e., a quite severe episode, involving stresses that would be "akin to those caused by impact in a traffic accident or achieved by jumping from a height of around 5 m" (p. 431).
(2) Long-term fatigue of joint/lubrication, also which can accelerate development of OA. Anderson suggests this type of loading has usually more to do with posture than lifting and is probably more tied to mobility of a joint as opposed to "degree of effort."

A few well-controlled studies have helped confirm the role of *acute* trauma. The supposed and much-argued influence of *chronic* trauma is at the heart of the debate discussed throughout this and the succeeding chapter.

In one longitudinal study of 86 patients with ligamentous injuries (mostly from football or downhill skiing), after 4–5 years, 42% showed chondromalacia of the patella, and 52% had some radiological abnormality (Warren and Marshall, 1978). A similar longitudinal, prospective study of knee OA following menisectomy traced patients at 4.5 and again at 14.5 years (Jorgensen et al., 1987). Indications of knee instability in the post-operative patient sample rose from 10% (at 4.5 years) to 36% (at 14.5 years). In addition, radiographic changes increased over the same period from 45% to 89%. Using the most rigorous radiographic criteria for defining OA, 8% of the patient sample developed knee OA, and the degeneration more frequently followed lateral as opposed to medial menisectomy. (Note: See Chapter 2 for a discussion of the radiographic criteria for diagnosing OA as well as information relating to general epidemiological methodologies.)

Prior fracture involvement may also increase the risk of athletes developing OA, especially when ligamentous instability or abnormal joint motion occur (Panush and Lane, 1994). Overall, however, the link between trauma and secondary OA is not yet clearly enough established to specify the risk, but it does appear that in younger individuals, post-traumatic OA does not develop, unless the fracture involves the epiphysis (Wright, 1990). As discussed in the prior chapter, and again further below, the age of onset of such injury as well as other types of mechanical stress, is an important consideration.

GENERAL EPIDEMIOLOGICAL STUDIES OF ACTIVITY AND OA

As noted, there is a considerable literature relating to a variety of presumably stressful activities and their relationship to the development of OA. Such studies have focused on several different forms of repetitive stresses encountered in occupational or in athletic contexts and have investigated OA involvement in numerous joints. Clearly, there is little consensus on the significance of the findings, particularly as results are highly mixed. Nevertheless, numerous studies, some quite well controlled, have found positive correlations of OA and various activities.

Studies Reporting Positive Correlations of Activity and Increased OA

Some of the earliest systematic research, using large samples and appropriate statistical controls, were the ground-breaking studies by

J.H. Kellgren, J.S. Lawrence and their associates (Kellgren and Lawrence, 1952; Kellgren, 1961; Kellgren et al., 1964; Lawrence et al., 1966; Lawrence, 1969). In particular, this research concerned the prevalence of OA in industrial settings, focusing on large samples of miners and dockworkers (compared to large control groups of more sedentary workers); for the studies of miners, total samples were approximately 7,000 individuals. Assessment involved questionnaires, clinical exams, and extensive radiography. The strongest correlations were found for spinal disease, especially disk degeneration (Note; not OA, strictly speaking; see Chapter 2). Nevertheless, for peripheral involvement, there was slightly higher prevalence in the knees of miners than in controls, but most of the involvement was slight. The authors concluded that these slight degenerative changes seen in the knee were most likely the result of prior injury and increased body weight, but *not due to kneeling*. (Interestingly, while the work of Kellgren and Lawrence regarding spinal involvement is often cited in publications written by paleopathologists, the outcomes of the research actually relating to OA are often ignored; see below for further discussion.)

Anderson et al. (1962) in another sample of Scottish miners (data including only reports of complaints) found similar results, with the highest incidence of complaints, mostly again apparently due to disk degeneration, among those miners working directly at the coal face. Similarly, Partridge and Duthie (1968) also found higher incidence of back complaints in dockworkers. This latter study also investigated peripheral OA and found a higher prevalence in dockworkers, compared to office workers, for multiple-joint involvement, especially of the hand. For single joint involvement, however, the incidence was similar in the two groups of workers.

In a more recent study, and one specifically looking at peripheral OA, Lindberg and Montgomery (1987) found among Swedish shipyard workers a somewhat higher incidence of knee OA in those workers performing "heavy labor" for more than 30 years (compared to white-collar workers; $p < 0.05$). Finally, Videman et al. (1990) in a more restricted study, using a relatively small sample of cadaver materials, also reported higher spinal involvement (disk degeneration, VOP, and OA) among individuals reported (by family interviews) to have been engaged in "heavy work."

It must be noted, however, that most of this research (excepting Lindberg and Montgomery, 1987) on miners and dockworkers have concentrated on spinal involvement, especially self-reporting of back pain. Given the considerable ambiguities and strong age-correlation demonstrated for degenerative spinal disease (see Chapter 2), these results are of limited utility.

The most comprehensive *general* research on the relationship of OA to occupation has come from analysis of knee OA from HANES I data (Anderson and Felson, 1988). In this analysis of a very large population sample ($N > 5{,}000$), occupation was classified using broad (Department of Labor) categories, assessing particularly the amount of knee-bending; all subjects were radiographed for degenerative knee changes, and possible confounding variables (age, sex, race, weight, educational level, socioeconomic status) were controlled. However, occupation was classified only as to current status, and thus no data on prior occupations, length of exposure, or age of onset were available. Moreover, presence of prior injury was not included.

Nevertheless, for a large sample, the controls used represent some of the most systematic approaches yet attempted at this scale, and the results are thus informative. Among males in the 55–64 age group the investigators found an odds ratio of 2.45 in males participating in more strenuous occupations and an even higher odds ratio (3.49) in this same age group among more strenuously employed females. The Department of Labor occupational categories also contained some information on "strength" demands, but this factor was not significantly correlated with OA in males (but was so in females; odds ratio = 3.13).

From the demographic profile of affected individuals, Anderson and Felson suggested that duration of exposure may be important, with degenerative changes appearing only after a long period of repetitive occupational exposure. They thus concluded "a substantial proportion of osteoarthritis of the knee is associated with occupational overuse" (Anderson and Felson, 1988: 188). It will be most illuminating to see whether the HANES II data also follow similar patterns (as of this writing, analysis of occupational correlations for HANES II has not yet been reported).

Another broad retrospective population study of occupational influence on OA was done in Sweden and included census data on more than 250,000 individuals (Vingård, Alfredsson, Goldie et al., 1991). Hip and knee OA were assessed from comprehensive medical records, and occupation, based on exposure to mechanical forces, was broadly categorized as to "high" and "low" workload. Results showed higher incidence of hip OA in males among farmers (discussed further below), construction workers, firefighters, and food processors. Among females, house cleaners had higher rates of knee OA and mail carriers a higher incidence of hip OA. Here, as with the U.S. (HANES I) analyses, careful controls were attempted, also including in the Swedish study, partitioning the sample by geographic area. The authors concluded: "The findings support the hypothesis that heavy physical work load contributes to osteoarthrosis of the hip and knee" (Vingård, Alfredsson, Goldie et al., 1991: 1025).

The investigators were, however, quite cautious in not drawing *specific* hypotheses regarding functional demands and clearly recognized the multifactorial nature of OA: "Physical overuse might be *one* of the aetiological factors" [emphasis added] (Vingård, Alfredsson, Goldie et al., 1991: 1030).

While the results of such large retrospective studies as those just discussed provide much insight, for a progressive disorder such as OA, longitudinal data are yet more informative. The largest and most long-term analysis of this kind is part of the Framingham study (Felson et al., 1991). Indeed, except for some more restricted studies in sports settings (especially regarding runners) and mostly short-term follow-up analyses following surgical patients, the authors of the Framingham OA analysis correctly claim their results as "the first longitudinal population study examining the relationship between occupational physical demands and OA" (Felson et al., 1991: 1589).

In the Framingham research, assessment included periodic clinical exams over a number of years as well as intermittent radiographs. For diagnostic purposes, OA was determined radiographically and included only more unambiguous features (i.e., narrowing of joint space).

The analysis failed to find a correlation of occupation and OA in females, but consistent occupational data for women were limited. Among men, however, a consistent correlation was found, especially relating to jobs involving strenuous labor *combined* with knee-bending. Moreover, knee OA was more frequently expressed in individuals who were continuously, as compared to intermittently, employed in strenuous occupations. The authors concluded, "Thus among men, occupation which combines knee bending and physical demands may be an important cause of idiopathic OA" (Felson et al., 1991: 1587).

Although the apparent correlation of general occupational stress among the Framingham males supports certain (general) aspects of the stress hypothesis (and these data are, in fact, probably the best that do), the situation is not as clear-cut as first impression might suggest. A later study (Hannan et al., 1993), conducted, in fact, by some of the same researchers on the very same sample, assessed more general levels of physical activity, determined longitudinally through questionnaire, and its relationship to radiographic knee OA. Here, in addition to work-related tasks, leisure-related activities were also addressed.

However, and somewhat disturbingly, the results are at direct odds with the strictly occupationally related findings just discussed. Conversely, for general activity, the authors conclude:

> There is no increase in the risk of knee OA with increasing physical activity in either men or women (Hannan et al., 1993: 704).

> In summary, although knee OA is often presented as a wear-and-tear disease, our data do not suggest that high levels of habitual, usually weight bearing activity increases rates of knee OA (p. 708).

Why the contradictory findings from the same data set? The authors of the second study (Hannan et al., 1993) suggested that the occupational data (Felson et al., 1991) might reflect more accurately activities leading to overuse, while habitual (more general) activities may be more non-specific (and might include a variety of activities not promoting overuse). Moreover, it could be added that the more general study on activity might reflect usage later in life, while the occupational data might more accurately include habitual stresses perhaps starting earlier. Although, even here with the most rigorous longitudinal controls yet attempted, good information on duration of use and age of onset is still nearly entirely lacking.

As emphasized, the Framingham longitudinal data and rigorous methodological protocols provide the most controlled approach currently available. Nevertheless, the contradictory results which have emerged thus far are a sobering reminder that even with superior clinical and epidemiological documentation, the complex phenomena leading to OA are difficult to control and interpret meaningfully. The lesson for skeletal biologists working on archaeological samples is obvious.

While not as rigorously longitudinal (and prospective) in approach as the Framingham research, a few clinical follow-up studies have contributed to understanding of some possible activity-related features of OA. Kohatsu and Schurman (1990) examined a small sample of patients ($N=46$) requiring knee replacement (matched with controls) and found increased likelihood of such severe knee OA in individuals who were obese (3–5 times risk), who had engaged in "heavy work" (2–3 times risk), or, particularly, who had a history of prior knee injury (5 times risk). In a similar, but larger, study of a male Swedish arthroplastic patient sample ($N=239$), Vingård, Hogstedt, Alfredsson et al. (1991) determined from work history questionnaires that risk was significantly increased in individuals who had experienced high static work loads (odds ratio $=2.42$) and even more so in those who regularly performed heavy lifting (O.R. $=3.31$). In this retrospective analysis, long-term exposure appeared to be especially important. The same patient sample was also investigated for history of sports activity (Vingård et al., 1993), and again positive correlation of hip OA was found with activity (assessed as hours of exposure to sports of all kinds). Indeed, for those individuals with the most frequent exposure, as compared to matched controls, the odds ratio of developing hip OA was 4.5. Some, mostly anecdotal, data on type of sport was also recorded, and from these data, track and field and racquet sports appeared to be the most hazardous. When the risk of increased sports activity was assessed in combination with occupational history (with high physical loads from both), there was a striking odds ratio of 8.5 of developing hip OA (Vingård et al., 1993; Olsen et al., 1994).

Similar results were also reported by these same researchers in a study of a female patient sample which had undergone total hip replacement surgery (Vingård et al., 1997). In women with severe hip disease an association was found with higher exposure to physically demanding tasks, at work, at home, as well as in sports activities. In fact, the highest risks were noted in those individuals "who jumped or moved between levels, who frequently climbed stairs, *and* who had physically demanding tasks outside occupational life" (p. 293) [emphasis added].

These retrospective studies suffer from inherent biases, most especially the study population is self-selected and adequate controls are difficult, if not impossible, to establish. Other retrospective investigations, using questionnaires or hospital records, are similarly limited. Still, they provide some insight into the relationship of activity and OA. Jacobsson et al. (1987) found that in a small sample ($N=85$) of males with severe hip OA, compared to controls, affected individuals had been more frequently exposed to heavy labor, heavy lifting, and frequent tractor driving (perhaps, reinforcing other results obtained for farmers; see below). Kujala et al. (1994) reviewed Finnish hospital admissions records of elite athletes with OA of the hip, knee, or ankle (compared to controls) and found a slightly increased risk of required hospital care among the athletes. Similarly, Kujala et al. (1995) compared knee OA among 117 male former athletes. In this investigation various sports activities were classified and corresponding knee OA prevalences were found as follows:

weight lifters (31%)
soccer players (29%)
runners (14%)
shooters (3%)

This study also separately assessed tibio-femoral as compared to patellofemoral involvement. Interestingly, soccer players had the most tibiofemoral disease, while weight-lifters displayed the most patellofemoral involvement. Moreover, the authors suggested that among soccer players previous injury was the most important contributing factor, but among weight lifters body mass (associated with kneeling/squatting) was more influential.

Another recent retrospective study of female former elite athletes in England also found similar patterns of hip and knee OA. In this radiological survey of 67 former runners and 14 tennis players, higher prevalence of OA, as compared to controls, was found at all sites, but most particularly, for osteophyte development in the knee (tibiofemoral and patellofemoral compartments). The authors concluded that "Weight-bearing sports activity in women is associated with a 2–3 fold increased risk of radiologic OA (particularly the presence of osteophytes) of the knees and hips" (Spector, Harris, Hart et al., 1996: 988).

English patients, reporting knee pain and showing radiographic signs of knee OA (Cooper et al., 1994), in yet another retrospective study, also showed association with increased occupational stress (as compared to controls). From questionnaire data, lifetime occupational exposure was assessed, including data on eight specific activities (squatting, kneeling, stair-climbing, heavy-lifting, walking, standing, sitting, and driving). Results showed strong correlation with regular squatting (O.R. = 6.9), kneeling (O.R. = 3.4), and stair-climbing (O.R. = 2.7). Cooper and colleagues also concluded that heavy lifting, while not independently increasing risk, "appears to augment the risks associated with kneeling and squatting" (p. 92). This study also separately evaluated the major anatomical compartments of the knee and found the strongest occupational associations in the medial tibiofemoral compartment.

In a final general study of occupational associations with OA, Stenlund (1993) focused on acromioclavicular involvement among construction workers (and collaterally assessed history of sports participation). Results showed that bricklayers and rockblasters had higher incidence of radiographic OA than foremen, and particularly so as associated with number of years working and amount of lifting (but apparently among the rockblasters not from vibration of pneumatic tools; for further discussion, see below). These findings, in this traditional cross-sectional investigation, are particularly tentative, as a dated radiographic method of assessment (Collins, 1950) was employed.

General Epidemiological Studies Finding Negative Correlations with Activity

A systematic overview of the relevant literature, especially those reports in the last decade, would easily find almost as many *negative* correlations of activity with the development of OA. These more general studies, discussed below, employ equally well-controlled and systematic methodologies as those discussed above which have found positive correlations. Both those general surveys which have found positive associations, as well as those that have failed to do so, are summarized in Table 3-1. As noted, the pioneering work by Kellgren, Lawrence, and associates is often cited by osteologists, most usually relating to the positive association of back complaints and presumed activities of miners and dockworkers. However, the results relating to activity and peripheral OA were more disappointing. For example, in the best controlled study (of knee OA) in which specific mining and dockworker activities were assessed, Lawrence (1955) found no association with kneeling, stooping, distance walked at work, heavy lifting, or duration of work in wet conditions. Excepting a few instances of secondary OA following prior injury, Lawrence (1955: 260) concluded: "We have found no factor in mining which appears causally related to osteoarthritis of the knee."

Likewise, Anderson and Duthie (1963) did not find a correlation among dockworkers for incidence of complaints (for the back or limbs) as related to "heaviness of work;" and Lindberg and Danielsson (1984) also found essentially the same prevalence of hip OA in Swedish shipyard workers as in (white collar) controls. Anderson (1974) in a general review of English occupational data did find some slight increase in number of back complaints with "heavier" work, but again found no evidence suggesting increased risk of limb OA. Sairanen et al. (1981) in a study of Finnish lumberjacks (compared to office workers) surprisingly found no correlation of activity with back complaints, but did find a slight increase in frequency on X-ray of disk degeneration (which, however, was *not* related to length of exposure). Perhaps, even more surprisingly, the prevalence of radiographic hip OA was actually *less* in lumberjacks than in controls, while knee OA was only slightly higher in those doing felling work (5% compared to 3%, and not statistically significant). From such radiographic findings, Moskowitz (1993: 1739) concluded: "In summary, there is no evidence suggesting that exercise is deleterious to normal joints, when the criterion used is loss of roentgenographic joint space."

In a more general epidemiological survey, also in Finland, Julkunen et al. (1981) failed to find a correlation between OA or

Table 3-1. General epidemiological surveys relating to Activity and OA

Study group/Activity	Joint(s) evaluated	Reference
Studies finding positive association:		
General (HANES I)/ by occupation (knee-bending)	Knee	Anderson and Felson (1988)
General (Framingham)/ by occupation (knee-bending)	Knee	Felson et al. (1991)
General survey/ "high static work loads;" "heavy lifting"	Hip	Vingård, Hogstedt, Alfredsson et al. (1991)
General (patients w/ pain); various activities (includ. squatting, kneeling, etc.)	Knee (all compartments)	Cooper et al. (1994)
Farmers, construction workers, firefighters (males)	Knee, Hip	Vingård, Alfredsson, Goldie et al. (1991)
Mail carriers, cleaners (females)	Knee, Hip Knee	" "
General/sports activities (all kinds)	Hip	Vingård et al. (1993)
General/occupational/ home/sports (females)	Hip	Vingård et al. (1997)
General (New Haven)/ handedness only	Hand	Achinson et al. (1970)
Athletes (elite)/various sports	Hip, Knee, Ankle	Kujala et al. (1994)
Post-operative/ "heavy work"	Knee	Kohatsu and Schurman (1990)
Outpatient/"heavy mechan. load occupations"	Hip	Roach et al. (1994)
Former athletes(male)/ various sports	Knee (Tib/Fem; Pat/Fem)	Kujala et al. (1994; 1995)
Former athletes (female)/ various sports	Hip, Knee (Tib/Fem; Pat/Fem)	Spector, Harris, Hart et al. (1996)
Cadaver/general occupation	Spine (disk degen., VOP, OA)	Videman et al. (1990)
Patients w/ severe disease/"heavy labor, heavy lifting"	Hip	Jacobsson et al. (1987)

Table 3-1. *(Continued)*

Study group/Activity	Joint(s) evaluated	Reference
Coal miners/complaint reporting only	Disk disease	Anderson et al. (1963)
Miners	Hip, Knee, Shoulder	Schlomka et al. (1955)
Construction workers	Acromio-clavicular	Stenlund (1993)
Shipyard workers	Knee	Lindberg and Montgomery (1987)
Studies failing to find association:		
General (Framingham)*/ by general activity	Knee	Hannan et al. (1993)
General survey/ occupation and sports activities	Knee	Julkunen et al. (1981)
General survey/ manual workers	Limbs	Anderson (1974)
General survey/by occupation	Knee	Schouten et al. (1992)
Miners, dockworkers	Knee	Lawrence (1955)
Dockworkers	Spine, Limbs	Anderson and Duthie (1963)
Physical education teachers (female)	Knee	White et al. (1993)
Male athletes (veteran)	Spine	Healy et al. (1996)
Weavers and other occupations	Hand	Waldron and Cox (1989)
Athletes (runners—Stanford)/ handedness by occupation	Hand	Lane et al. (1989)
Aging survey/various morphological indicators (mineral density, forearm circum.)	Hand	Hochberg et al. (1991)
Studies with mixed results:		
Miners (compared to office workers)	Disk degeneration, Knee	Kellgren (1961)
Dockworkers (compared to office workers)	Spine, Elbow, Hand, Knee	Partridge and Duthie (1968)

*Same study group as in Felson et al. (1991).

"soft tissue rheumatism" and working conditions, although, interestingly, they did find a higher incidence in those individuals engaged more actively earlier in life in sports (and thus possibly indicating an etiological factor related to age of onset).

Three more recent studies have all also failed to find evidence to support the stress hypothesis. In a general population survey in the Netherlands, Schouten et al. (1992) did, not surprisingly, find correlations of knee OA with age and weight. They, however, did not find any statistically significant correlations with gender, meniscectomy, uric acid concentration, smoking or, occupation (activity), "except possibly standing."

Similarly, in another partially longitudinal study of physical education teachers, compared to controls, White et al. (1993) found among the more active group lower rates of knee OA, more lumbar spine involvement, and a similar frequency of hip OA. At a 12 year follow-up the physical education teachers actually displayed *less* joint pain and less stiffness than seen among the controls.

In another study of older male athletes engaged in "lifelong" sports participation the incidence of spinal involvement, as determined by MRI for disk herniation, spondylosis, and spinal stenosis, was no more frequent than in less actively engaged population samples (Healy et al., 1996).

Lastly, in their systematic and innovative study of the Spitalfields skeletal materials, Waldron and Cox (1989) did not find any correlation of occupation with OA. They focused primarily on any association between those individuals who were documented to have been weavers with degenerative hand changes. However, here, as elsewhere in the skeleton, no occupational pattern was found. "Further analysis failed to show a relation between occupation and osteoarthritis at any site" (Waldron and Cox, 1989: 420). Even at the gross level of grouping manual *vs.* non-manual workers, "there was no evidence that manual work *per se* was an aetiological factor" (p. 422), leading the authors to conclude that the main result "seems at variance with the generally held view that the likelihood of developing osteoarthritis is directly related to the physical effort involved in work" (p. 422).

SPECIFIC STUDIES OF OCCUPATIONAL AND SPORTS-RELATED STRESS

In addition to the more generalized attempts to correlate activity with OA, an impressive variety of studies have been done seeking to link particular activities with OA in various joints. This research is especially relevant to osteologists, as such analyses attempt to isolate *specific* activities and relate them to their possible influence on degenerative disease. Thus, these more focused epidemiological studies are analogous to the attempts frequently made by skeletal biologists to relate degenerative skeletal lesions to specific ancient activities. Indeed, this same clinical/epidemiological literature is oftentimes cited by osteologists as providing substantial support for the stress hypothesis in general and osteological interpretations in particular (although, as noted above, such references are frequently highly selective).

Investigations of Hand Involvement

The most complete and often-cited study of occupational influences on hand OA was carried out by Hadler et al. (1978). In this survey of long-term (>20 years employed) textile workers, three different kinds of workers were identified: burlers, winders, and spinners. Evaluation was done on both hands by clinical exam as well as radiographically. Results showed a greater preponderance of degenerative changes in the right compared to the left hand as well as some interesting correlations apparently related to the *type* of task performed (e.g., more OA in the wrist of winders; more involvement of digits 2 and 3 in burlers and spinners). These correlations led the researchers to conclude that "highly significant task-related differences were demonstrated. These task-related differences in the structure and function of the hands were consistent with the pattern of usage" (Hadler et al., 1978: 210).

Another strong aspect of the study was the sophisticated statistical tests (including multivariate methods) employed to evaluate significance. Nevertheless, the small sample size ($N=64$) and complexities of etiopathogenesis, relevant especially to hand OA, make these results quite tentative, a view also voiced by the principal investigators: "We urge caution in extrapolating these findings until they are confirmed in similarly designed studies in other populations"(Hadler et al., 1978: 219).

In fact, some other research of hand involvement does at least partly corroborate the findings of Hadler and associates. For example, the general epidemiological New Haven survey of a quite large sample (with >1400 radiographs taken of both hands) found a higher frequency of right hand involvement, as well as, not unexpectedly, higher prevalence in women. Here, as in the majority of other studies of hand OA, the greatest involvement was seen in DIPs.

A more recent study of a relatively small sample ($N=36$) of rock climbers, Bollen and Wright (1994) found a higher prevalence (on X-ray) of both subchondral cysts and osteophytes among climbers, as compared to controls. Interestingly, and relevant to attempts to correlate activity with bone density/geometry (see Chapter 7) as well as enthesophytes (see Chapter 5), among the rock climbers, increased cortical thickness (of middle phalanges) as well as thickening of the distal ends of the fibrous portions of the flexor fibrous sheaths was observed. Moreover, Nakamura and colleagues (1993) found a higher prevalence of Heberden's nodes in Japanese school cooks, as compared to women presumably engaged in less demanding hand-related tasks.

Nevertheless, other epidemiological findings relating to hand involvement do not support the admittedly tentative findings noted above. Lane et al. (1989) in a systematic radiographic study did not find a higher prevalence of OA related to handedness. Moreover, OA was *very* common in all individuals surveyed (some changes seen in 133 of 134 individuals surveyed); accordingly 95% of those individuals with hand OA were classified into "non-physical" occupations.

In addition, an earlier survey of cotton textile workers (Lawrence, 1961) did not find any strong correlation of activity with increased hand OA, and as noted above, Waldron and Cox (1989) also did not find a relationship of hand OA and an occupational history of weaving.

The various studies of hand involvement are clearly contradictory, but the balance of evidence suggests the hand is not a particularly good indicator of activity-related factors. Indeed, given that other, more systemic influences probably predominate in the etiopathogenesis of hand OA (see Chapter 2), the failure to establish consistent relationships is not surprising. Other factors may also be at work here. Glick and Parhami (1978) found few cases of unilateral hand involvement (again, not surprising, considering the likely role of genetic factors), and they further suggested that frostbite might initiate erosive lesions of the hands (an influence potentially important in population surveys of the Inuit). In addition, as in other joints, age is strongly correlated with the progression of hand OA (Kallman et al., 1990). But, more surprisingly, other factors, apparently significant in other joints, seem to play less of a role in hand OA. Hochberg et al. (1991) did not find a relationship of hand OA with body mass (obesity), bone mineral density, body fat distribution, grip strength, or forearm circumference.

Thus, while some systemic factors (age and genetic influences, the latter as related to generalized OA) are clearly highly significant in the etiopathogenesis of hand OA, others (obesity) may not be.

Furthermore, the role of localized mechanical factors (correlated with activity) in the hand is not well-established, nor does it seem likely to be.

Pneumatic and Other Mechanical Tool Users

The mechanical stress placed on the upper limb by heavy (and vibrating) mechanical tools, especially pneumatic drills, might be expected to lead to degenerative disease. Certainly, if the impulse loading hypothesis as developed by Radin and associates (Radin and Paul, 1971; Radin et al., 1972; see Chapter 2), were accurate, users of such implements *should* show some consistent pattern of involvement related to exposure.

Early studies, conducted in Europe (Fischer, 1932; Linde, 1932; Vossenaar, 1936) initially seemed to confirm the relationship of pneumatic tool operation and OA (especially of the elbow). The relationship appeared most clearly in those contexts where *duration of use* was controlled (Fischer, 1932; and later as confirmed by Lawrence, 1955 and Roche et al., 1961). Further, in addition, to the elbow, Kouba (1967) found some association of pneumatic tool use and OA of the shoulder and hand; Schumacher et al. (1972) found a possible association with OA of the wrist; and Bovenzi et al. (1980) also found a correlation with involvement of the wrist, hand, and shoulder.

As might be expected, results are not entirely consistent. An early study (Hunter et al., 1945), well-controlled and using a large sample ($N=286$), did not find an association of OA with pneumatic tools: "there was no convincing evidence that they [arthritic changes] can be brought about by use of pneumatic tools" (Hunter et al., 1945: 16).

Along these same lines, in a more recent investigation, Stenlund et al. (1992) found an increased incidence of acromio-clavicular joint OA in operators of pneumatic tools, but only when *also* associated with other activity risk factors such as heavy lifting. In this latter, more controlled study, when these other factors were separately evaluated, vibration alone was not found to contribute to increased OA.

Moreover, in the most systematic review of this specific literature, Burke et al. (1977) critique much of the earlier work as being mostly of an anecdotal nature, and thus lacking appropriate controls. In their own quite small ($N=34$) but intensive study, employing clinical exams as well as radiographs, these researchers could not establish a strong relationship of pneumatic drill use with OA of the cervical spines, shoulder, elbow, or hand.

Surveys of occupationally related OA risks of other heavy equipment use have also produced mixed results. Mintz and Fraga (1973) in a very small sample ($N=3$) observed a modest relationship of elbow involvement among foundry workers using tongs for lifting (while twisting). Among a sample of symptomatic ($N=35$) lumberjacks using chainsaws for 7–20 years, however, Kumlin et al. (1973) found almost no diagnostic evidence of hand OA (surprising, when considering these individuals presented with symptoms). Likewise, as noted above, the lack of association of increased OA in the back and lower limb of Finnish tree fellers (Sairanen et al., 1981) agrees with these results, although this latter investigation did not evaluate upper limb involvement.

Overall, the research concerning the effects of repetitive mechanical equipment use, especially pneumatic tool use, on upper limb involvement has yielded contradictory results. Some, mostly early, research showed fairly consistent associations, but more controlled research has failed to confirm this relationship, or has demonstrated that mechanical stresses are not primarily due to impulse loading (thus failing to support the hypothesis advanced by Radin and colleagues). Nevertheless, the balance of evidence does indicate a modestly strong relationship, especially in those situations where long-term and probably severe mechanical loading was involved (and also quite possibly requiring an early age of onset). Such a pattern is well in keeping with results observed osteologically in archaeological populations, most notably among the Inuit (see Chapters 2 and 4 for further discussion). These epidemiological investigations into the effects of mechanical equipment are some of the *best* data available and could give slight encouragement to those seeking to make behavior-based interpretations of OA in the past. These results are, however, hardly convincing. Commenting on these mixed and often negative findings, Panush and Lane (1994:84) note: "wear and tear may indeed predispose to OA, but this notion *should be considered tentative and not accepted uncritically*" [emphasis added].

Farmers

Probably the most consistent occupational association with OA (in this case, of the hip) involves farmers. Given that agricultural tasks are frequently both habitual and occupational as well as a "family" activity (i.e., participation by children as well), the analogy with pre- and proto-historic populations is, no doubt, better than most contemporary occupations, where activities are engaged in intermittently, usually only as adults. Thus, the results relating to OA in farmers are intriguing.

From general occupational data collected in France, Louyot and Savin (1966) found considerably more hip OA among agricultural workers (34%) compared to non-agricultural workers (7.8%). Likewise, Vingård et al. (1991b) also found a higher incidence of hip OA in farmers in individuals participating in "lower workload" occupations. In a much more controlled British study, Croft et al. (1992) confirmed these findings, especially among those who had farmed for more than 10 years. Lastly, in a retrospective Swedish study, Thelin (1990) reported that, again in long-term farmers (>10 years working), there was a 3.2 increased risk factor of hip replacement (compared to the general population).

The most detailed evaluation of possible contributory factors comes from the British report, in which Croft and colleagues attempted to control for specific farming-related activities (including heavy lifting, vibration from heavy machinery, walking over rough ground). No clear correlation with any specific activity, however, could be established, although the authors suggested "heavy lifting" may have been most significant. Still, as shown here in contemporary studies with reasonably precise controls, the association of OA with *specific* activities is extremely difficult to demonstrate (a point not to be lost on osteologists).

A further complication, indicating multifactorial causation, has recently been raised by Croft (1996). He noted that the samples from Jamaica (Bremner et al., 1968) and Nigeria (Ali-Gombe et al., 1996) were largely drawn from rural agriculturists; yet, both groups showed a *low* prevalence of hip OA.

Perhaps the most intriguing aspects of the pattern of hip OA involvement among farmers relate to duration of use, and most especially, *age of onset*. Both Thelin (1990) and Croft et al. (1992) report that OA is most frequent in individuals who farmed for more than 10 years, many of whom, no doubt, began some tasks as children. Croft et al. recognize the potential influence of an early age of onset of stress-related activity when they suggested:

> Another question as yet unanswered is whether risk relates particularly to activities at an early age. Many of our subjects started farm work in their early teens, when the hip joint is not fully developed. The hip may be particularly vulnerable to trauma or physical stress at this stage of life (Croft et al., 1992: 1271).

A similar hypothesis was proposed by Murray and Duncan (1975) as related to early sports activity and later increased prevalence of hip OA (discussed in Chapter 2). It appears that if hip OA is to appear other than as secondarily to a major traumatic causation or as idiopathically (and very late in life), it results from stress begun *early* in life. Repetitive activities, thus, can influence hip OA, but this joint in particular seems to be highly age-specific in its accommodation to such mechanical influences.

Curiously, osteological data for hip OA do not mirror the findings relating to involvement among contemporary farmers. Perhaps, however, this apparent dissimilarity might relate more to the types of lesions (particularly eburnation) frequently utilized by osteologists to diagnose OA (and its consequent relatively infrequent prevalence in archaeological samples; see Chapters 2 and 4 for further discussion). Of course, it is also possible that hip OA is underrepresented archaeologically due to the fragile nature of the proximal femur and acetabulum. Thus, differential preservation may give the *appearance* of a lower prevalence than was, in fact, actually the case among a prehistoric group.

Baseball Pitching and Other Sports-Related Stresses on the Upper Limb

Another possibly fruitful area of contemporary analogy with activities commonly practiced in the past can be gleaned from the sports literature, especially that relating to the effects of baseball pitching. The deleterious effects of pitching have been investigated in adults (college and professional athletes) and were recognized as early as 1941 (Bennett, 1941). However, degenerative involvement has been described most especially in children, producing a syndrome often referred to as "Little League elbow" (Adams, 1965). The ramifications of such severe, repetitive *and* early onset stress may well be especially relevant to understanding the pattern of degenerative elbow disease in archaeological populations (particularly among the Inuit). It must be noted as well that those results observed in adult pitchers also are relevant, as the majority of these athletes began throwing as children; thus, their joints reflect cumulative stresses incurred over perhaps decades of use.

What has been called Little League elbow actually involves various degenerative changes of the elbow joint. Adams (1965) characterized several types of lesions that he observed in young pitchers, including:

(1) medial epicondyle epiphysitis (i.e., tendinitis)
(2) more severe avulsion and fragmentation of medial epicondyle
(3) involvement of radial head and capitulum (particularly vulnerable if epiphysis is ununited), including traumatic osteochondritis dissecans

To these types of lesions, Kern et al. (1988) added the possibility of loose bodies in the joint, and others have noted that the radial head may fuse late or fail to fuse (King et al., 1969; Pappas, 1982); O'Neill and Micheli (1988), however, postulated that, with extreme use, the proximal radial epiphysis may actually be induced to close early.

King et al. (1969) found in a sample of 50 professional baseball players marked asymmetry in musculature (especially in the forearm) and frequent flexion contraction, such that 50% of the pitchers were unable to completely extend their elbows. They also noted usually mild valgus deformities (seen in 30% of pitchers) and medial traction spurs (small exostoses) of the ulnar notch (in 44% of the sample).

Much of the pain reported by pitchers (again, professionals) was attributed by Wilson et al. (1983) to the development of osteophytes on the posterior and particularly posteromedial aspect of the olecranon process, causing impingement of the anterior wall of the olecranon fossa. Finally, O'Neill and Micheli (1988) suggested that involvement may also include overgrowth of the radial head.

The majority of researchers consider the lateral lesions (to the capitulum and radial head) to be the more severe (Larsen et al., 1976) and they can, in the most severe cases, lead to fibrillation/exfoliation of the cartilage (Adams, 1965). With continued throwing, the condition can progress to degeneration of the joint surfaces and resulting joint incongruity (Pappas, 1982). Interestingly, this array of severe lateral degenerative changes denotes a very similar pathology to what Angel initially termed "atlatl elbow," and such severe elbow changes have been well documented among the Inuit (Ortner, 1968; Jurmain, 1978; Merbs, 1983; see Chapter 4).

The biomechanical stresses of pitching which can produce pathological changes have been studied by Albright et al. (1978) and in more detail by Gainor et al. (1980). Albright and associates observed, mostly anecdotally, that a sidearm delivery appeared to place more stress on the elbow, an observation also suggested by Gainor et al. (in the latter, supported by video and computer analysis, in which axial loading was seen to produce extremely high rotary torque). Moreover, the lateral compartment appeared to be most subject to compressive loads, while the medial structures were subjected most to torsional ones.

Thus, clearly a variety of degenerative elbow changes *may* result from pitching, but the risk factors (probabilities) have not yet been well assessed. Adams (1965) found some evidence of degenerative involvement in every individual in his sample of 80 Little Leaguers; moreover, the degree of changes was in direct proportion to the amount and type of throwing. As noted above, King et al. (1969) also documented in professional pitchers frequent asymmetries and contraction deformities, and Tullos and King (1973) reported that 50% of all professional pitchers they surveyed had at some point in their careers been prevented from pitching due to elbow and/or shoulder symptoms.

In a large systematic survey of little league pitchers ($N=166$), Larson et al. (1976) found only minor symptomatology and slight radiographic changes to the pitching arm; the majority of minor changes were found in the medial compartment, with only 5% of individuals showing some changes to the lateral compartment (and here included cystic lesions, increased density, flattening and irregularity of joint surface, and irregularity of the epiphyseal line). Larson and colleagues argue that the medial changes are "frequent and benign," but lateral lesions are considerably more severe in nature.

One explanation of why the more severe elbow changes observed by Larson et al. (1976) were less common than those reported by Adams (1965) may be the ages of their respective samples. Larson and colleagues restricted their survey to 11–12 year olds (the majority pitching just 1–3 years). Adams' sample included individuals up to 14 years old, and the average duration of pitching in his sample was three years.

In addition to the obvious role of duration of use, age of onset may well be a parallel risk factor in elbow OA. It appears that the more severe lateral involvement takes somewhat longer to be manifested than the more benign medial changes, and these more debilitating effects might not become clinically apparent until years following the initial damage. In particular, if in young athletes the cartilage is permanently damaged at the osteochondral junction, or more fundamentally at the tidemark, the joint repair capacities might be compromised, leading to an early adult onset of severe OA (a similar situation perhaps present in *some* archaeological populations).

Clearly, the need for prospective, long-term studies of young baseball pitchers as they mature is obvious (but yet to be accomplished). Retrospective analyses of college and professional pitchers are also instructive, but they are highly biased, self-selected samples. On the one hand, these elite athletes are "survivors" of years of joint abuse, and they may have endured due to some initial (i.e., at least partly genotypic) advantage. On the other hand, at this advanced level of performance, they regularly subject their joints to peak levels of stress. Thus, the representativeness of such self-selected elite athletes is open to question, and an issue that impacts all investigations of such a nature.

In addition to the well-described degenerative changes seen in the elbow of baseball pitchers, some pathological involvement has been noted in the shoulder. This condition, suggested by O'Neill and Micheli (1988) to be termed "Little League shoulder," also appears to be much influenced by duration and age of onset. Jobe, in fact, suggested that in professional pitchers there is a much higher incidence of impingement syndrome of the shoulder in those individuals who began pitching as children (personal communication, cited in O'Neil and Micheli, 1988). Shoulder involvement, however, is probably not as clearly related to pitching as is elbow OA, and Albright et al. (1978) comment that elbow changes were more common and symptoms seemed to develop earlier than those in the shoulder (arguing perhaps that the elbow is more at risk of early injury than the shoulder).

Upper appendage OA developing in sports contexts has been best documented in baseball pitchers, but other (less severe) changes have also been observed in other athletes. Most notably is the condition termed "tennis elbow," typically involving strictly soft tissue changes around the lateral humeral epicondyle. Moreover, occasionally, the olecranon fossa or the trochlea may be involved, and hypertrophic spurs may develop, the latter sometimes fracturing and producing loose bodies in the joint (Priest et al., 1974). Nevertheless, in the majority of these individuals the changes are mostly benign and probably rarely involve the underlying bone. In fact, most patients presenting with "tennis elbow" are not tennis players, the condition usually heals spontaneously within 12 months, and, furthermore, it is strongly correlated with age (Garden, 1961).

Lastly, a rare but intriguing pathological elbow condition has been reported in javelin throwers (Miller, 1960). The condition, sometimes dubbed "javelin thrower's elbow," does not appear to be OA, as it does not produce internal joint changes. Indeed, the condition in its most severe form is a stress fracture involving the tip of the olecranon. In its more chronic manifestation it results from inflammation of the medial ligament as it courses over the medial humeral epicondyle (and less commonly over the ulnar attachment of the ligament). Radiographs, however, appear normal (Miller, 1960), so it is doubtful that any clear osteological signs would be observable. Almost certainly, the condition does not appear to produce elbow OA.

Nevertheless, such observations *may* be of interest to osteologists. The research by Priest et al. (1974) on tennis elbow was initially focused on delineating degenerative elbow disease, but this same work has become a central foundation for skeletal biologists investigating the role of activity in shaping bone geometry (see Chapter 7). Indeed, the opportunities to use more widely similar studies in this manner are considerable, and still largely untapped. Moreover, several other epidemiological studies of OA in occupational and sports contexts provide a resource with great potential for those attempting to establish the utility of so-called enthesopathies (and here too, one generally ignored) (see Chapter 5).

Soccer and American Football Players

The investigations of activity-related OA of the upper appendage have, to this point, predominantly focused on baseball pitchers. Likewise, studies of lower limb OA involvement have emphasized evaluation of soccer players and (American) football players (and more recently, runners; see below).

Several studies of lower appendage OA in soccer players have been done (Solonen, 1966; Adams, 1976; 1979; Klunder et al., 1980; Chantraine, 1985; Deacon et al., 1997), but the results are mixed. Solonen (1966) radiographically examined the hips, knees, and ankles of 36 soccer players and found a particularly high prevalence of OA in the talocrural joint (97% incidence of X-ray changes in soccer players, compared to 20% in controls), but most of these individuals had suffered severe prior injuries. Likewise, Adams (1976) in a study of 67 professional and former professional soccer performers found but rare knee involvement, and again most cases were correlated with prior injury. In a follow-up study of ankle changes, sharp osteophytes were sometimes seen, but only very rare clearly diagnostic features (i.e., narrowing of joint space) were observed. Klunder et al. (1980) examined veteran (retired) soccer players (average age = 36 years) who had formerly played, on average, for more than 20 years. The results were again mixed, as radiographic evidence showed a higher incidence of OA in the hip (49% compared to 26% in controls), but not of the knee. Moreover, only a minority of those with radiographic hip changes reported any symptoms, and a good proportion had a history of prior injury. Chantraine (1985) also evaluated (for knee OA) veteran soccer players who had played on average for close to 20 years. Unlike the studies mentioned above, however, Chantraine did find a higher prevalence of knee OA in the athletes (lacking specific controls, comparisons were made with the general population). However, once again, prior injury and surgical treatment (menisectomy) were important risk factors. Moreover, as in other studies, most cases of radiographic "disease" were asymptomatic. Finally, Deacon et al. (1997) surveyed 50 retired Australian elite soccer players for prevalence of knee OA as compared to controls. Similarly to the studies discussed above, they observed a significantly increased risk among the former soccer players, most specially for those with prior injury. Thus, taken together, the evidence for an association of soccer with OA of the lower appendage is not especially compelling.

A few studies have also been conducted on American-style football players. Vincellete et al. (1972) found a high proportion of radiographic changes in the ankle and foot of professional football players (up to 90% involved). However, most of these radiographic changes included periarticular alterations and were thus not diagnostic of OA (but could relate to investigations of enthesophytes). Moretz et al. (1984) found quite common radiographic knee OA among former college players, including many who reported prior serious injury. As with soccer athletes, prior injury is particularly common in football players (Rall et al., 1964). Felson (1988) concluded that prior major injury appears to be a significant risk factor for knee OA in sports, and most especially so among American-style football players.

Runners

Over the last decade, among the most systematic studies of OA in athletes are those on runners. Probably the best-controlled of these investigations are part of the long-term longitudinal studies conducted at Stanford University by Lane and associates (Lane et al., 1987; Lane et al., 1990; Lane et al., 1993; Lane, 1996). In addition, a number of other excellent investigations have recently been completed, and the results consistently have failed to demonstrate any link between running and an increased risk of OA.

The Stanford study involved prospective ascertainment by clinical examination and radiography of the knee initially in 498 long distance runners, compared to carefully matched controls (Lane et al., 1987). Results showed no increase in OA among runners; in fact, runners actually showed less musculoskeletal disability than that found in controls. This difference was found, even after controlling for sex, age, and occupation. Commenting on this *negative* association of OA and running, Eichner (1989: 147) noted "... there is no conclusive evidence that running causes the disease; in fact, running may actually slow the functional aspects of musculoskeletal aging." Eichner later concluded: "All things considered, then, the question for the future is not 'Does running cause osteoarthritis?' but rather, 'Does sloth cause osteoarthritis'" (Eichner, 1989: 154).

In a later more detailed study, employing both standard radiography (of lumbar spine and knee) as well as CT scans (of L1), in a sample of 34 runners, the athletes were found to maintain more bone density than controls (Lane et al., 1990). The runners did show somewhat more osteophyte ("spur") development, but Lane et al. do not recognize osteophytes *alone* as indicative of OA (see further discussion in Chapter 2). Finally, another subsample of older runners (average age = 63 years) also failed to show any association of running activity with knee OA, even after controlling for age, occupation, and level of education (Lane et al., 1993). These findings led the investigators to conclude: "In summary, running did not accelerate the development of radiographic or clinical OA of the knee" (Lane et al., 1993: 461).

Konradsen et al. (1990) in a study of veteran Danish long-distance runners, many running for 40 years or more, found no differences in hip, knee, or ankle involvement, when compared to controls matched for age, weight, and occupation. More specifically, no degenerative differences were seen relating to joint alignment, range of motion, complaints of pain, or, on X-ray, in thickness of articular cartilage, grade of degeneration, or osteophytes.

Three other investigations also evaluated OA in veteran runners, the first of the hip (in Finnish champion runners) (Puranen et al., 1986). The second study analyzed knee and foot involvement in long-distance or marathon runners (Panush et al., 1986). Again, in neither study was there any evidence found of increased risk of degenerative disease related to running. Finally, Sohn and Micheli (1983) surveyed (by questionnaire) a large sample ($N = 504$) of former college cross-country runners compared to controls. Here too, although less controlled than the above studies, there was no increase in prevalence of OA, nor was there any link established with amount of mileage or years running.

One study of knee OA in runners (McDermott and Freyne, 1983) did find in a small ($N = 20$) self-selected sample (those presenting with pain) a greater degree of radiographic knee involvement in those individuals who had run for longer periods of time. However, most of the symptoms, as well as radiographic signs, appeared to be most correlated with prior injury, deformity, or ligamentous laxity. Marti et al. (1989) in a study of elite athletes also found more OA of the hip in runners ($N = 27$) than in a sample of bobsledders ($N = 9$) or 'normal' controls ($N = 23$).

In conclusion, then, the great majority of data on runners, even for those athletes participating for most of a lifetime, show no indication of increased OA related to activity (if anything, the best-controlled studies indicate the opposite, i.e., that running deters musculoskeletal deterioration). Indeed, in one recent review of 12 different studies involving a combined sample of 1597 athletes in which runners averaged 33 miles/week over an average of 16 years, the general finding was no increase in OA of lower limb joints in comparison with more sedentary controls (Panush and Inzinna, 1994). In some respects these results are a surprise. Running exerts considerable, and obviously highly repetitive, biomechanical force, estimated for the ground reaction force at 250–300% of body weight. Indeed, in a 70 kg runner, moving at 1175 steps per mile, an estimated 220 tons of force would be absorbed for every mile run! (Herring and Nilson, 1987).

This consistent lack of correlation of running and OA also further questions the impulse loading hypothesis proposed by Radin and colleagues (see Chapter 2). Moreover, the failure to find a consistent association of OA with such long-term repetitive stress as that experienced by runners also questions many of the basic assumptions commonly made by osteologists in their speculative interpretations of past behavior.

Miscellaneous Studies of Occupational and Sports-Related Activities

A considerable variety of other reports have been published suggesting links of various activities with OA. Some of these investigations have collected fairly systematic data, while the majority are primarily anecdotal. Those studies presenting reliable data, as well as the less systematic research (discussed above) are summarized in Table 3-2. The following review begins with the more comprehensive reports and proceeds to those more anecdotal in nature.

One interesting study of a quite large sample of sport and military parachutists ($N = 221$) was assessed through questionnaire by Murray-Leslie et al. (1977). Of this sample, 104 of the parachutists were X-rayed, providing data on the knee and some information on the ankle as well. There was no systematic control group used, except general epidemiological comparisons, but the authors were still able to conclude "that parachutists as a group do not show an increased prevalence of radiological osteoarthritis of the knee or ankle" (Murray-Leslie et al., 1977: 327).

A recent investigation of professional sailors ("seafayers;" $N = 299$) has reported quite high prevalence of knee OA (51%) (Pearce et al., 1996). Of interest here, as well, was observation of frequent leg deformity (genu varus or "bow legs") in this sample.

Table 3-2. Specific epidemiological studies of occupational and sports-related activities and OA

Study group/Activity	Joint(s) evaluated	Reference
Studies finding positive association:		
Textile workers	Hand	Hadler et al. (1978)
Pneumatic drill users	Shoulder, Elbow, MC joints	Fischer (1932)
Pneumatic tool users	Elbow	Linde (1932)
Pneumatic drill users (miners)	Elbow	Vossenaar (1936)
Pneumatic drill users (miners)	Elbow	Lawrence (1955)
Pneumatic tool users	Wrist	Schumacher et al. (1972)
Pneumatic tool users (shipyard workers)	Shoulder, Wrist Hand	Bovenzi et al. (1980)
Foundry workers (using tongs)	Elbow	Mintz and Fraga (1973)
School Cooks	Hand	Nakamura et al. (1993)
Sailors ("seafarers")	Knee	Pearce et al. (1996)
Baseball pitchers (Little League)	Elbow	Adams (1965)
Baseball pitchers (professional)	Upper limb asymmetries (not OA)	King et al. (1969)
Baseball pitchers (professional)	Shoulder, Elbow	Tullos and King (1973)
Baseball pitchers (Little League)	Elbow	Larson et al. (1976)
Baseball pitchers	Elbow	Hansen (1982)
Soccer players	Ankle	Brodelius (1961)
Soccer players	Ankle	Solonen (1966)
Soccer players	Knee	Chantraine (1985)
Soccer players	Knee	Deacon et al. (1997)
Amer. football players	Knee	Rall et al. (1964)
Amer. football players	Ankle, Foot	Vincelette et al. (1972)
Amer. football players	Knee	Moretz et al. (1984)
Runners (presenting with pain)	Knee	McDermott and Freyne (1983)
Runners (elite)	Hip	Marti et al. (1989)
Ballet dancers	Ankle	Brodelius (1961)
Ballet dancers	Cervical spine, Hip Knee, Ankle	Washington (1978)
Ballet dancers	Cervical spine, Hip Knee	Ende and Wickstrom (1982)
Wrestlers	Cervical spine, Elbow, Knee	Layani et al. (1960)

Table 3-2. *(Continued)*

Study group/Activity	Joint(s) evaluated	Reference
Weight-lifters	Spine	Kotani et al. (1970)
Weight-lifters	Spine	Aggrawal et al. (1979)
Swimmers (breastrokers)	Knee	Stulberg et al. (1980)
Rock climbers	Hand	Bollen and Wright (1994)
Farmers	Hip	Louyot and Savin (1966)
Farmers	Hip	Thelin (1990)
Farmers	Hip	Croft et al. (1992)
Studies failing to find association:		
Cotton (textile) workers	Hand	Lawrence (1961)
Pneumatic tool users	Elbow	Hunter et al. (1945)
Pneumatic tool users	Cervical spine, Shoulder, Elbow, Hand	Burke et al. (1977)
Chainsaw users (lumberjacks)	Hand	Kumlin et al. (1973)
Lumberjacks	Knee, Hip	Sairanen et al. (1981)
Soccer players	Knee	Adams (1976)
Soccer players	Ankle	Adams (1979)
Parachutists	Knee, Ankle	Murray-Leslie et al. (1977)
Ballet dancers	Hip, Knee, Ankle	Schneider et al. (1974)
Ballet dancers	Ankle, Foot	Ambre and Nilsson (1978)
Weight-lifters	Acromio-clavicular, Shoulder, Elbow, Wrist, Hip, Knee	Fitzgerald and McClatchie (1980)
Karate instructors	Hand	Crosby (1985)
Porters	Neck (Cervical spine)	Levy (1968)
Porters	Neck (Cervical spine)	Scher (1978)
Runners (elite)	Hip	Puranen et al. (1975)
Runners (long-distance)	Knee	Sohn and Micheli (1985)
Runners (long distance)	Hip, Knee, Foot	Panush et al. (1986)
Runners (long-distance)	Knee	Lane et al. (1987; 1990; 1993)
Runners (long-distance)	Hip, Knee, Ankle	Konradsen et al. (1990)
Studies Finding Mixed Results:		
Soccer players	Hip (positive assoc.)	Klunder et al. (1980)
Soccer players	Knee (no assoc.)	" " "

Moreover, two studies of professional ballet dancers found no clear association of activity and increased prevalence of lower appendage OA. Both studies used fairly small samples, one including 52 dancers (only 28 radiographically surveyed for the hip, knee, and ankle) (Schneider et al., 1974), and the other investigated the ankle and foot in 20 ballet performers (Ambre and Nilsson, 1978). In all cases degenerative changes were surprisingly infrequent and only slight in degree. An earlier, less systematic, analysis was done by Brodelius (1961), who reported a high prevalence of talar joint involvement in 16 ballet dancers.

A quite complete radiographic survey of peripheral involvement in weight-lifters was carried out by Fitzgerald and McClatchie (1980). In their analysis of 25 experienced weight-lifters, no clear increase in prevalence of OA was found in the acromioclavicular joint, shoulder, or knee. Another more restricted study of the lumbar spine (Aggrawal et al., 1979) did, however, report a higher incidence of spinal involvement in weight-lifters.

Young swimmers who incorrectly perform the breaststroke can develop a condition called "breastroker's knee" (Stulberg et al., 1980). Apparently, an improper whipkick can produce knee tenderness, particularly on the medial facet of the patella (observed in 18 of 23 swimmers reporting pain). Even more interestingly, those swimmers who had swum in such a fashion for more than eight years ($N=4$), all showed evidence of patellofemoral OA.

Two informative studies have been done, both in South Africa, of possible neck involvement related to carrying heavy loads on the head (so-called "porter's neck"). One investigation (Levy, 1968) included a survey of a hospital sample and the other (Scher, 1978) was more anecdotal in nature. Interestingly, in neither study, in which some individuals habitually carried loads as heavy as 200 pounds, was any increase in cervical involvement found. Those few cases of neck disease that were found were most apparently the result of acute injury, in which the porter slips or falls while carrying a heavy load.

Other suggestions and case studies are considerably more anecdotal. Such reports include metacarpophalangeal involvement in a baker (McDonald and Marino, 1990); metacarpophalangeal and DIP changes in pianists (Bard et al., 1984); and patellofemoral involvement in cyclists (Bagneres, 1967). Even more anecdotally, Hellman et al. (1983) mention possible complications in the shoulder of polo players, elbow of handball players, hip of mountain climbers, and the ankle of cross-country motorcyclists.

Taking speculative whimsy even further, Dahl et al. (1981) suggested two new syndromes ("musher's knee" and "hooker's elbow"), and Mebane (1981) proposed an affliction he termed "dog-walker's elbow." Both of these last contributions appeared as letters in the *New England Journal of Medicine*, usually an august periodical, but one that tolerates flights of fancy in its non-refereed correspondence section. Despite the obvious limitations of these latter suggestions, one can occasionally find in the osteological literature reference to these "sources" as though they were serious, scholarly research. Perhaps, osteologists, in devising their own imaginative scenarios, can become too uncritical of others.

EARLY ONSET OF STRESS

An important point raised above (and also discussed in Chapter 2) concerns the potential differential effects of the *age of onset* of biomechanically stressful activities. Other than the suggestions made relative to the onset of hip disease among farmers, most contemporary occupational data obviously are not the most appropriate source for understanding this phenomenon. Nevertheless, the further analysis of agricultural workers in a variety of settings in different cultures, including migrant workers in the U.S., offer considerable potential for greater elucidation, especially if such studies are designed prospectively.

More recently, however, the growing sports literature has provided considerable insight into the potential differential influence of early onset of mechanical stress. Williams (1979) called attention to the deleterious effects of "overuse" in young athletes, particularly the consequences of excessive training for too long a period of time, in which tissue changes (resulting from exercise) do not recover prior to the next bout of exercise.

Micheli and Klein (1991: 6) observed that "Children, it seems, are not only susceptible to overuse injuries, but may in fact be at increased risk as compared with their adult counterparts."

The mechanics involved (as discussed in Chapter 2) might involve early damage to the osteochondral junction and/or the tidemark, producing such early pathological changes as osteochondritis dissecans:

> . . . a repetitive injury to the subchondral bone and overlying cartilage, it reflected the particular sensitivity of the pediatric joint surface to sheer stress (Micheli and Klein, 1991: 6–7).

> Biomechanical and clinical evidence suggests that growth cartilage, especially that of the articular surface, is less resistant to repetitive microtrauma than the mature adult counterpart (p. 7).

In fact, early onset of such changes as chondromalacia and osteochondritis dissecans might well later in life be manifested as OA, a conclusion (as noted in Chapter 2) also suggested some time ago by Duncan (1979). To date, sufficient data on pediatric arthropathies are still limited, but some support for this hypothesis is accumulating.

Bruns and Kilma (1993), for example, examined osteochondritis in the knees of adolescent boy and girl patients ($N=97$) following surgical treatment. Among this admittedly self-selected sample, these patients had histories of intensive sports activities and/or trauma. In addition, there also was a correlation of medial condyle lesions and varus malalignment of the knee, suggesting constitutional factors might be important here. In addition, Williams (1979) anecdotally observed elbow osteochondritis in girl gymnasts, as well as an increasing number of cases of foot traction epiphysitis in young soccer players.

Young baseball pitchers are also at risk, a point emphasized above, and given some support as related to elbow lesions (Adams, 1965) as well as those of the shoulder (Hansen, 1982; Jobe, as cited in O'Neill and Micheli, 1988). Likewise, the hip may also be prone to early modification from sports activity, perhaps leading to development of OA much later in life (Murray and Duncan, 1971).

Finally, various traumatic/degenerative lesions of the back also appear to display clearly different age-related risks. For example, in individuals presenting with low back pain, adolescents display far more spondylolysis than adults, who, alternatively, had significantly more disk degeneration and OA (Micheli and Wood, 1995). The effects of repetitive biomechanical stress on the developing spine and resulting initiation of spondylolysis have been carefully documented by Merbs (1995; 1996a,b), and will be discussed further in Chapter 5.

As the available data suggest, age of onset of mechanical loading *might* be an important consideration in understanding the development of OA. However, the necessary data adequately to test this hypothesis are not yet at hand, a point well-made by McKeag (1992: 476):

> ...early participants in sports at a very intense and frequent level may increase the risk of osteoarthritis in future life, but until prospective control studies of long-term effects can be done, this can remain nothing more than an interesting observation.

CONCLUSIONS

Despite years of intensive investigations, the available data regarding the effects of activity (occupationally and/or sports-related) on OA are conflicting. Indeed, regarding the inconsistency of results, McKeag went on to comment:

"Explanations for the patterns just described tend to defy explanation based on our present knowledge of the causes of osteoarthritis" (McKeag, 1992: 480). One immediate issue, as raised by the quote above is of even more general methodological concern: "There are virtually no prospective controlled studies of the long-term effects of exercise on the musculoskeletal system" (Panush and Brown, 1987: 59).

Panush and Inzinna further elaborated on these same concerns as relating to the sports data:

> However, most of these reports were anecdotal, retrospective, and uncontrolled. These data can best be interpreted as suggesting but not proving a possible relationship between exercise activities and DJD. It can also be suggested that, in many of these instances joints may have been damaged by being subjected to abnormal and/or unusual stresses. (Panush and Inzinna, 1994: 3.)

Such lack of rigorous longitudinal data is clearly a handicap, but recent publication of the Framingham and Stanford results have begun to redress this problem. Obviously, however, much more remains to be done (Lane, 1991).

This need for better-controlled investigations was further emphasized by Maetzel et al. (1997) in their recent overview of clinical studies of purported occupational associations with OA in the hip and knee. From their survey of 28 years of clinical publication, they found only seven studies fully meeting their standards of methodological rigor (controlled comparisons, diagnostic criteria, etc.). Thus these authors could make but few systematic conclusions; they did note, however, that there was "a strong positive relationship between work related knee bending exposure and knee OA. The evidence between work related exposure, farming in particular, and hip OA is consistently positive but weak" (p. 1599).

Of course, in addition to mechanical stress, there are a variety of confounding influences (as emphasized in Chapter 2) including age, obesity, bone density, prior injury, other arthropathies, etc., and Waldron (1994a) appropriately reemphasized a complex model of OA etiopathogenesis. Clearly, then, present understanding of occupation and sports data do *not* support an overly simplistic explanation of OA causation:

> The traditional notion was that cumulative effects of weightbearing, strain, and/or stress damaged the articular cartilage and resulted in DJD. However, this is now considered to be an oversimplification. Biomechanical

factors are an important but not exclusive nor necessary contributor to DJD (Panush and Inzinna, 1994: 2).

In addition to prospective studies, and considering *some* role of mechanical factors, more precise data on the degree and types of mechanical factors acting on joints relative to different sports and occupational activities are greatly needed (Mankin et al., 1986); this same point was also much emphasized in osteological interpretation by Woods (1995):

> It is now clear that simple use of a joint does not equate with loading that necessarily leads to osteoarthritis development, especially if porosity and eburnation are not distinguished. Conversely, the presence of osteoarthritis in a joint does not equate to excessive use of that joint (Woods, 1995: 115).

If interpretations and potential confounders were not difficult already, recent publications have suggested further complications. For example, Hannan et al. (1992) found in their analysis of HANES I data that educational level influenced self-reporting of knee OA, even when controlling for previous injury, race, obesity, *and* occupation. Of course, it is problematic that such sociological variables actually affect joint damage (seen on X-ray or on gross by osteologists). Nevertheless, another consideration raised recently by Bergmann et al. (1995) might well be important in evaluating archaeological samples. Using hip implant instruments these investigators found that, compared to walking and running barefoot, wearing shoes increased loads by up to 50%. Moreover, the effects were found regardless of type of shoe, although very hard soles were apparently the most disadvantageous. These results are especially significant for reconstructing behavior (and consequences) in past populations, particularly as this research emphasized that gait stability is the most important factor influencing joint loading.

From all the confusing and conflicting information, it may appear no conclusions are currently possible. Certainly, an appreciation of *complexity* is in order. Nevertheless, two recent reviews (Hoffman, 1993; Panush and Lane, 1994) reached very similar conclusions, namely:

(1) Normal joints in people of all ages may tolerate long-term and vigorous exercise without accelerating development of OA.
(2) Individuals may be at increased risk who have underlying conditions (e.g., muscular weakness, imbalance, neurological abnormalities, anatomical variances) *and* who also engage in significant exercise (occupations) that stress relevant joint(s).

(3) Individuals with prior injury are also at increased risk, especially when engaging in repetitive stressful activity.

To these points could also be added some other possibly useful general conclusions:

(4) Some joints (elbow and hip particularly, as compared to other joints) appear to be under differential risk, given the age of the onset of mechanical loading; early injury and/or modification of joint mechanics can produce OA changes later in life.
(5) Variable types of repetitive stress *may* be of particular importance for initiating degenerative changes in different joints. For example, the knee appears most prone to activities involving repetitive bending, while the spine and hip appear more at risk as the result of heavy lifting.
(6) Beyond these generalities, the association of OA with *specific* activities is not clearly supported in contemporary contexts by either the occupational or sports literature. Further, the implications for and limitations on osteological interpretations are obvious.

At the most general level, what can be gleaned from contemporary data is that influences on OA are multifactorial, and, thus, osteologists should always approach their interpretations with considerable caution. Waldron (1994a) reminds anthropologists of the inherent pitfalls:

> Since we know that occupation is *not* the sole cause of osteoarthritis, there cannot be any likelihood of being able to deduce the former from the latter (Waldron, 1994a: 96).
>
> The only way in which an occupation could accurately be determined in a *single* skeleton would be if the appearance of the disease at a single site, or a combination of sites, was unique to those following a single occupation. And we know that this is not the case (p. 97) [emphasis added].

Chapter 4

Osteoarthritis: Anthropological Interpretations

> There is a perfectly understandable drive to make the most of what little evidence survives in the skeleton and this sometimes has the effect of overwhelming the critical faculties (Waldron, 1994a: 98).

INTRODUCTION

Skeletal biologists, most specifically, paleopathologists, have been among the most enthusiastic supporters of the stress hypothesis, and many have thus sought to make a variety of behaviorally based interpretations from the skeletal pattern of OA. As discussed in the prior two chapters, and as further reinforced by the quote by Waldron above, this desire is quite understandable. Rogers et al. (1987) give further voice to this motivation, laying the blame more at the feet of archaeologists, rather than osteologists (who should more appropriately shoulder it):

> Because archaeologists naturally want to present as complete a picture as they can of their finds and to "bring the bones to life" for the non-specialist reader, the danger of over-interpretation may sometimes be irresistible and lead to over-simplification (Rogers et al., 1987: 93).

Osteologists, in fact, bring a host of biases with them to the interpretation of past behavior, perhaps as best illustrated in their application of skeletal OA in inferring ancient activity patterns. Indeed, one could ask: "Why is osteoarthritis so frequently used to reconstruct behavior (i.e., 'activity') in past populations?" Several considerations bear on this approach and carry inherent biases with them. First, OA is clearly a very common skeletal disease, probably surpassed only by dental disease in frequency as a pathological condition in human skeletal material (Jurmain, 1990; 1991b). Secondly, it is relatively easy to diagnose in skeletal material by simple gross examination. Thirdly, as detailed in Chapter 2, in contemporary populations it is a frequent and sometimes debilitating condition, with a rich clinical literature evaluating manifold aspects of the condition. And, lastly, of course, there is the assumption, one might say, "the hope," that among archaeological samples OA can provide a window to understand behavior, specifically, patterns of activity. This last motivation, as Rogers et al. (1987) suggest can be "irresistible" as it promises in Waldron's words such an "alluring prospect" (Waldron, 1994a).

The willing, sometimes over-willing, temptation to make more of skeletal data than is scientifically prudent is, no doubt, the most fundamental bias relating to interpretation of skeletal OA, as well as the other types of lesions documented in this book. Moreover, with OA, particularly, and leading directly from the more general biases listed above, other misinterpretations can quickly follow. For example, osteologists frequently view OA as a condition that is irreversible and inevitably progressive in nature. This over-simplification of the pathogenesis of OA most likely derives from a data-collection perspective that is necessarily cross-sectional. From such data, it will obviously *appear* that, with age (and/or biomechanical stress), OA becomes increasingly common and more severe. The limitation of such a limited approach is failure to appreciate the considerable capabilities of joints, including bone and articular cartilage, for repair (see Chapter 2). Only from longitudinal studies can such a perspective be gained, a source obviously unavailable to skeletal biologists. Merbs (1983: 2) makes this point clearly, in noting that osteological studies are "not a substitute for careful longitudinal studies of living organisms."

Merbs, in fact, in his systematic analysis of the Saldermiut Inuit and monograph detailing the results (1983) remains, at a general level, cautious in making interpretations. For example, he reminds us of the two abysses faced by paleopathologists, as initially described by Johnson (1966), namely, the "nonspecificity of etiology" and the "nonspecificity of diagnostic morphologic criteria."

Given such major limitations imposed upon skeletal biologists, Merbs is rather pessimistic about the approach's likely contribution to understanding the etiology of OA (or most other skeletal conditions). Bennike (1985: 123) is also extremely careful in arguing that skeletal data "do not contribute directly to a clear understanding of factors responsible for osteoarthritis." Merbs and Bennike are quite correct in emphasizing a conservative approach, especially in making behavioral interpretations (a degree of caution not always strictly followed even by those of us making such warnings); however, in some respects, they may be overly pessimistic in their attitude relating to the potential of osteology to contribute to clinical understanding of disease causation. Indeed, Merbs' own work on spondylolysis (1995; 1996a,b) is a beautiful example of how careful osteological research can advance a general understanding of certain conditions (see Chapter 5 for further discussion). Moreover, regarding OA specifically, skeletal biology has made and continues to make positive contributions to understanding its etiopathogenesis (e.g., Rogers et al., 1990; Jurmain and Kilgore, 1995).

Another bias osteologists almost inevitably bring with them is very obvious, that is, an over-emphasis on bone. Without question, bone *can* be, and frequently is, a major component in the pathogenesis of OA. This is not to say, however, that bone is the *primary*, or even usually the most significant, tissue in the full pathogenesis of the disease. Nevertheless, osteologists are understandably most concerned with bone changes, and they interpret clinical data through this lens. Accordingly, great emphasis is frequently placed on the research of Radin (Radin et al., 1972; Radin, 1975; 1976), as exemplified in some of my early publications (Jurmain, 1977a) as well as that of many other researchers (e.g., Pickering, 1979; Merbs, 1983; Walker and Hollimon, 1989; Nagy, 1996). What is less than obvious in such a narrow approach is the clear multi-factorial etiology of OA and its manifestly multi-focal pathogenesis. Indeed, in more recent publications (Radin and Rose, 1986; Mankin et al., 1986), Radin, himself, has grown more circumspect regarding the primacy of impulse loading/bone involvement in the etiopathogenesis of OA (see Chapter 2).

Clearly, then, osteologists have a variety of unique perspectives, some of which have potential to contribute meaningfully to understanding OA and other skeletal diseases. Nevertheless, numerous biases are also inherently tied to these perspectives, and are best recognized and contained (at least as far as is reasonable). It is also worthy of mention that some skeletal biologists have for some time recognized the strengths and weaknesses of osteological approaches, most especially as they apply to interpretation of OA. Most notably, in the United States the work of Bridges and in the United Kingdom that of Rogers and Waldron have highlighted many of the same concerns raised here. Most especially, these researchers have addressed critical issues relating to the methodology of evaluating OA in skeletal materials.

METHODOLOGY

The types of bone lesions that have been evaluated as part of OA are discussed in Chapter 2. In recent years some further attempts have been made to more fully systematize osteological criteria of OA diagnosis (Rogers et al., 1987; Waldron and Rogers, 1991; Rogers and Waldron, 1995). According to Rogers et al. (1987: 185), the following four types of degenerative bony lesions characterize OA:

(1) formation of true, marginal osteophyte
(2) subchondral bone reaction (eburnation, sclerosis, cysts)
(3) pitting of the joint surface
(4) and, in severe cases, alteration of joint contours.

Rogers and associates emphasize that osteophytes *alone* are not a sufficient diagnostic criterion, and, thus, must be accompanied by at least one of the other changes to allow a diagnosis of OA. "Unless they [osteophytes] are accompanied by certain signs of disease (eburnation, sclerosis, or fusion, for example) they should not be regarded as pathological and should be scored separately" (Rogers et al., 1987: 192).

This conservative interpretation of osteophytes has been suggested by others and has also been advocated here (see Chapter 2 for further discussion). Recently, Buikstra and Ubelaker (1994) have advocated scoring osteophytes, porosity, and eburnation, each separately. Moreover, in addition to degree (e.g., for eburnation whether grooving is evident), extent is separately evaluated; in this set of recommended standards extent is evaluated on the basis of amount of the respective joint area that is involved (e.g., extent of circumference affected by most severe expression of osteophyte development).

The need for a greater degree of consensus regarding scoring criteria is quite apparent, and is probably the main explanation why consistency between different observers has thus far proven so poor. In a brief study conducted by Waldron and Rogers (1991) at the 1990 European Paleopathology Association meetings, the observations of 30 different participants were compared, ranging from beginners to "experts." There proved to be a discouragingly high degree of interobserver variation in scoring, even to the point of recognizing whether any degenerative changes were present. Moreover, the disagreements were as pronounced among the experts as among the beginners. While far from truly consistent, the criterion with the greatest degree of agreement was presence of eburnation.

Bridges (1993) also addressed the issue of methodological consistency in scoring OA changes and showed how differing criteria, as well as differing methods of treatment (i.e., whether or not involvement was scored for individual bones or as composite joint scores), significantly impacted results. In one damning set of comparisons Bridges contrasted results from three different studies of VOP, all as determined in the same collection (Indian Knoll). Reported frequencies here varied from 39%, to 60%, to 72%. Bridges also made the important point that differing criteria (e.g., whether emphasizing eburnation or osteophytes) will not just affect overall scores within joints, but even the *relative* ranking of comparisons between joints. For example, if eburnation is given more weight, then the knee would show proportionally more OA relative to that seen in the shoulder or elbow. Finally, what *segments* are assessed within joints affects both totals and relative rankings. In this regard, whether the acromio-clavicular segment is assessed as part of total shoulder OA clearly impacts results, as does separate evaluation of compartments within the knee (patellofemoral *vs.* tibiofemoral) and in the elbow (trochlear-ulnar *vs.* capitulum-radial).

Not until notably more consistency can be attained will it be possible to make productive comparisons across studies conducted by different observers. For this reason, Bridges in her comprehensive review of OA in the New World (1992) only makes very general comparisons (emphasizing ranking, i.e., which joints show greatest incidence as reported in different groups). Even here, results proved so inconsistent as to make comparisons of apparently very little utility. Perhaps, the only way to bring greater rigor (and with it, we hope, greater consistency) is to either *simplify* scoring procedures (and include only eburnation, as argued in Chapter 2) or to elaborate them (with much more detailed recording of regional involvement, as in Woods, 1995 and Nagy, 1996).

Some of these considerations have already been discussed in Chapter 2. A point raised there also is that *age*, and, obviously, sex as well, must be systematically controlled to make results in any way comparable. Accordingly, initial sample sizes must be quite large (a point also emphasized by Bennike, 1985) to allow such stratification, a limitation usually recognized by osteologists (but also frequently finessed in the service of necessity rather than rigor). The particular requirements of sample stratification will be returned to below in discussion of specific osteological approaches, and some of the more controlled samples will also be evaluated and compared (and are also partially listed in Appendix A).

As a final point, it should be noted that no approach, no matter how apparently systematic, is immune to the types of problems discussed above. My own scoring technique, emphasizing specific criteria within each joint (e.g., 16 for the elbow, 17 for the knee) attempted to more consistently ensure repeatability. Still, a colleague employing exactly the same techniques (Pierce, 1987) and having some considerable prior experience working with me, reported markedly elevated frequencies in two North American samples (from even those seen among the Inuit).

Pierce's results were most dramatically distinct for the Averbuch (Tennessee) material. Are these differences mostly 'real,' that is, bioculturally based epidemiological contrasts between the populations, or are they more a product of inter-observer variation in implementation of the same methodology? Perhaps, the most frustrating aspect of this dilemma is that there is no way to distinguish between the two alternatives. Nevertheless, my impression is that much of the variance does result from inter-observer differences in application of a purportedly precise, but complex, methodology. The basic message for scoring criteria is, no doubt: "simpler is better."

STUDIES OF FOSSIL HOMINIDS AND NONHUMAN PRIMATES

Probably the best known of the reports relating to OA among fossil hominids relates to the La Chapelle skeleton and the analysis done by Straus and Cave (1957). In their reanalysis of the famous La Chapelle skeleton, Straus and Cave suggest Boule misinterpreted the spinal arthritis, so to produce an ape-like posture. However, as Trinkaus (1985) has more recently argued, the spinal arthritis in La Chapelle (including some VOP and some moderate apophyseal OA of C5/C6, T1/T2, and T10/T11) was not all that extreme and Boule did recognize it, but chose to ignore it in his more generally biased interpretation. Other apparent osteoarthritic changes seen in the La Chapelle individual include bilateral eburnation of the humeral heads, a rare finding in *any* hominid.

Perhaps the most consistent analyses of OA in fossil hominids were further studies of Neandertals, and these were also done by Trinkaus (1983). In his detailed study of the Shanidar remains, Trinkaus reports on some degenerative changes in five individuals (Shanidar 1–5), mostly vertebral involvement. In addition, Shanidar 1 showed changes in the lower appendage (knee, foot, and especially ankles), which Trinkaus suggests were probably secondary to severe trauma in the individual. Shanidar 3 also had severe foot/ankle involvement, here argued by Trinkaus as "due to mechanical stress." Moreover, he suggested the slight involvement of the left wrist in Shanidar 4 was probably "use-related."

OA has been evaluated in a wide variety of animals, both fossil and extant (Rothschild, 1990; 1993; Rothschild and Rothschild, 1994). More specifically, considerable work on OA in nonhuman primates has been done, much of it quite recently, and is comprehensively reviewed by Lovell (1990; 1991).

One systematic comparative analysis (Rothschild and Woods, 1992) found in a sample of 267 prosimian and 1250 anthropoid skeletons quite low rates of OA. In captive animals rates of involvement were less than 5% in both prosimians and anthropoids, and, interestingly, were even lower in free ranging animals (0.8% in prosimians and 0.9% in anthropoids). In addition, there was an intriguing difference in pattern, with more hip lesions in captive animals, but more knee involvement among free-ranging individuals.

In another study, focusing on great apes (Lovell, 1991), rates of OA were also found to be quite low in both the spine and in peripheral joints. The highest rates of involvement were for VOP among mountain gorillas, where prevalence approached 20%.

In some of my own research on African great apes, results similar to that observed by Rothschild and Woods (1992) as well as by Lovell (1991) were found. In the small, but extremely well-documented sample of chimpanzees from Gombe National Park virtually no VOP was found among 11 adults surveyed (total sample of vertebral body surfaces = 474) (Jurmain, 1989; in preparation). Moreover, the pattern of no vertebral body involvement was seen even in the oldest individuals, in one case, a female thought to be > 40 years of age at death (Figure 4-1).

In a larger comparative sample of gorillas and chimpanzee skeletons (Jurmain, 1997), similar patterns of degenerative involvement have emerged. For VOP less than 1% of elements were involved in the combined samples (> 2500 surfaces evaluated). Likewise, for peripheral joints in chimpanzees only about 1% of joints ($N = 556$) were affected, and in gorillas about 2% of joints ($N = 511$) were involved (Jurmain, in preparation) (see Table 4-1).

The somewhat higher incidence of VOP in gorillas compared to chimpanzees has been explained (Lovell, 1991) as the result of greater body weight in gorillas. Perhaps, of even greater interest, are comparisons of the patterning of VOP/OA involvement in great apes as compared to that seen in humans. Consistently, great apes are dramatically less involved in the spine than are humans. It is easy to conclude that the differences *may* relate to the altered biomechanical demands

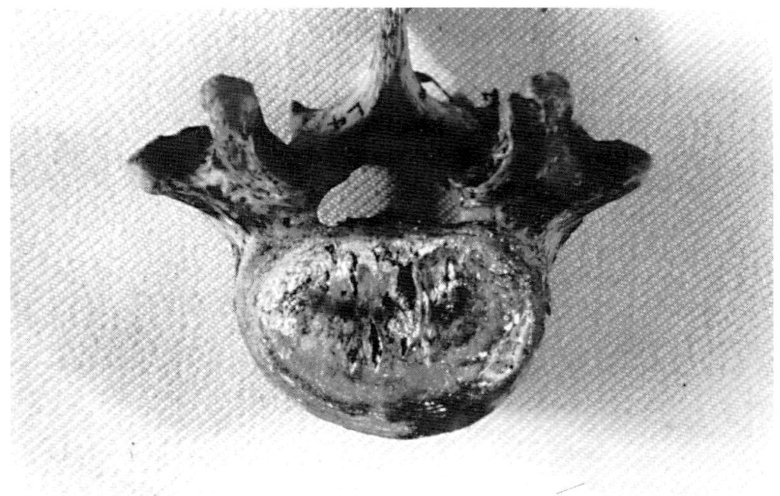

Figure 4-1 Fourth lumbar vertebra. Adult female chimpanzee (Flo), aged 40+ years. Note absence of vertebral osteophytosis. Photograph by the author.

concomitant with bipedality. However, such a conclusion may, in fact, be too easy. The pattern of spinal degenerative disease among Old World monkeys argues for more complexity in interpretation of spinal disease among primates. An osteological analysis of baboons (Bramblett, 1967) and radiological survey of macaques (DeRousseau, 1988) both found a high incidence of VOP in these quadrupedal monkeys. Moreover, macaques show fairly high frequency of knee OA involvement (Carlson et al., 1996). Thus, bipedality *alone* apparently does not simply account for the rates of VOP seen in humans. Perhaps, apes with considerable selection for stability of the lower back, are the exception among catarrhines. Clearly, more data on a wide variety of nonhuman primate species are needed.

HUMAN OSTEOLOGICAL STUDIES OF SPINAL INVOLVEMENT

For a number of years, the greatest osteological focus on degenerative disease concerned spinal involvement. For example, Elliot-Smith and Wood-Jones as early as 1910 described "spondylitis" in ancient Nubians as "very common." Some later work focused on contemporary samples, either dissecting room materials such as the Todd Collection

Table 4-1 Summary of degenerative involvement in African ape samples*

OA—peripheral (TMJ and four major joints)

Gombe chimpanzees (*Pan troglodytes schweinfurthii*)
 Total joints scored = 107
 Individuals = 11
 Joints affected = 2 (both, knees)

Powell-Cotton chimpanzees (*Pan troglodytes troglodytes*)
 Total joints scored = 556
 Individuals = 70
 Individuals involved (mod–sev) = 4 (0.057)
 Joints involved = 8 (0.014)
 Joints affected: 1 shoulder; 2 hips; 5 knees (only one severe—left hip)

Powell-Cotton gorillas (*Gorilla gorilla gorilla*)
 Total joints scored = 511
 Individuals = 65
 Individuals involved (mod–sev) = 5 (0.07)
 Joints involved = 11 (0.022)
 Joints affected: 3 shoulders; 4 hips; 4 knees
 (four severe—3 knees, 1 shoulder)

Spinal involvement:

	VOP	Apophyseal OA
Gombe chimpanzees		
Individuals:	11	8
Surfaces:	474	1020
N involved:	0	8 (0.008)
Individuals involved:	0	2
Powell-Cotton chimpanzees		
Individuals:	57	57
Surfaces:	2520	5292
N involved:	11	8
Freq. involved:	0.004	0.002
Powell-Cotton gorillas		
Individuals:	52	52
Surfaces:	2265	4896
N involved:	82	44
Freq. involved:	0.036	0.009

* Except for Gombe results (Jurmain, 1989), data, unpublished (in preparation).

(Willis, 1924; Nathan, 1962) or Terry Collection (Stewart, 1947; 1958; Roche, 1957), or Korean War dead (Stewart, 1958). Several early studies of archaeological populations concentrated both on New World samples (Hooton, 1930; Stewart, 1947; Goldstein, 1957; Chapman, 1962; 1964; 1972; Anderson, 1963) as well as Old World materials (Inglemark et al., 1959; Brothwell, 1961; Jonck, 1961; Bourke, 1967; 1969; Acsádi and Nemeskéri, 1970; Wells, 1972; Swedborg, 1974). This last study, by Swedborg, for its time, and still stands as such, was among the largest ($N=1126$) and most rigorous analyses of spinal degenerative disease (both VOP and apophyseal OA).

More recent studies of spinal degenerative disease, post-dating 1980, include Gejvall's (1983) report on a Swedish series; Merbs (1983) detailed investigation of the Saldermiut Inuit; Bennike's (1985) highly systematic analysis of prehistoric and historic Danish material; Walker and Hollimon's (1989) report on Southern California Indians, and my own analysis of Central California Indians (Jurmain, 1990). Four recent quite systematic studies are those by Bridges (1994), Lovell (1994), Kilgore (1990; n.d.), and Knüsel et al. (1997) (Figure 4-2).

In the first of these recent works, Bridges (1994) investigated both VOP and OA in a relatively large series ($N=125$) from Northwest Alabama. As in numerous other studies, she found the highest incidence of VOP in the lumbar segment and suggested this is a quite

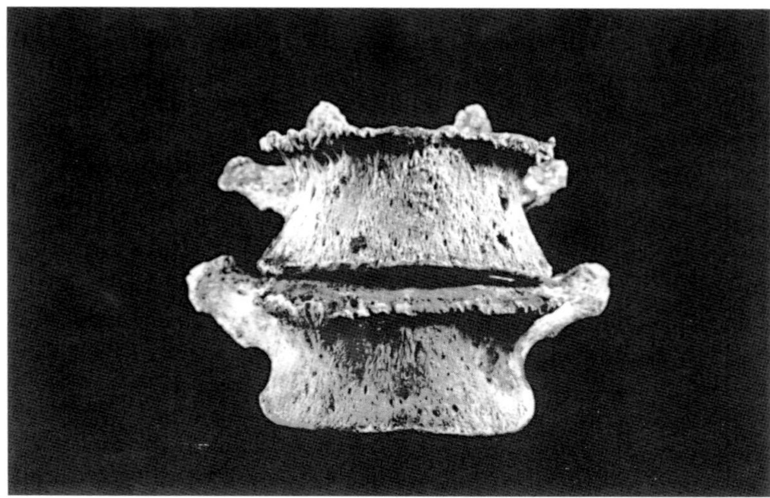

Figure 4-2 Severe vertebral osteophytosis, lumbar vertebrae. Kulubnarti, Sudanese Nubia. Photograph courtesy, Lynn Kilgore.

typical pattern (see below). Bridges also noted that comparisons with other studies were frequently not feasible, as prevalence for individual segments was not usually reported completely. In one comparison she was able to make (with my California data, Jurmain, 1990) she found lower frequency of involvement in the Alabama series than that reported for Central California. Bridges, however, was appropriately concerned with methodological issues, and thus she suggested that the apparent inter-population variance could easily be as much attributed to differences in methodology and data presentation as to real differences in the respective samples (points discussed above). Finally, Bridges also noted that in her investigation, following lumbar incidence, cervical VOP was "unexpectedly" high, and suggested this heightened cervical involvement could be due to activity (see below).

Lovell (1994) analyzed a somewhat smaller and unfortunately much fragmented sample from Harappa (Pakistan). Her analysis was admirably detailed in specifying the types of lesions (i.e., eburnation or lipping) found in apophyseal expression of OA. Moreover, she distinguished in vertebral bodies typical VOP from "focal pitting." However, due to the fragmentary nature of her sample, age control was not possible, nor, obviously, was systematic analysis of the pattern of involvement within individuals.

In a controlled biomechanical study of a quite large sample ($N=138$) from Sudanese Nubia, Kilgore (1990; n.d.) found similar general patterns of both VOP and OA to that reported elsewhere (see below). She thus argued for a multi-factorial explanation for the etiopathogenesis of degenerative spinal disease, emphasizing the effects of age, weight-bearing, and trauma, and only secondarily, of activity. Finally, Knüsel and colleagues (1997) analyzed both VOP and apophyseal involvement in a 13th and 14th Century sample ($N=81$) from northern England. Likewise, these results mirrored those of the other studies discussed above.

Common Patterns of Spinal Involvement

Probably the most relevant and certainly the most *consistent* finding relating to degenerative spinal disease is the regular patterning of involvement. The highest rates of VOP are generally found in the lumbar region, followed by the cervical, and the thoracic vertebrae are most typically the least affected spinal elements.

Consistent results from a variety of studies have supported this uniform pattern of VOP involvement (Roney, 1966; Nathan, 1962; Anderson, 1963; Larsen, 1982; Thould and Thould, 1983; Kilgore,

1984; 1990; Pickering, 1984; Jurmain, 1990; Bridges, 1992; 1994; Lovell, 1994). A few studies, however, have reported a slightly different order of peak involvement, most commonly with the cervical segment incidence just slightly exceeding that found in lumbar elements (Stewart, 1958; Guinness-Hey, 1980; Merbs, 1983; Walker and Hollimon, 1989) (Note: the thoracic segment is uniformly always the least involved vertebral region).

The biomechanics related to bipedal locomotion and the resulting curvatures of the spinal column are most likely the explanation for the regular patterning of VOP, a point made some time ago by Stewart (1966) and much reinforced by Merbs (1983: 169):

> The distribution of vertebral osteophytosis, at least in a general sense, thus seems to reflect a simple mechanical principle, the greater distance the center of the disk from the line of weight transmission, the greater the stress placed on it and the greater the likelihood of degeneration occurring.

In general, for VOP, across varied studies (Nathan, 1962; Swedborg, 1975; Merbs, 1983; Kilgore, 1984; Jurmain, 1990; Bridges, 1994; Jurmain and Kilgore, 1995; Knüsel et al., 1997), the peaks of involvement are at:

C4–C7 (especially, C5/C6)
T7–T9
L1–L5 (especially L3–L5)

Consequently, for VOP, the upper thoracic vertebrae are always the least affected elements. In addition, some data suggest a slightly greater degree of VOP on the right side as compared to the left. Following Nathan's (1962) suggestion, Merbs (1983) also emphasized this asymmetry might be influenced by possible osteogenic stimulation caused by the aorta.

Apophyseal OA also follows a regular pattern of involvement, but one slightly different from that seen for VOP (and here, too, suggestive of biomechanical factors). For apophyseal OA, as observed by several researchers (Inglemark et al., 1959; Nathan, 1962; Merbs, 1983; Kilgore, 1984; Bridges, 1994; Knüsel et al., 1997), peaks of involvement occur at:

C2–C5
C7/T1
T9/T10
L5/S1 (most especially)

It should be noted that for osteological data the relative patterning of apophyseal involvement does not stand out as dramatically as that for VOP. This difference in resolution is the result of the relative infrequency of apophyseal OA as ascertained in human skeletal material, almost always with incidence by element less than 5% (the only exception being L5/S1) (Jurmain, 1990).

What emerges from the data on degenerative spinal disease, especially those relating to VOP, is a clear-cut pattern, presumably related to *general* and fairly uniform biomechanical stresses. These stresses reflect most immediately direct compression loading as influenced by curvature of the spine. Thus, as Merbs (1983) suggested, and as reinforced by others (Kilgore, 1990; n.d.; Bridges, 1994; Jurmain and Kilgore, 1995), degenerative spinal disease does not reflect *specific* activities, but is more the product of "species specific biomechanical parameters imposed by weight bearing and locomotion in an upright posture" (Jurmain and Kilgore, 1995: 448). Bridges (1994: 93) makes exactly the same point: "Much of the patterning seen in arthritis in this and other samples is due to stresses imposed by spinal curvature and weight-bearing due to our erect posture."

Spinal Involvement and Interpretations of Activity

Despite the limitations noted above, osteologists have still attempted to draw some behavioral conclusions from the patterning of degenerative spinal disease. Merbs (1983), for example, was obviously quite focused in his monograph on "activity-induced pathology," but for spinal disease he judiciously kept his conclusions conservative. Merbs, as mentioned above, has emphasized the observed similarities in spinal involvement as likely "characteristic of *Homo sapiens* as a whole" and also suggested that some differences in pattern between populations

> may well be related to differences in activity patterns, possibly the way things are carried on the head or with a tumpline, the way heavy objects are picked up from the ground, or even the habitual posture, *but such activities would be difficult to identify precisely* [emphasis added] (Merbs, 1983: 168).

An explanation of the patterning of cervical OA as possibly related to carrying objects on the head, or by use of a tumpline, has also been proposed by Kilgore (1984) and recently much emphasized by both Lovell (1994) and Bridges (1994). Considering such a possibility, Gejvall (1983), in assessing a well-documented Swedish historical sample, however, does not think the cervical VOP he observed was due to carrying burdens on the head. Bridges noted further that the pattern of cervical involvement she observed in her Northwest

Alabama series was characteristic of Eastern Woodland groups and argued it might well reflect burden carrying or "some other activity."

Lovell (1994) also emphasized the likely influence of carrying burdens on the head and appropriately cited relevant clinical sources (Levy, 1978; Scher, 1978, discussed in Chapter 3). She, however, did not draw the same conclusions from this admittedly restricted comparative base, but she did note the current lack of good clinical analogies. Lovell (as well as Bridges) was justifiably cautious, in not drawing too specific conclusions regarding the role of activity on inducing degenerative spinal disease. Clearly, more clinical data are urgently required, and, in their absence, continued restraint is urged.

Another recent analysis of a small series of prehistoric ($N=18$) and historic ($N=21$) skeletons from Nebraska has also sought to draw behavioral interpretations from the patterning of spinal involvement (Reinhard et al., 1994). In testing possible differences between prehistoric and historic samples in VOP involvement, no osteological differences were found (possibly limited by sample size), and Reinhard and colleagues appropriately suggested most of the VOP observed probably was tied to the influence of aging. Yet, in comparison to OA elsewhere in the skeleton, the authors cited another of their publications (Sandness and Reinhard, 1992) and argued "osteophytosis is a more sensitive indicator of activity-induced stresses." In the face of accumulating evidence to the contrary, such a conclusion is surprising. Further, in the same report, and despite their own caveats, the authors apparently could not resist drawing *some* behaviorally based interpretations. They thus claim that the incidence of VOP in the historic sample (11/21 individuals) compared to that in the historic material (13/18 individuals) "indicates the spines of historic people were under greater activity-related stress" (Reinhard et al., 1994: 67).

Despite some occasional lapses, as indicated above, the current trend in the interpretation of spinal degenerative disease is towards a much more conservative approach. Such a perspective is well reflected in the recent study by Knüsel and colleagues (1997). After controlling their sample for age, lack of clear patterning of vertebral involvement led the authors to conclude that "the vertebral column may not be an ideal structure to study markers of occupational stress" (Knüsel et al., 1997: 481).

Another behaviorally based interpretation of OA in the spinal column related asymmetries of apophyseal involvement, either in the cervical and/or the upper thoracic segments. Both Merbs (1983) and Bridges (1994) suggested these asymmetries may have been produced by the differential influence of handedness. This is an interesting suggestion, but, again, one still lacking supporting clinical evidence.

Without some fairly systematic independent line of evidence, distinct from that of osteological data alone, it is difficult to see how such claims can be falsifiable. In fact, much of apophyseal involvement is highly asymmetric, and more than likely, results from trauma, rather than such chronic wear-and-tear influences as would be produced by differential handedness. Of course, trauma could be linked to culturally patterned (and highly interesting) types of behavior. However, and always remembering the unrelenting constraint of rigor, how does one get at it?

ANTHROPOLOGICAL STUDIES OF PERIPHERAL OA

While many of the early osteological investigations of degenerative disease concentrated on spinal involvement, a number of studies also reported some findings related to OA of the peripheral skeleton. Some of the earliest investigations included those by Elliot-Smith and Wood-Jones (1910) and Ruffer and Rietti (1912) of Egyptian materials; Hrdlicka's review (1914) of Peruvian skeletal materials; Hooton's work (1930) on the Pecos Pueblo collection; Krogman's analysis (1940) of Iranian skeletons; Snow's detailed investigation of the Adena and Indian Knoll collections (Webb and Snow, 1945; Snow, 1948); Goldstein's (1957) brief analysis of Texas materials; Brothwell's (1961) quite detailed report of ancient and modern British materials; Anderson's (1963) systematic analysis of an Iroquois ossuary; Miles' study (1966) of a fairly large sample from Mesa Verde; and finally, Roney's (1966) description of OA in a small sample from California.

Prior to the mid-1970s probably the greatest influence relating to interpretation of OA, especially *behavioral* interpretations come from the published works of J. Lawrence Angel in the U.S. (Angel, 1966; 1971) and Calvin Wells (1962; 1963; 1964; 1965; 1972) in the U.K. (A more detailed discussion of their individual contributions can be found in Chapter 1).

In the late 1970s I began to publish my own findings (Jurmain, 1977a,b; 1978; 1980) relating to patterns of peripheral OA in the shoulder, elbow, hip and knee among modern Americans (Terry Collection), Pueblo Indians, and Alaskan Inuit. Some of these results (those relating to contemporary samples) are discussed in Chapter 2. Regarding the two archaeological samples, the results can be summarized by a few specific conclusions:

(1) OA of all joints was quite rare in Pueblo Indians.
(2) Compared to the Pueblo Indian sample, OA was consistently more frequent among the Inuit (and most dramatically so in the elbow).

(3) Several patterns of within-group variation (correlation with age in different joints, right/left correlation, factor analyses) are suggestive of *possible* biomechanical influences (these specific analyses were carried out primarily on the better-documented and more complete sample of contemporary Americans from the Terry Collection, and the resultant interpretations are discussed in Chapter 2).

Beginning about 1980, there was a marked increase in research relating to peripheral OA by osteologists. There were several reasons explaining this greater interest and intensified research focus. Firstly, and most obviously, a growing number of professionals entered the field, and an appreciable portion of these new Ph.D.'s were trained in paleopathology. Secondly, the growth of the Paleopathology Association itself (founded in 1973) helped foster increased interest in a variety of topics relating to ancient diseases; more specifically, a special symposium sponsored by the Paleopathology Association in 1978 and published in 1979 (Cockburn et al., 1979) generated yet further interest. Thirdly, as some of the more widely circulated articles of the 1970s became better known, a growing perspective emerged among osteologists that osteoarthritis was a highly useful avenue by which to understand behavior (i.e., activity patterns) in the past.

Given the intensive and multidisciplinary approaches of osteoarchaeological interpretation which took hold in the 1980s, the potential offered by evaluation of osteoarthritis was viewed optimistically, and, indeed, proved irresistible. Among the best of such systematic approaches to utilizing skeletal biology in a broader biocultural interpretive framework was the multidisciplinary research led by Larsen, mostly in the Southeast (Larsen, 1982; 1987; 1990), as well as that by Walker, mostly in California (Walker and Hollimon, 1989; Walker and Lambert, 1989) (see below for further discussion, as relating specifically to OA).

Probably the most influential of the publications stimulating this relative burst of behavior-oriented research concerned the elbow and derived from research at the Smithsonian (National Museum of Natural History). Angel's highly influential 1966 publication on the California Indian remains from Tranquillity has, of course, been widely cited by two generations of skeletal biologists. The immediate attraction was both in the concept, a presumably clear behavioral link to a skeletal lesion, as well as an eminently memorable name for it: "atlatl elbow" (Figure 4-3).

What, perhaps, is not as well recognized is that Angel was reporting results for a general skeletal analysis of a small, poorly preserved

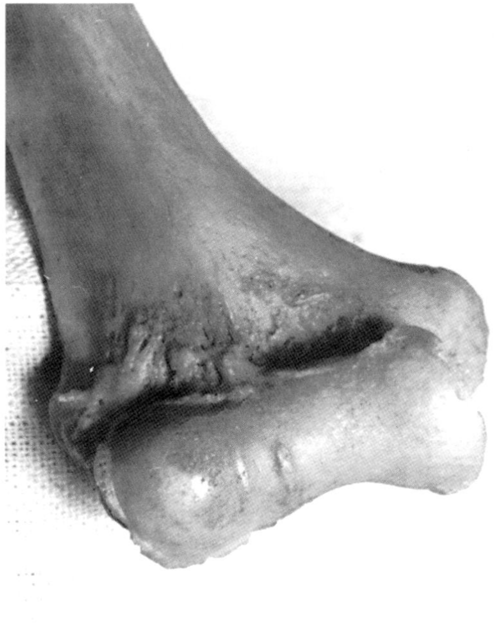

Figure 4-3 Severe osteoarthritis of elbow; distal humerus (capitulum) showing eburnation. Such involvement has been termed "atlatl elbow." Terry Collection, U.S. National Museum of Natural History, male, 71 years. Photograph by the author.

sample ($N = 30$, with only 7 adults complete enough for analysis). Accordingly, osteoarthritis was mentioned primarily in passing. Nevertheless, of the 13 elbows that were adequately preserved, 6 showed marked OA changes (most with eburnation).

Clearly stimulated by the intriguing suggestion made by Angel, Ortner (1968) followed up this initial report with a more detailed study of his own. In this research, Ortner concentrated specifically on elbow OA and compared the prevalence and pattern of involvement between two archaeological populations: Peruvians and Alaskan Inuits. The results were quite striking in contrast of prevalence (Peruvians, 5%; Inuits, 18%). Moreover, the *pattern* of lesions, especially among the Inuit was most intriguing, with most of the severe changes isolated to the lateral (capitulum-radial) compartment.

Ortner interpreted the pattern of elbow degeneration from a functional perspective, emphasizing the mixed flexion-extension as well as rotary motion found in the lateral elbow compartment. Following on this research, and, indeed, at the direct suggestion of Ortner, I further investigated the distribution of OA involvement in all four major peripheral joints, and, again, found the patterning in the elbow, especially among the Inuit, the most distinctive (Jurmain, 1977a; 1978).

Merbs also found exactly the same pattern in his research of Inuit from Greenland which he published in 1983. In this study, of the relatively small Saldermiut skeletal sample, Merbs creatively displayed more precisely the pattern of lesions using simple joint anatomical maps. His approach has more recently been elaborated by both Woods (1995) and Nagy (1996).

Beginning with Angel's insight and developing through Ortner's more focused studies, the elbow eventually assumed a central place in osteological interpretations of OA, indeed, to a far greater degree than it is utilized in contemporary epidemiological research (see Chapter 2). That a particularly serious manifestation of elbow OA should have influenced osteologists so dramatically is interesting, and also has led to a persistent bias. In particular, the most distinctive and reproducible pattern was not found in Angel's California series (or others from there), but among the Inuit (as shown by Ortner, Jurmain, and Merbs). However, there is no reason to suspect that the Inuit are a typical population, not necessarily genetically, nor dietarily, and certainly not behaviorally.

In fact, from a behavioral (activity) perspective, in contemporary circumstances, the best analog with Inuit extreme upper appendage use might be found in baseball pitchers (see Chapter 3). Thus, what initially (in the 1970s and 1980s) appeared as a fairly easy link between OA lesions and a particular behavior might apply *only* to the specific archaeological case from which the model was derived. That it has proven far more difficult to establish such clear associations in other populations, and using other joint areas, while disconcerting, should thus not be a surprise.

Merbs' investigation of Saldermiut activity patterns, indeed, emphasized OA involvement, as reflected throughout the skeleton (spinal disease, TMJ, shoulder, elbow, wrist, hand, costovertebral joints, hip, knee, ankle, and foot). The analysis was admirably intensive, but, unfortunately, the sample (for adults, $N = 91$) was too small for really adequate stratification (by both age and sex together). In fact, Merbs does provide detailed data showing differences by sex (a major emphasis of his study), but does not attempt any age control. As noted previously, for a condition as age-influenced as OA, this lack of control produces a major constraint on interpretation, and, most particularly, on inter-population comparisons.

Other, less ambitious, studies of OA in New World archaeological populations include Martin et al.'s (1979) work on the Dickson Mound material; Pickering's (1979) pilot study and later expanded investigation (1984) of a Late Woodland sample from Illinois; Tainter's (1980) analysis of a Middle Woodland sample from the Illinois Valley, Larsen's incisive review (1982) of biological changes among the St. Catherine's island population; Bradtmiller's (1983) focused interpretation of the effects of horseback riding among the Arikara; Miller's (1985) study of elbow involvement in Arizona; Pierce's (1987) detailed comparison of major joint involvement at Averbuch (late Mississippian) and Indian Knoll (see Appendix A); Walker and Hollimon's (1989) comprehensive review of patterns among Santa Barbara Channel Indians, and Bridges (1991) definitive study of patterns of OA involvement in Northwest Alabama.

It should be noted that much of this same literature (as well as a host of other studies) was comprehensively reviewed by Bridges (1992). In this highly useful review and evaluation of available data (mostly from New World contexts), Bridges selected 25 samples for comparisons (including those which reported joint involvement systematically). Bridges drew several interesting (although not always encouraging) conclusions:

(1) In archaeological samples, the elbow is relatively more frequently affected than in contemporary populations (in complete accord with my own findings—Jurmain, 1977a; 1978).
(2) Either the knee (in 17 of 25 studies) or the elbow (ranked 1st or 2nd in 15 studies) is the most frequently involved of the major joints.
(3) There was no consistent relationship in pattern of involvement related to subsistence economy (i.e., hunter-gatherers *vs.* agriculturists) (see below).
(4) Hand and foot involvement were rarely reported systematically, but when they were, the hand frequently showed high levels of involvement (Pfeiffer, 1977; Larsen, 1982; Merbs, 1983; Walker and Hollimon, 1989; Jurmain, 1990).
(5) Asymmetry of involvement was most apparent in the knee, followed by the elbow (Note: my results suggest the opposite ordering). Bridges also made the telling criticism that even when such bilateral data were presented, statistical tests rarely were applied.

(6) Patterns of sexual dimorphism were quite mixed between studies. Occasionally, males were more frequently involved (particularly so among the Inuit). However, in most of the well-controlled studies, where statistical testing was performed, no significant sex differences in OA prevalence were found.

Not quite as much intensive recent research concerning OA patterning has been done in the Old World, but generally similar results have been found in a variety of studies including: Angel's (1971) influential work on the Lerna (Greece) material; Edynack's (1976) creative analysis of a Hungarian cemetery sample; Griffin et al.'s (1979) study of British and Norse archaeological samples; Rogers et al.'s (1981) detailed study of Romano-British, Saxon, and Medieval English remains; Thould and Thould's (1983) brief analysis of a large and significant series from Poundbury (England); Kilgore's (1984) controlled and detailed study of peripheral OA in Sudanese Nubians (see Appendix A); Bennike's (1985) impressively thorough review of OA in Danish remains (see Appendix A); Waldron's (1992) systematic investigation of a large collection from a Black Death cemetery in London; and Robb's (1994) comprehensive analysis of various "skeletal signs of activity" among skeletons from the Italian Iron Age.

Finally, there have been also some recent studies of populations from elsewhere, including a very large series from Hawaii quickly summarized by Pietrusewsky and Douglas (1994) and Webb's (1995) systematic review of degenerative disease among Native Australians.

Many of these studies of OA patterning are admirable, collecting systematically considerable information on degenerative skeletal patterning. In several instances, behavioral interpretations were also drawn, occasionally stretching the limits of reasonable inference (see below for discussion). In addition, many studies suffer from lack of adequate age control, a point also raised by Bridges (1992). In those cases where more systematic sample stratification was practiced, and the results are more compatible with my own results, specific data are included in Appendix A. These data are provided as a potential resource for future comparative analyses.

Some investigators have sought to deal with potential age effects by initially testing their samples for age bias, *prior* to interpretation of OA patterning. For example, Larsen (1982) and Walker and Hollimon (1989) each statistically tested their respective samples for possible age bias, and failing conclusively to find such bias, proceeded to analyze comparative aspects of OA, as if they were age-independent. The problem here is a reversal of what typically is considered the null hypothesis. The failure to demonstrate differential age bias between samples might easily be a product of sampling error. It is thus not justifiable to assume age-related effects are not influencing the results. Similarly, Tainter (1980) attempted to control for age influences by limiting his comparative samples to individuals over age 35 (given the prevalence distribution of OA *after* age 35, this methodology is also not sufficient; see Chapter 2). The most conservative approach demands that the null hypothesis be: There is no difference in OA prevalence between respective samples (be they male/female or comparisons of different populations), even when controlling for age. Before proceeding with more elaborate interpretations, this basic null hypothesis must, obviously, first be falsified. Such a cautious methodology further demands large, well-controlled samples, but there is simply no adequate substitute for such an approach.

RELATIONSHIP OF PATTERNS OF OA AND SUBSISTENCE ECONOMY

An interesting focus attempted by skeletal biologists, especially during the past two decades, has been to seek information on biological correlates of the transition to agriculture. Among the most commonly used skeletal features investigated for such evidence has been OA. Cockburn et al. (1979) posited a fairly traditional opinion, that, following development of agriculture, stress generally increased (due to "digging, hoeing, back-breaking stooping," etc.). As a result, these investigators suggested the prevalence of OA should have increased among early agriculturists, as compared to their hunter-gatherer precursors.

However, more systematic analyses have failed to confirm this prediction, or, for that matter, establish *any* clear pattern of OA related to subsistence economy. In a comprehensive group effort (symposium and subsequent publication), Cohen and Armelagos requested participants to investigate systematically a variety of potential biological indicators relating to the agricultural transition. Among those indicators explicitly addressed was OA. In the edited volume summarizing the results (Cohen and Armelagos, 1984a) eight articles presented some data on OA prevalence (and a ninth, Kennedy, 1984, refers to it in very general terms for South Asian samples).

Among the participants presenting some data on OA were the contributions by: Meiklejohn et al. for western Europe (Mesolithic/Neolithic); Larsen for the Georgia coast; Cook for the Lower Illinois

Valley; Goodman et al. for the Dickson Mounds (Illinois); Cassidy for Kentucky; Rose et al. for the Lower Mississippi Valley; Allison for Chile; and Rathbun for Southwest Asia (Iran and Iraq).

Owing to inherent constraints, most of the samples employed were quite small, and age controls were not systematic in most of the investigations (with the exception of Cassidy). Larsen, as in the 1982 study of much of the same materials (mentioned above), again controlled for age in an *a priori* fashion. Thus, he argued that the age distribution in his two samples (pre-agriculturists and agriculturists) did not differ, and, in terms of statistical significance, the entire samples apparently did not. Still, for data he presented, among the pre-agriculturists 29.9% of the sample were aged 35+, while among the agriculturists only 11.3% survived to age 35 (It is thus not at all certain whether OA differences between the samples might have been strongly influenced by their differing age distributions).

Without more systematic standardization, it is difficult to conclude much from the varying reports, and the symposium editors (Cohen and Armelagos, 1984b) also could not identify any real consensus concerning the patterning of OA as it related to subsistence economy. Nevertheless, there is a persistent tacit assumption running through the papers that OA *should* somehow reflect activities (and would do so, if only the data were adequate). Some representative comments include:

> Individuals who habitually engage in activities which put strain on the joint systems are more likely to show degeneration (Goodman et al., 1984b: 35).

Mieklejohn et al. argued that the differences in their European samples showed:

> heavier biomechanical stress in the Mesolithic than in the Neolithic (1984: 83).
>
> On the other hand, there is a reduction in functional demand on the postcranial skeleton as is indicated by the decreased frequency of degenerative joint disease (Larsen, 1984: 379).
>
> Degenerative pathology is inferred to be due to increased physical-work stress or interpersonal strife (Goodman et al., 1984a: 300).
>
> the reduction in degenerative disease indicates a less rigorous life style (Rose et al., 1984: 417).
>
> The relatively high frequency and severity of degenerative change can be attributed to physical stress and not to age alone (Rathbun, 1984: 156).

Rathbun, however, did add a partial caveat, recognizing some degree of complexity as well as the potential influence of age of onset,

> No simple pattern or explanation should be invoked, since osteoarthritis changes occur early in life compared to modern standards (p. 159).

Such behavioral assumptions have more recently been subjected to increased critical evaluation, a topic discussed in detail in Chapter 3, and elaborated below. In further research to establish some kind of consistent relationship of OA and the agricultural transition, even the best efforts have continued to be frustrated. Most notably, Larsen and colleagues, especially for data from the southeastern U.S., have marshaled an impressive array of methodological approaches to address the biological nature of the hunting-gathering/agricultural transition (Larsen, 1982; 1984; Larsen and Ruff, 1984; Larsen, 1995). Some of the evidence (e.g., dental data) has proved highly informative. Regrettably, skeletal data on OA have not generated anything approaching consistent results. For example, Larsen initially suggested that a decrease in OA with agriculture in coastal Georgia indicated a less active life style, but in his more recent thorough overall review (1995) of biological correlates relating to the agricultural transition, he observed, more generally, an increase in mechanical loading accompanying agriculture. In the end, he concluded, however, that OA provided no conclusive evidence of any general trends.

This lack of a consistent pattern relating to subsistence economy was also noted by Cohen (1989) who suggested "skeletal evidence of physical stress (arthritic and trauma) does not show clear trends through time" (Cohen, 1989: 117). Despite such highly inconsistent results, however, Cohen did not appear seriously to consider alternatives to activity-related explanations of OA patterning, and, in fact, continued to argue in favor of the stress hypothesis: "High rates of physical labor are recorded in the skeleton in the form of degenerative joint disease" (Cohen, 1989: 109).

Lastly, Bridges (1991; 1992) in her detailed and balanced discussion of the entire topic of OA and subsistence economy found basically no correlation (a finding directly contradictory to some of her biomechanical data; see Chapter 7 for further discussion). The results, concerning OA specifically, proved so discouraging, in fact, that Bridges questioned the utility of using OA at all to reconstruct activity. As a possible alternative, she suggested that OA might not reflect so much consistent types of loading, as traditionally implied by the stress hypothesis, but rather intensive/infrequent activities. She does not, however, suggest how one might test this hypothesis; and certainly from skeletal data *alone*, rigorous testing would not be feasible.

OA AND BEHAVIORAL INTERPRETATIONS

As suggested by many of the quotes and other discussion above, for many osteologists the central component of their OA investigations is ultimately to reconstruct behavior; indeed, it is frequently the *reason* for such research in the first place. Yet, as this book emphasizes, many of the assumptions underlying this perspective are inherently biased and are, for the present, essentially unproven. Thus, given the motivation, "alluring prospects," irresistible temptations, etc., anthropologists have devised a host of imaginative scenarios attempting to link osteoarthritic changes to a variety of activities.

Opinions, of course, to some degree do vary, and in recent years greater critical evaluation of the validity of using OA to reconstruct ancient activities has been voiced by numerous researchers (discussed below). Nevertheless, over the last 30 years the overwhelming consensus has been the widely held view that OA can be used to reconstruct past activity patterns, both at a general level and, for many zealous practitioners, even for quite specific contexts.

As noted, Wells and Angel both were among the earliest to emphasize behavioral interpretations of OA. Wells (1964), for example, in commenting on a possible explanation of cervical involvement in Nubian skeletons, presumably due to carrying burdens on the head, was actually quite restrained: "It is a neat explanation, but rather too facile to be satisfying because cervical disease also occurs in peoples who never carried burdens in this way" (Wells, 1964: 64).

Still, as was more characteristic of Wells' approach, he was drawn more strongly to a general functional explanation of OA, concluding regarding its etiology that "injury, especially repeated episodes of minor stress is the most important single factor in determining the joints it will attack" (Wells, 1964: 60). In fact, Wells goes on to make a variety of highly speculative suggestions regarding particular joint involvement, including that of the first metatarsal in Macedonian soldiers (due to "marching in poor footwear") and in the wrist and thumb in west and southern Africa (due to hoeing hard soils).

As discussed above, Angel's (1966) investigation and naming of "atlatl elbow" in a small California Indian sample was to have a wide influence. Moreover, as two of the involved individuals were thought to be females, Angel suggested "seed grinding" as another possible cause of severe elbow disease.

In the 1970s the trend to emphasize behavioral interpretations continued, as in Edynack's survey of Medieval Hungarian skeletal material, most notably, her suggestion of pelvic and limb involvement being the result of horseback riding, an inference also proposed for the Arikara by Bradtmiller (1983). I also was publishing my first results in 1977, arguing:

> Given enough time, the cumulative effects of biological aging and other systemic agents will eventually produce some degenerative disease in all joints, but the crucial factor determining expression of degenerative disease in severe form at an early age is the presence of chronic, severe functional stress (Jurmain, 1977a: 364).

In the same year, regarding knee involvement specifically among the Inuit, I further suggested,

> The greater frequency of degenerative involvement apparently is in large part directly related to the kind and amount of functionally induced stress (Jurmain, 1977b: 8).

In this interpretation of knee involvement, and in succeeding publications dealing with elbow OA (Jurmain, 1978), I not only emphasized general functional components of OA etiology, but also attempted to draw numerous *specific* behavioral conclusions from the pattern of involvement (discussed further below and also included with other such imaginative suggestions in Table 4-2).

Table 4-2 Some suggested* behavioral associations inferred from skeletal OA in Inuit samples (after Jurmain, 1978; Merbs, 1983)

Suggested activities producing elbow disease (Jurmain, 1978)

	Bilateral	Unilateral
Males:	Use of ice prod sled driving and pushing rowing	Harpooning, lancing bola throwing; bow and arrow use butchering
Females:	Sled driving and pushing hide preparation	Bola throwing

Suggested activities and peripheral OA (Merbs, 1983)

Joint area involved	Activity
TMJ	Hide preparation (females)
Acromio-clavicular	Kayaking; harpooning (males)
Shoulder (gleno-humeral)	Paddling; bow and arrow use (males)
Elbow	Kayaking (males); throwing (males)
Elbow	Scraping skins (females)
Wrist	Kayaking (males)
Wrist	Scraping skins (females)
Interphalangeal	Sewing (females)

* These inferences were merely suggestions at the time of their publication. They are most certainly *not* meant to be taken literally. Indeed, such specificity is generally not warranted when using osteological data (see discussion in text).

The organizers of the Paleopathology Association symposium on osteoarthritis also were somewhat drawn to behaviorally based interpretations, and thus suggested: "The pathology so produced in a joint may be useful to indicate the individual's way of life" (Cockburn et al., 1979: 74). This conclusion indeed was more cautious than many of my own inferences at that time, and these authors went on further to suggest how one might approach *testing* such a behavioral hypothesis: "The interpretation of the lesions can sometimes be checked for accuracy by comparative studies with peoples living in similar circumstances today" (p. 74). Cockburn and colleagues clearly acknowledged that it is not always possible to find appropriate clinical models, but their relative caution and emphasis on contemporary analogy is similar to that argued here (see Chapter 3).

Tainter (1980), in his behaviorally focused study (especially relating to status) of an Illinois Valley skeletal series, left little doubt regarding his position: "... any variation in such motion patterns that are related to specialized roles should be reflected in the distribution of degenerative involvement in skeletal material" (Tainter, 1980: 309).

Larsen (1982), echoing some of the same perspectives mentioned above, concluded similarly,

> Although degenerative joint disease is complex and influenced by both systemic and mechanical factors, the differences between human populations can be best interpreted in the light of *mechanical stress associated with lifeway* [emphasis added] (Larsen, 1982: 225).

Larsen surmised that the prevalence of OA was most likely related to the (total) time spent in subsistence-related work, a view at odds with Bridges' (1992) suggestion, and one no longer strictly advocated by Larsen.

In addition to the systematic and, understandably, broadly influential work of Larsen, at this same time publication of Merbs' detailed monograph (1983) was also to have a wide impact. While Merbs in his initial discussion (cited above) remained prudent in his approach, the primary thrust of his work, as made clear in its title, and as seen most obviously in his conclusions, was to make specific associations of pathological changes (especially OA) with behavior. Curiously, in dealing with interpretations of OA, Merbs relied less heavily on clinical support than he did in his ground-breaking research on spondylolysis (Merbs, 1995; 1996) (see Chapter 5). As a result, Merbs' conclusions relating to OA accounted more superficially for its multi-factorial etiology, leading him more irresistibly to simple behavioral interpretations (see Table 4-2).

Clearly following the impetus of Merbs (as well as others), Miller (1985) in her interpretation of elbow OA in a small Arizona sample unambiguously stated her position: "The pattern and frequency of degenerative disease of the elbow in the series form Nuvakwewtaqa leaves little doubt that activity patterns are implicated in the etiological process" (Miller, 1985: 395).

In their detailed review of a large series ($N = 967$) from Southern California, Walker and Hollimon (1989) defined a moderately cautious general view:

> The association between osteoarthritis and habitual movements means that the everyday activities of prehistoric humans can to some extent be reconstructed through studying the distribution of osteoarthritis of the skeleton (Walker and Hollimon, 1989: 171).

Nevertheless, in their attempts (and eagerness) to identify specific skeletal markers relating to subsistence economy and cultural transitions in the past, these investigators were drawn, as have so many others, including myself, into making explanations of how specific behaviors actually *produced* specific OA changes.

As discussed in some detail above, Bridges (1992) has probably attempted the most comprehensive review of various paleoepidemiological studies of OA. She also addressed several important methodological issues and remained judiciously cautious regarding overly specific behavioral interpretations. Since, in her opinion, long-term behaviors do not seem to produce easily interpretable skeletal changes, Bridges alternatively posited that "High intensity infrequent forces may injure the joint while low-level habitual activities may not" (Bridges, 1992: 87).

In recent years the tendency to interpret OA using an explicit behavioral focus has continued. Larsen and Ruff (1994: 31) assumed that OA was due to "wear and tear of articular joints due to activity." Likewise, Pietrusewsky and Douglas (1994: 186) concluded that OA changes reflected "for the most part biomechanical wear and tear and functional stress." Finally, Larsen (1995: 200) succinctly summarized this behavioral perspective:

> Because degenerative changes in joints result largely from physical demands occurring over the course of an individual's lifetime, their prevalence in past populations provides an important perspective on activity in both living and extinct societies.

Not all skeletal biologists have been as eager to utilize OA evidence to reconstruct behavior, and some of these more critical perspectives have been articulated for some time. For examples, Ortner's (1968) biomechanical analysis of elbow OA was clearly quite functionally oriented.

However, his careful depiction of specific functional components within the elbow allowed more precise testing of his specific hypotheses. Moreover, Ortner realized that, at a more fundamental level, such functional interpretations required controlled studies to test even the most basic assumptions. "Whether the difference is genetically determined or results from a culturally patterned adaptation to a society's natural environment must await studies which control for these two variables" (Ortner, 1968: 145).

Rogers et al. (1981) were also quite careful in their interpretations of degenerative disease (and other arthropathies) as found in ancient British skeletal remains. They particularly were cognizant of the possible confusion which could be generated by populations also containing individuals presenting with DISH. These researchers thus suggested that some aspects of the skeletal patterns they found (particularly shoulder involvement among Saxons) *may* relate to 'occupation;' but, as the involvement mostly includes exuberant osteophyte, they concluded the conditions "both may be part of an inherited tendency to form excess bone" (Rogers et al., 1981: 1670).

Perhaps the most incisive comment on limitations of the traditional osteological interpretation of OA patterning was made by Bennike.

> It is otherwise interesting to study the many different explanations offered to account for a high incidence of osteoarthritis in connection with skelet[al] studies. Skeleton finds from very different cultures may be involved, and no matter what group be represented—hunters, riders, farmers, men or women—it is apparently always possible to find a reason for the pattern obtained (Bennike, 1985: 137).

Bennike's own sample of prehistoric and historic Danish remains was initially quite large. However, she rigorously controlled her evidence for sex, age, and cultural period, leading to subsamples which often were fairly small. Bennike, in addition to commenting ruefully regarding the interpretive fancies of others, maintained rigor in her own work. Thus, even given some tempting hints of pattern (in her small subsamples), she concluded: "There is thus insufficient basis for discussing the results further or drawing conclusions from the tendencies observed" (Bennike, 1985: 139).

As noted, Bridges also has intensively investigated degenerative disease from a behaviorally oriented perspective (Bridges, 1990; 1991; 1992; 1993; 1994). She has, however, been generally prudent in drawing specific conclusions from the patterning of OA changes. For example, in her comparative analysis of upper appendage involvement, she was not able to demonstrate any pattern apparently linked to weapon use. In other words, if the males in her sample had "atlatl elbow," she concluded it was "not caused by atlatl use." In further comparisons, and with additional reflection, Bridges has grown even more cautious, especially regarding the specificity of OA patterning.

> However, in most cases it is not possible to associate specific types of arthritis with particular behaviors, precisely because so many kinds of activities or injuries can result in joint trauma. Increased arthritis may be a sign of a change in the level or type of physical activity, but it is not generally possible to specify which activity or combination of activities caused joint deterioration (Bridges, 1992: 82).

Finally, Webb (1995) in his comprehensive review of paleopathology in Australian skeletal material exercised considerable restraint in his interpretation of OA, arguing:

> The functional stress hypothesis has been used widely to make predictive statements concerning human behavioral patterns in past societies, especially hunter-gatherer and early farming communities. This logical approach has been recently questioned, however, in terms of its value, particularly if it cannot be supported by ethnohistorical or ethnographic data (Jurmain, 1990). In other words, *one should not expect to be able to say too much about specific activities from the pattern and frequency of arthritis* [emphasis added] (Webb, 1995: 173–174).

OA Involvement and Specific Activities: Historical Contexts

Some of the more exaggerated claims relating to potential behavioral influences, as inferred from skeletal patterning of OA, have come from recent studies of historical/contemporary materials. Perhaps, since in these contexts there is usually some historical documentation relating to occupation and so forth, osteologists feel more compelled to use this information creatively. It thus seems there is an even greater willingness to link *something* observed in the skeleton to behavioral information obtained from historical sources. However, and quite obviously, the quality of (independent) behavioral data almost always is not up to the task (and are certainly much more limited than the clinical sources reviewed in Chapter 3). Accordingly, the hypotheses derived from this work are no more testable than ones proposed from more ancient contexts; thus, the reasoning is usually equally circular.

Lai and Lovell (1992) attempted to relate various skeletal markers seen in three historic male skeletons from Canada who presumably engaged in fur trading. In order to bolster their hypothesis, a remarkable array of lesions was interpreted, including enthesophytes, Schmorl's nodes, etc. (for further discussion, see Chapter 5). For degenerative disease, lesions evaluated, and uniformly attributed to behavioral causation, included VOP (and vertebral compression), spinal OA, and peripheral joint OA (shoulder, elbow, hip and knee). In most cases, however, identification was restricted to osteophyte development.

The care and detail of such a thorough osteological investigation are admirable. However, building an argument of weakly supported data, even apparently copious amounts of such data, does not strengthen one's case. Such attempts are analogous to building a house on a weak foundation; further attempts to reinforce the structure by adding more rooms, all with equally shaky underpinnings, will not strengthen the edifice. In the end, the whole structure falls at its weakest point.

Reinhard and colleagues (1994) also sought to relate specific behaviors to OA lesions found in human skeletons from Nebraska. Although their sample was initially reasonably large ($N=92$), only 37 adults were available from the historic period collections. The authors stated their perspective clearly: "If, extrapolation from ethnographic and historical writing are correct, these alterations in life experience should have left *recognizable signatures* on the skeleton" [emphasis added] (Reinhard et al., 1994: 63).

In part, the authors were referring here to possible evidence of inflammatory disease, and such historical documentation may indeed provide useful clues for diagnosis of some lesions (at least narrowing the range of reasonable possibilities in a differential diagnosis). However, for OA, with a highly non-specific etiology, the logic is backwards. Socio-behavioral sources predict *nothing* about skeletal involvement. Only the skeletons themselves can provide the hard data, and the lesions must be linked directly and *independently* to *known* (well-documented) behavioral environments (i.e., best done in clinical contexts).

Larsen et al. (1995) undertook a similar study of historic cemetery remains from Illinois. Perhaps, because of greater temptation to make interesting behavioral associations in such material, the researchers appeared more ambitious in this regard than is typical of their work on more ancient remains (e.g., Larsen 1982; Larsen and Ruff, 1994). They thus concluded:

> The Cross series exhibits a number of pathological conditions that are related to activity, including degenerative joint disease (also called osteoarthritis), non-articular joint changes resulting from posture, and fractures arising from trauma (Larsen et al., 1995: 147).

The evidence to support these claims, regardless of obvious theoretical considerations, is far from overwhelming, as this was a very small collection. The authors did note that, of the 11 adults (>35 years), eight showed signs of some OA. However, such a prevalence is quite typical of most human populations, especially when characterized by osteophyte formation, so it is difficult to see how such evidence argues for a behavioral interpretation.

A final example of imaginative inference was recently published by Hershkowitz et al. (1996). From the Todd Collection of dissecting room specimens these authors identified an individual who presumably was a boxer, yet no evidence of independent documentation was presented. Thus, the reader is left to assume the only evidence to support this occupational attribution derived from the skeleton, inferred from such amorphous lesions as "damage" of the humeral lesser trochanter (tubercle?). Hershkowitz and associates interpreted such skeletal evidence as unambiguous in its behavioral implications, suggesting the lesions,

> possibly resulted from repeated forceful medial rotation of the humerus during punching while training and boxing (p. 177).

and they thus speculated:

> When we apply other information, e.g., lesions of the humeral lesser trochanter, it becomes clear that we are not dealing with [a] deliberate accidental isolated case of violence, but rather with an individual who was engaged in frequent hand-to-hand combat (Hershkowitz et al., 1996: 178).

As a counterpoint to such bold inference, stands Walker's (1997) recent systematic investigation of clinical ramifications on the skeleton of boxing, particularly as relating to trauma (see Chapter 6 for further discussion).

It would thus seem that the accumulating literature (both clinical and anthropological) on OA has done little to dent enthusiasm regarding behavioral interpretations as derived from skeletal involvement. Quite the opposite trend, in fact, seems apparent. Indeed, in many cases, the assumptions regarding the theoretical bases of such conjectures are taken as established. While in more controlled paleoepidemiological surveys, obvious controls regarding sampling and rigor of hypothesis testing are taken as basic, in some contemporary research, occupation (and other specific behavioral conclusions) are thought reasonable to infer from one (or a handful) of skeletons. What such an approach actually *proves*, is not at all clear.

ADEQUACY OF ETHNOGRAPHIC DATA FOR BEHAVIORAL INFERENCE

If the historical documents relating to recent finds are usually inadequate regarding *specificity* of behavior, to establish adequately relationships with skeletal lesions, what then of ethnohistorical accounts? These, of course, are almost universally even more meager in the types of specifics that a skeletal biologist requires. For example, in my own research I concentrated heavily on Inuit behavioral patterns, making a variety of conjectural, but, ultimately, highly superficial behavioral inferences (see Table 4-2). Inuit peoples are, in fact, among the *better* documented archaeological groups, and some reasonably detailed ethnographies were accumulated while some individuals still practiced a traditional lifestyle (Boas, 1888; Steensby, 1910; Birkett-Smith, 1929; Griffin, 1930; Nelson, 1969; Spencer, 1969).

Still, the resolution of detail regarding specific behaviors was frustratingly insufficient. One could glean very general patterns (use of harpoon, rowing, ice fishing, sled use, etc.). to obtain "relevant" behavioral details from the available (but imprecise) literature; it was thus a simple matter for me to cull scattered glimmers of evidence from otherwise rich ethnographies (all the easier, in fact, when one was looking for certain *expected* types of activities).

Merbs' (1983) Inuit study group provided, in many ways, superior data to the Inuit samples I used. His collection, from just one village, thus represented a more cohesive sample, and ethnographic details were collected on the village shortly prior to its extinction (in 1902–1903). Some of the individuals in his sample probably died in that last winter of 1902–1903, and were thus probably among those individuals observed by visitors to the village. Merbs was thus entirely correct in asserting his enthusiasm for the tremendous potential offered by the materials (skeletal and documentary).

Nevertheless, even in such relatively ideal circumstances, much specific information was lacking. For example, Merbs was keenly interested in types of rowing behavior, particularly as pertaining to upper appendage degenerative patterning. However, it was not possible to document clearly whether the Saldermiut Inuit used kayaks or umiaks (an important point, as the oars, rowing stroke, and concomitant biomechanical stresses vary). Thus, even with such feasible conditions, to try to associate a groups' prior behavior with what is preserved in the skeleton leaves many gaps. As we reach further into the past, with ever more tenuous forms of documentation, the gaps grow larger.

Therefore, what is generally not ascertainable in archaeological contexts includes such data as:

(1) *who* practiced what behaviors?
(2) for *how long* were the behaviors practiced; in the course of a day, a year, a lifetime?

(3) *at what age* did individuals begin the behavior?
(4) how much variability was there among individuals in 'occupational' specialization?
(5) how exactly were specific behaviors practiced (be it spear-throwing, rowing, etc.); i.e., what biomechanical loads were placed on joints?

From an epidemiological viewpoint, what skeletal biologists dealing with archaeological collections, even presumably well-documented ones, lack is any precise information concerning: amplitude of stress; age of onset of stress; and duration of stress (note further, large samples are rare, and samples with good age controls, even rarer). It would be hard, even in the best of circumstances, to infer even general levels of activity strictly from skeletal patterns of OA. Attempting to reconstruct *specific* activities in such contexts is scientifically unjustified.

CONCLUSIONS

Much of the data presented here are conflicting, and the overall picture of the utility of OA as a behavioral ("activity") indicator is discouraging at best. What then can be done? Two researchers, personally familiar with both clinical contexts and skeletal manifestations of OA, have made some possibly useful suggestions regarding understanding of *overall* patterns of involvement. Waldron (1994a) suggested that looking at the combination of different joints affected in different groups might be informative, and Rothschild (1995) emphasized also that understanding the composite picture of population involvement might provide more information than looking at individuals (an approach used productively by Rothschild and associates in investigations of systemic arthropathy; Rothschild et al., 1988). This active interplay between clinical research and osteological investigation most certainly offers the most productive avenue for increasing future understanding of skeletal manifestations of OA. Such a useful interaction has also been employed successfully by Rogers and colleagues (e.g., Rogers and Dieppe, 1993).

Nevertheless, the patterns will not likely prove easy to interpret. While, as Waldron suggested, the overall patterning of multiple joint involvement might be most useful, comparative results thus far obtained are not encouraging. Bridges (1992) in fact, could find very little consistency among various skeletal investigations, except that the elbow tends to be relatively more affected in archaeological samples than in contemporary ones.

Bridges (1992; 1994) made the excellent point that different methods of scoring are probably contributing to some of this inconsistency.

This is a view with which I completely agree. Thus, probably the most urgent current need is to have more widely employed systematic methods of scoring skeletal OA. To this end, it has been suggested here that osteophytes not be used in such evaluations, but rather diagnosis should depend on identification of eburnation. At a minimum, such a methodology would greatly assist in promoting inter-observer reliability. Further, as discussed in Chapter 2, and following the standards recommended by Buikstra and Ubelaker (1994), separate scoring criteria should be utilized for eburnation, surface pitting (i.e., porosity), and osteophytes.

Beyond some internal controls related to methodological issues, at a broader etiological level, skeletal biologists must come to recognize that, even under the most ideal of circumstances, biomechanical factors (which would correlate with activity) play only *one* part in initiating the disease, and perhaps only a small part. The multifactorial nature of OA is not simply a challenge to overcome, so to distill out the "useful" pattern, it is a fundamental complicating reality that itself conveys real information—and may, in the end, frustrate all attempts at simplification.

There are certainly unique components to OA patterning in *some* populations. Perhaps, in fact, the Inuit are the most unusual, in the sense that the prevalence and distribution of OA in their joints differs markedly from that seen in most other groups (and, most especially, in the elbow). Looking at the world of skeletal OA through the fascinating, but ultimately narrow, window of Inuit elbow disease, has served in my research (and more broadly as well) to overly emphasize the value of behavioral interpretation. In many cases, the resulting enthusiasm for such an approach has overwhelmed judicious science and produced simplistic, untestable scenarios.

Thus, most behavioral reconstructions are inherently circular, since there is no independent means by which to test the hypotheses proposed, other than superficial analogs selectively gleaned from woefully imprecise sources. What do we really *know* of the behaviors, especially, activities, of past populations? From archaeological and ethnohistorical accounts, usually, the information is frustratingly incomplete. But the best source to more fully complete the picture is not necessarily from skeletal data, at least not skeletal involvement relating to OA. Twenty years ago I was much more optimistic regarding the potential of such an approach, for reconstructing both general and specific activity patterns. Now, it seems that what we have learned from skeletons is unreliable and what we can surmise from clinical sources is mostly discouraging. If we are confidently to reconstruct activity patterns in most ancient populations, it appears we will have to look elsewhere.

Chapter 5

Enthesopathies and Other Osteological Indicators of Activity

In fact, researchers have had a difficult time conclusively demonstrating that any specific activity is the cause of a particular morphological feature. In, part this may be because bones are influenced by a wide variety of activities; their structure represents a balance between often competing forces (Bridges, 1996b: 117).

INTRODUCTION

In addition to osteoarthritis, several other osseous markers have been suggested in recent years to result from physical activity, and in some cases, they have been argued to relate to quite *specific* activities. In particular, various osteologists have postulated that remodeling changes at tendinous and ligamentous insertions, that is, so-called "enthesopathies," are potentially good indicators of physical activity. In addition, other such changes as Schmorl's nodes, stress fractures (including, especially, spondylolysis), parturition scars, and development of auditory exostoses have all been proposed as indicators of activity. Some of these conditions, notably spondylolysis and auditory exostoses, each appear to develop as a consequence of a now well-established etiopathogenesis. Especially, in the case of spondylolysis, detailed correlations with clinical manifestations, as thoroughly researched by Merbs, have provided convincing evidence, and thus added considerable rigor to such osteological ventures.

However, the majority of these other osteological manifestations have received but tenuous clinical support. Nevertheless, in recent years, uncritical and oftentimes exaggerated behavioral reconstructions utilizing these bone "markers" have proliferated in the anthropological literature. Frequently, possibly informative, although, admittedly, tentative, clinical data are simply overlooked. Perhaps, even more disturbingly, in much contemporary osteological research, the requirement of any substantive independent documentation is tacitly dismissed; the results, of course, are highly reinforcing, but reflect largely circular arguments.

ENTHESOPATHIES

Enthesopathies are skeletal manifestations associated with tendinous or ligamentous insertions. Some researchers (Stirland, 1991) draw a distinction between tendinous insertions (entheses) and ligamentous attachments (syndemoses). However, most clinical sources (Resnick and Niwayama, 1983; Doherty and Dieppe, 1986) refer to both types of lesions as enthesopathies (or simply as "entheses").

> In the broad sense an enthesis is the site of insertion of a tendon, ligament, or articular capsule into bone. An alteration at this site is most accurately termed *enthesopathy* and can accompany many disorders including traumatic, inflammatory, endocrine, and metabolic conditions (Resnick and Niwayama, 1983: 1).

At such insertion sites, the biomechanical stresses are particularly demanding, as two very different types of materials, with highly different elastic properties (as measured by Young's modulus), are joined. Currey (1984) pointed out that, in standard engineering designs, rarely are two such different types of materials joined *directly* with each other:

> For engineers the problem of attaching two materials of very different Young's modulus may be considerable. Usually the materials cannot be bonded to each other, because the strains each material undergoes near the interface are very different and lead to large stress discontinuities. These discontinuities in turn make adhesion difficult. Often, rather complex knots or other fastenings must be produced. In fact, there are few examples in engineering technology where a low-modulus material that is bearing a high tensile stress is attached to a high-modulus material. One example is attachment of ropes to boats (Currey, 1984: 191).

Even though tendons, as compared to bone, show elastic properties differing by an order of magnitude, the physiological joining of the two tissues is effectively accomplished in biological systems through gradual integration of variable tissue zones:

(1) tendon or ligament
(2) unmineralized firbrocartilage

(3) mineralized firbrocartilage
(4) bone
(Noyes et al., 1974; Resnick and Niwayama, 1983; Mankin and Radin, 1985).

The transition between bone and tendon or ligament is greatly facilitated and the forces evenly distributed, since "the fibers of the ligament or tendon become more compact, then cartilaginous, and finally calcified before entering osseous structure" (Resnick and Niwayama, 1983: 1). The transition of tissue structure is also characterized by changes in fiber orientation. In tendons, the orientation of connective tissue fibers differ from that in most areas of bone. On the surface of bone, especially, the ("intrinsic") fibers are usually roughly parallel to the bone. The ("extrinsic") penetrating tendon/ligament fibers often are at a large angle, and where they merge, the mineralized fibers are often referred to as "Sharpey's fibers." In this way, the low modulus tendon penetrates into the high modulus bone, but transforms to high modulus tissue *very* quickly (Currey, 1984).

Etiology

With but a few exceptions, clinicians have not addressed the topic of enthesopathy very intensively. The reasons are quite obvious. Such lesions are hardly life-threatening and, indeed, rarely produce identifiable symptoms. Indeed, except for some secondary involvement associated with spondyloarthropathy and DISH, they are largely overlooked clinically. Moreover, the changes are subtle, and even in related studies employing some radiography, the lesions do not image consistently (thus remaining almost "invisible" to the clinician).

In fact, in a recent Medline search for enthesopathy references encompassing 1985–1996, only 24 sources were found. Of these, eight related to spondyloarthropathy, one each to trauma/inflammation, leprosy, Vitamin D-resistant rickets, and shape factors, five to treatment, and six had some aspect of activity included. However, all of these latter sources related specifically to "tennis elbow," and only one such source for the entire decade was in English (the other five were published either in German or Serbo-Croatian). It would thus seem that the topic of enthesopathy has not exactly fired the imagination of our clinical colleagues.

Nevertheless, over the last several years there is still some useful source material. While not addressing enthesopathy separately, several studies have investigated more general aspects of ligament/tendon injury and repair. Experimental data, for example in dogs (Laros et al., 1971) and rhesus monkeys (Noyes et al., 1974), have shown that near-total immobilization weakened ligaments. However, in these studies the controls were not adequate to assess relative effects of *normal ranges* of activity on the tissues. Noyes and colleagues thus made very cautious conclusions and clearly point to a major gap in our knowledge:

> Thus, conclusions on the beneficial effect of exercise on strength of ligament units must be qualified as to the type of activity or exercise involved.

Even though it frequently is assumed that exercise presumably increases ligament strength,

> the reported increase in ligament strength above control values as a result of increased activity may represent instead the prevention or reduction of the disuse effect of caging. Inasmuch as ligament units respond to mechanical stimuli, it *may* be possible to increase the strength of bone and ligament structures above normal by exercise and other activity-related procedures; however, to our knowledge, this has *not been experimentally or clinically documented* (Noyes et al., 1964: 1416) [emphasis added].

In a more detailed evaluation if the biomechanical/biochemical effects of activity/immobilization in both rats and dogs, Tipton et al. (1975) also collected data on ligament water content, collagen concentration, and elastic stiffness. Interestingly, none of these variables seemed to be influenced by exercise "in normal animals." Exercise, however, was found to be important in facilitating repair following acute injury (in agreement with the experimental studies mentioned above).

Further, Tipton and colleagues investigated the effects of hormones (ICSH, testosterone, TSH, thyroxin, ACTH, and growth hormone) on repair of ligaments following surgery. Of note, ICSH and testosterone *increased* the repair process, and resulting tissue growth, while the other hormones decreased repair strength. In humans, as well as other mammals, the relative effect of testosterone on bone turnover as well as response of the tendon or ligament/bone junction may well be a major influencing factor; it thus also may produce considerable sexual dimorphism in accommodation to "normal" exercise (see below).

Resnick and Niwayama (1983) comprehensively reviewed a variety of factors which can influence enthesopathy, including: inflammatory rheumatic conditions (especially, seronegative spondyloarthropathies such as ankylosing spondylitis); DISH, degenerative disease (i.e., osteoarthritis); endocrine disorders; CPPD; and trauma. Doherty and Dieppe (1986) also noted the possible association of enthesial involvement in CPPD (tendon insertion calcification, especially Achilles, triceps, and quadriceps obturator), and Shaibani et al. (1993) suggested possible secondary involvement in a number of conditions (especially spondyloarthropathy). Indeed, given all the possible conditions mentioned (Resnick and Niwayama list more than 20), the relative silence regarding effects of activity is noteworthy. It should also be noted that Stirland (1991) emphasized as well the likely complex etiology of enthesopathy, making the timely reminder that "The problem is once again one of specificity" (p. 41).

The types of lesions most commonly addressed by osteologists are hypertrophic changes at usually well-defined attachment points, what are sometimes called "enthesophytes." However, other types of reactions can also occur at tendinous/osseous junctions which *might* be informative. For example, Michael and Holder (1985) define a condition they term the "soleus syndrome" which manifests as a mild periosteal reaction (periostitis) on the tibia "corresponding to the medial origin of the soleus muscle and its fascial attachments" (p. 87). What is potentially useful here is that the condition may be related to activity, especially when the Achilles tendon is stressed while the heel is pronated. The resulting symptomatic condition is often seen is athletes as "shin splints." This portion of the tibia frequently manifests mild periosteal changes corresponding presumably to what Kennedy (1989) terms "rough patches." Such lesions are usually interpreted as being the result of inflammation (which they are); still, the *initial* etiology might well be a form of chronic trauma. Whether, however, such lesions can be even generally correlated with activity is problematic.

From the above discussion, it is clear that enthesophyte etiology is multiple and complex, and it still remains poorly understood. Indeed, Mankin and colleagues commented: "Little is known about ligament and tendon healing after trauma (even less is known following degeneration)" (Mankin et al., 1986: 1148). Nevertheless, some osteologists (e.g., Hawkey and Merbs, 1995) have posited the assumption that enthesophyte formation is stimulated by *increased* vascularity (which is, in turn, stimulated by exercise). Yet, in studies of the rotator cuff (supraspinator insertion involvement) (Rathbun and McNab, 1970) and the Achilles tendon insertion (Clement et al., 1984), the investigators, alternatively, suggested that exercise might actually *reduce* vascularity and cause degeneration (rather than hypertrophy). It is not that the views expressed by osteologists are necessarily incorrect, but that simplistic statements tend to gloss the complexities of the process (as well as de-emphasize the accompanying lack of understanding of the basic mechanisms involved).

Kennedy (1989) addressed some of these issues, in part, when he commented:

> In human beings these markers are not tested experimentally. When animals are used in experiments involving relationships of muscle to bone, parallels with markers of occupational stress in humans are not obvious. Therefore, occupational activities must be inferred from clinical records in industrial and athletic medicine, ethnographic accounts, and the archaeological and historical records (Kennedy, 1989: 137).

Certainly, the occupational and sports literature could offer some insight (as discussed below), but the data currently available are so anecdotal as to be only of limited utility. Nevertheless, as emphasized later in this chapter, this type of documentation probably offers the only verifiable (and independent) means of testing the "activity hypothesis" to explain enthesopathy. Where I disagree, however, is in Kennedy's suggestion of using ethnographic accounts as well as the archaeological and historical records to infer "occupational activities." As strongly emphasized in Chapter 4 (in discussion of OA), these sources are notoriously lacking in the specificity required to test adequately such behavioral hypotheses. The danger, of course, in using such unsubstantiated sources is that the argument becomes irresistibly circular in nature.

Kennedy goes on further to note the limitations and possible directions of osteological research: "In short, anthropological understanding of the *causes* of the markers and their correct *identification* await further investigation by skeletal biologists" (Kennedy, 1989: 154) [emphasis added]. It can be assumed that "anthropological understanding" refers to that gained by osteologists, and, as far as "identification" is concerned, greater standardization in the recognition and recording of lesions is, of course, desirable. In fact, some recent progress has been made in this area (Buikstra and Ubelaker, 1994; Nagy and Merbs, 1995; Steen et al., 1996). However, in determination of causes (etiology), "investigation by skeletal biologists" is not likely to produce much further insight. Again, the problem is one of circularity.

Age Effects

An obvious confounding variable, so crucial in understanding the patterning of OA, is age, and it too has been shown to influence the development of enthesopathies. This is a point accentuated by Stirland (1991) in critiquing poorly controlled investigations. There are not, in fact, many well-documented clinical studies of entheses which control for age, but those that do clearly indicate strong age effects.

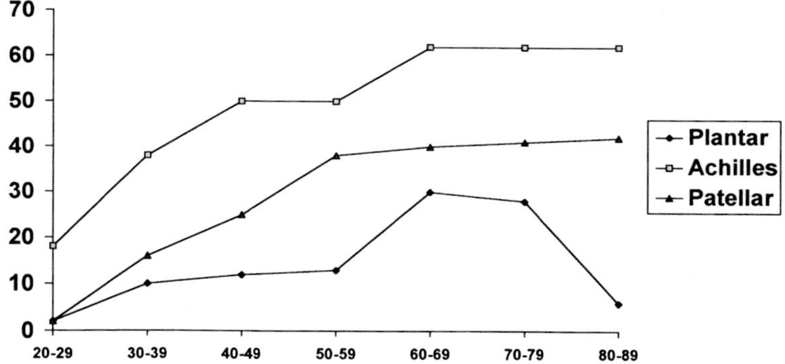

Figure 5-1 Age-related prevalence of enthesophyte formation: plantar, achilles, and patellar attachments (data from Shaibani et al., 1993).

For example, Resnick and Niwayama (1981) noted the role of age, and even more dramatically, Shaibani et al. (1993) provided controlled data showing an almost linear correlation of enthesophyte development with age. In this latter investigation, conducted on the reasonably well-documented Hamann-Todd collection, enthesophyte formation was plotted *vs.* age for three sites: Achilles, plantar calcaneus, and patellar (see Figure 5-1). Results showed a striking correlation with age for all these sites, increasing in frequency up to age 60, after which incidence plateaued. For any skeletal biologist, the meaning here is clear. Without rigorous age controls in analysis, reports on enthesial involvement are meaningless.

Likewise, it may be equally important to consider differential age effects influencing immature individuals, perhaps acting in a similar manner to the consequences of early onset of OA (see Chapter 3). Stanitski (1994: 94) concluded from sports-related injuries that,

> Many stress-related injuries in the skeletally immature athlete are junctional problems occurring at bone-muscle, muscle-tendon, or tendon-bone interfaces. Clinically they are commonly recognized conditions such as Osgood-Schlatter disease, Sever's disease, and Little League elbow.

Rates of Bone Turnover

In addition to age, another likely important contributing factor to enthesopathy is the rate of bone turnover. With respect to osteophyte development in OA, similar processes may also be operating (Duncan, 1983) (see Chapter 2). Indeed, it is highly probable that the *same* underlying systemic factors are at work on enthesial formation as those affecting osteophyte development, producing hypertrophic reactions in some individuals at certain skeletal sites. Identification of which systemic factors are influencing the phenotype is still far from certain. However, it seems likely that important genotypic factors are operating (most likely, regulatory gene(s)).

In fact, some recent findings lend support to this hypothesis concerning genotypic/systemic effects on bone formation at certain skeletal sites. As noted in Chapter 2, Uitterlinden et al. (1997) found variation in a vitamin D receptor locus with some variants correlated with increased incidence of radiographic OA. What is of particular interest to the present discussion is that the OA expression which was associated pertained *only* to osteophyte development. Thus, it is possible that such a genetic influence could also induce bone deposition elsewhere in the skeleton. Moreover, age, metabolic agents, endocrine status, and diet are all likely also to influence rates of bone turnover—and consequent appearance of such features as marginal osteophytes (in OA) and probably all lesions recognized as enthesopathies as well.

Excepting for the effects of age, detailed data on any of these factors, specifically as influencing enthesial involvement are still lacking. However, a preliminary study (Rogers and Waldron, 1989) has yielded interesting results. Specifically, in this investigation of 56 10th–15th Century individuals from Wells Cathedral, Rogers and Waldron found that 77% of those individuals with enthesial involvement also displayed osteophytosis. Moreover, 36% of the sample displayed not just entheses and osteophytes, but extra-spinal calcifications as well. Enthesophyte prevalence and severity was also most marked among males. Rogers and Waldron concluded that a probable entity they term "bone forming" underlay much of the variation in the appearance of hypertrophic bone lesions. "As well as occurring on its own 'bone forming' may well modify the expression of other bone diseases, especially joint disease, and may also modify new bone formation due to age or activity" (Rogers and Waldron, 1989: 170).

More recently, Rogers et al. (1997) have expanded the Wells' sample to include individuals from two other cemeteries (total $N=337$). Enthesophyte development was assessed at 14 ligament insertion sites and compared with osteophyte development around vertebral bodies and at 23 peripheral sites. As in the earlier study, a strong correlation was again found between osteophyte and enthesial involvement. Moreover, males were significantly more affected than females, and age also was correlated with increasing enthesial involvement. At the extreme end of the spectrum of individuals with the greatest osteophyte and enthesial involvement, the condition manifested as DISH. Rogers et al. argued, as in the prior investigation, that an underlying systemic predisposition for "bone forming" appears to be present in some individuals, both at and below the threshold of diagnostic DISH. These findings led these researchers to conclude: "This study suggests that the observed variation in bone formation could be due to differences in individual ability to form bone in response to stress *rather than difference in stress*" (Rogers et al., 1997: 90) [emphasis added].

It has, obviously, been known for some time that certain skeletal conditions which are characterized by bone hypertrophy, particularly ankylosing spondylitis and DISH, also commonly stimulate bone deposition at enthesial sites. The appearance of moderate-advanced enthesophyte development, in the absence of any other clear diagnostic criterion of DISH (e.g., substantial vertebral involvement) might, nevertheless, be but part of the same physiological process. Indeed, many individuals, especially older males, who display enthesial hypertrophy may simply be "subclinical" cases of DISH (as hypothesized by Rogers et al., 1997). Further, the common finding that males are more frequently involved than females for enthesopathies should, perhaps, not be all that surprising, in light of some experimental evidence showing the effects of testosterone on ligament repair (Tipton et al., 1975).

Finally, the pattern of involvement as shown in the Wells Cathedral series (Rogers and Waldron, 1989) also is suggestive of possible other factors. Many of these individuals were older males, who were apparently well fed (Rogers, personal communication). Thus, diet, in combination with age, endocrine status, and other metabolic factors, may also play a significant role in the etiology of enthesopathy; moreover, the effects may obviously differ among individuals as well as *between populations*.

Indeed, what is most obvious here is that we do not have a clear understanding at all of the etiology of enthesopathy. In addition to the factors known to exert influence on enthesial remodeling (age, sex, trauma), there are numerous others which potentially could act significantly, producing systemic bias both within and across population samples (see Figure 5-2).

"Activity" (i.e., chronic mechanical stress) may well be one of the primary influences producing the pattern of skeletal variation of enthesopathy. However, in light of the likely other confounders, the interpretations of such lesions as *indicative* of behavior (either generally or specifically) are currently insupportable.

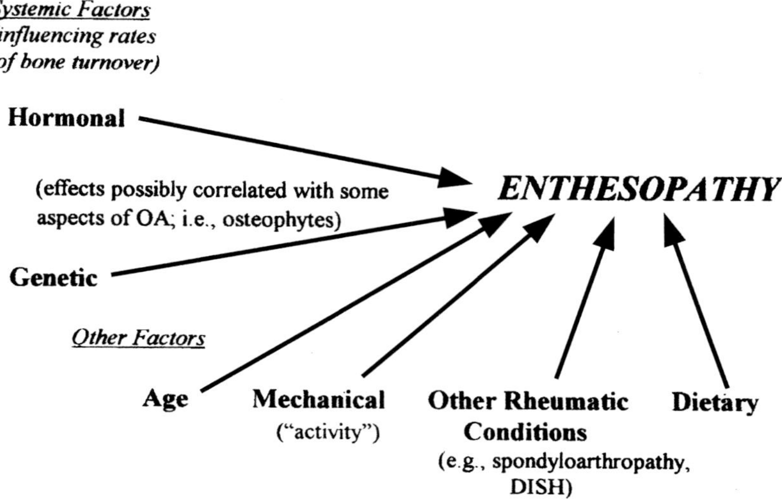

Figure 5-2 Etiolological factors influencing the development of enthesopathy.

OSTEOLOGICAL INTERPRETATIONS OF ENTHESOPATHY

Despite the weak substantiation for deriving behavioral interpretations, in the last several years numerous osteologists have drawn a remarkable array of simplistic conclusions regarding enthesial involvement. In fact, here, probably more so than any other area of contemporary skeletal/behavioral research, these perspectives reflect an unfortunate absence of scientific standards of verification and rigor.

Probably the best-known overall tabulation of enthesophyte interpretations (as part of a general review of "skeletal markers of occupational stress") was that by Kennedy (1989). Although Kennedy has been critiqued (e.g., Waldron, 1994a) for what is admittedly a remarkably eclectic list of "possible" occupational markers, Kennedy meant the list to be inclusive rather than evaluative. In this sense, it was a most scholarly contribution, and even a casual reading of the article would reveal Kennedy's real orientation: The contribution begins and ends with references to *Alice in Wonderland*. In fact, the brilliantly sarcastic verses from Carroll, used by Kennedy to introduce his article, are quoted again here (in Chapter 8) to conclude this work.

While considerable interest in what Kennedy calls "industrial medicine," has a long history, and is expertly summarized by him, osteological applications of enthesopathy have only recently become commonplace. Indeed, mostly in the last decade, researchers have increasingly sought to use such bone changes to make inferences regarding activity. As noted above, such research is frequently built upon the most fragile of theoretical foundations. The increasing popularity of this approach can be attributed to a number of factors. Firstly, as skepticism relating to the utility of OA in making behavioral interpretations has increased, skeletal biologists have sought other potential skeletal markers. Secondly, as earlier work (summarized below) has become more widely cited, further research has been stimulated, apparently in the mistaken impression that enthesopathy somehow "better" reflected activity than OA modifications. As emphasized above, such a conclusion is not, by any means, well supported. Indeed, more generally, the hypothesis that enthesial reactions relate to activity in any predictable fashion is still lacking scientific verification.

As early as 1968, Ortner mentioned some rotator cuff changes, particularly subscapularis attachment, in his study of degenerative elbow disease. While not the primary focus of this research, Ortner did find some association of the rotator cuff involvement with radio-capitular OA in a Peruvian sample (also of note, Ortner found such subscapularis involvement in 46% of individuals *without* elbow OA).

Gejvall (1983) identified nine cases (all males) in a Swedish archaeological sample showing what he termed "popliteal pitting" of femora. He suggested the lesion might relate to muscle insertion re-action of the gastrocnemius, and further asserted (rather superficially) the lesion might be due to canoe paddling or forceful spear throwing.

Also in 1983 Kennedy published what has proved to be an influential article relating changes in the ulnar supinator crest in skeletal samples from groups "known to have used spears, atlatls, and similar projectile weapons" (Kennedy, 1983: 871). It is, of course, quite a different matter to "know" an ancient society used some common weapons, probably true of a majority of societies antedating the Late Pleistocene, than to be able to determine actually which individuals did so (and thus reasonably infer any effect the behavior would have had on their skeletons). These are fundamental issues, and were also addressed for interpretations of OA (particularly as critiqued by Waldron; see Chapters 3 and 4). It thus remains unanswered whether those individuals who used spears regularly show any differences skeletally from those who did not (in the supinator crest or any other feature), Moreover, other than activity, what additional factors might influence supinator crest development? And, even if activity could be shown to be important, was there more than one activity which could lead to the lesion?

Kennedy alluded to some contemporary analogy, claiming "similar features appear in the right ulnae of living persons of both sexes who are habitually engaged in certain occupational and athletic activities involving angular displacement of the forearm" (Kennedy, 1983: 871). Yet, for clinical support, Kennedy offered only one source which investigated a specific activity: baseball pitching. This sports study (King et al., 1969) was mostly concerned with degenerative changes, but did mention, mostly anecdotally, the presence of medial "traction spurs" in the ulnar notch of 44% of pitchers.

What is surprising (and discouraging) is that over the last decade Kennedy's assertions have been frequently cited as "fact," i.e., that hypertrophy of the supinator crest is a skeletal marker of "spear throwing" or some similar activity. Perhaps, some of the broad misconception was fostered by Kennedy's unfortunate phrasing of his conclusions. Still, the oft-repeated claims are, at best, exaggerated. At worst, they represent unfamiliarity with original sources and unrestrained willingness (eagerness) to embellish otherwise questionable data.

Dutour's short contribution, published in 1986, has also been widely cited. In this research Dutour analyzed a small skeletal sample ($N=41$) from two sites in North Africa (Saharan region of Mali). Of the already small sample, many skeletons were fragmentary, yielding only 20 adults which could reliably be sexed. Obviously, in such a small collection, age controls were not possible, an important limitation in this work (and one pointed out by Stirland, 1991).

As with the 1983 report by Kennedy (and its ramifications), this contribution also has promulgated unsubstantiated conclusions, again, leading to repeated exaggerations by other workers. Dutour, for example, argued that enthesial reactions, most especially in the elbow, "may be tentatively correlated with javelin throwing, wood cutting, and archery" (p. 221). For support, he mentioned that "observations from sporting and occupational medicine indicate that specific enthesopathies are correlated with different activities" (p. 221). More specifically, he further argued that an osseous change to the medial epicondyle of an adult male:

> reflects an insertion pathology corresponding to hyperactivity of the muscles pronator teres, flexor carpi radialis, palmaris longis, flexor digitorum superficialis, and flexor carpi ulnaris or of the ligament. It is observed today in sporting medicine and generally affects javelin throwers or golf players (Dutour, 1986: 222).

Further, in two other cases showing changes to the olecranon process, Dutour concluded the relevant stress was produced,

> when the arm is horizontal and the elbow flexed. Under modern conditions, the lesion is observed in wood cutters, blacksmiths, and some baseball players. The pathology may be unilateral or bilateral, depending on the type of activity involved (p. 222).

What is troublesome here is that *none* of these claims of behavioral correlations, either general or specific ones, were supported by any clinical citations (although two others were each backed by one anecdotal source). Moreover, I am not aware of any consistent clinical research which has to date provided anything approaching scientific verification of such behavioral associations. That such weakly supported assertions were published by the *American Journal of Physical Anthropology* is somewhat surprising. Once published, of course, the work takes on a life of its own.

Lastly, Dutour, from the skeletal data themselves, reported some enthesial involvement in nine individuals (eight of whom were male). Dutour regarded the sex ratio as possibly "noteworthy," but, as shown above, such a male preponderance is completely expected (basically, regardless of the activity differential). Indeed, if activity were to be *assumed* as the major contributor to variation in incidence of such (enthesial) lesions, this approach would almost certainly grossly underestimate the physical workloads of females in the past. Such a conclusion would thus be biased on a number of counts.

In the last decade a number of other studies have built on this earlier research, much of it repeating prior claims, and, in so doing, providing a deceptive *appearance* of increasing validation. For example, Molleson (1989) attempted to relate bone changes seen in ancient Syria to activity (especially grinding on a quernstone). In this pursuit she evaluated enthesopathies as well as OA. For enthesial involvement, specifically, she concluded: "Symmetrical development of the deltoid and biceps muscles in these otherwise gracile people indicate that the upper arms were persistently involved in strenuous exercise" (Molleson, 1989: 369).

Likewise, Molleson concluded that the degenerative changes seen, both in the spine and peripheral joints, were related to activity. And, not just activity, but that specifically related to food preparation. Further, while a modestly different behavior (i.e., use of the mortar and pestle as compared to the use of a grinding stone) might account for the radial tuberosity changes observed here, Molleson argued this alternative behavioral explanation would not "lead to disk instability or the arthroses of the knees" (Molleson, 1989: 369).

Also published in the late 1980s were a series of publications using more recent, and presumably better historically documented skeletal materials, and this work gave further stimulus to using enthesial reactions to reconstruct activity and to make broader behavioral inferences. Many of these studies focused on presumably highly disadvantaged groups, most notably cemeteries associated with burials of slaves or those of former slaves (and their immediate descendants) (Kelley and Angel, 1987; Owsley et al., 1987; Angel et al., 1987). Kelley and Angel reported enthesial reactions including "muscle crests" associated with the deltoid, pectoral, teres major, and supinator. However, in this investigation of 120 individuals from 23 different slave cemeteries in Virginia and Maryland, no frequencies were detailed, and the sample was not controlled for either age or sex. Moreover, several other osteological indicators of uncertain validity (including Schmorl's nodes and parturition scars were employed in this analysis; see below).

Angel et al. (1987) did similar research on the sample from the 19th Century First African Baptist Church (FABC) cemetery in Philadelphia ($N=75$ adults). Enthesial data were again employed and included deltoid and pectoral reactions (humerus) as well as those of the supinator (ulna) and "adductor" (femur). These data were included within a table which also included "osteoarthritis of vertebrae" (presumably, VOP?). All of these osteological changes were simply attributed to "occupation." A major focus of this investigation was to compare patterns seen in this group of freed slaves, and their descendants, with those from a slave cemetery (specifically, Catoctin Furnace, Maryland). Again, however, none of the comparisons were controlled for age. Despite such limitations in sample size and relevant controls, Angel and colleagues (1987) derived a remarkable series of highly specific conclusions:

> The deltoid (for using an axe and lifting and lowering heavy weights) is not used as hard by the FABC people (p. 222).
>
> The pectoral is used hard by the FABC group (p. 222).

They further argued, regarding one individual with a number of enthesial reactions,

> We think she was a laundress (p. 222).
>
> Note how developed the *adductor tubercle crest* is in this group, possibly indicating horseback riding but rarely forming a rider's bone as in older cowboys (p. 223).

#64 has cervical breakdown, perhaps from carrying the laundry as a head load, and bending stress on lower vertebrae (p. 227).

Owsley et al. (1987) also did a detailed investigation of a slave cemetery from New Orleans ($N = 29$). In recording connective tissue ossifications and what they termed "myositis ossificans," the authors noted three individuals in particular (all older males) who showed such involvement "on almost all their bones." Owsley and colleagues thus concluded: "This general pattern reflects a very high level of physical activity and strain, which *undoubtedly* relates to occupational stress, probably as slaves" (Owsley et al., 1987: 191) [emphasis added].

These investigations of disadvantaged 19th Century African Americans were in many ways highly systematic, and reflect an enviable broad-based approach to biocultural interpretation of skeletal remains. Consistent data on mortality patterns, inflammatory reactions, indicators of metabolic stress (e.g., enamel hypoplasias), and bone density all contribute to a better understanding of the life stresses faced by these people. However, the data employed to reconstruct activity levels (OA, enthesial reactions, Schmorl's nodes) are consistently the weakest links in these otherwise systematic analyses. It is not a question of whether African Americans "worked hard;" clearly, many forms of documentation argue they did. It is here a question of whether skeletal data enable us to make such inferences, *independently* of other sources of information.

Other historically interesting archaeological contexts have also provided some fascinating hints relating to activities of past peoples. Most notably, Stirland's research on the *Mary Rose* remains has potential to yield some of the most systematic findings relating to behavioral influences on the skeleton. In her provisional report, Stirland (1991) listed several areas of enthesial involvement which she systematically scored. The most intriguing aspect of the *Mary Rose* sample is that, firstly, a relatively narrow range of occupations and associated activities would have been practiced by members of the ship's company, and these can be reasonably reconstructed from historical sources and tested further by contemporary experiment. Secondly, the provenience of osteological materials as associated with the ship's remains can be used further to narrow the likely occupations of some individuals (e.g., archers, common seamen). Thus, there is potentially considerably better resolution of behavioral documentation here than in the vast majority of archaeological samples and also a better likelihood of correlating at least some skeletons with probable "occupations."

Nevertheless, Stirland derived quite conservative inferences from these materials, stating:

> The amount and type of pathology and the degree of bony remodeling of various fibrous insertions in this sample has to be explained in terms of their environment. These would appear to be changes due to occupation in some members of the crew of the *Mary Rose*. Some of these are pathological and some morphological. The former may be due to work-related trauma, while the latter appear to be related to loading stresses. All, therefore, may be a consequence of occupation. The relationships between these changes and *specific* occupations, however, is another matter (Stirland, 1991: 45) [emphasis added].

From a contemporary context (the Hamann-Todd collection), Herskowitz et al. (1996) attempted to infer quite precise behaviors from skeletal changes in one individual (also discussed relative to OA in Chapter 4). Regarding muscle attachment involvement, the authors are quite cautious in making specific inferences and state such changes can be "produced by many types of strenuous work" (Herskowitz et al., 1996: 177). Curiously, however, they claim ligamentous modifications are "more informative," since they are "influenced directly by forces applied to the joint" (Herskowitz et al., 1996: 177).

From evidence of an enlarged conoid tubercle of the clavicle, Herskowitz and colleagues suggested the individual had tissue modifications to help insure stability of the shoulder. Such a conclusion is not unreasonable; however, the authors went further to argue these changes *specifically* reflected bone involvement related to boxing. Herskowitz et al. also included other bone changes, most of dubious validity. For example, they also asserted that the presence of "bony patches" on the upper limbs "are probably due to their defensive role, i.e., blocking the opponent's blows" (p. 177).

In other recent investigations of archaeological remains from both the Old and New Worlds, osteologists have, using especially enthesial reactions, continued to make interpretations of activity. Lai and Lovell (1992) evaluated a host of bone changes (including enthesopathies). In many cases, their behavioral attributions were highly specific. For example, hypertrophy of the supinator crests was related to "rowing or paddling;" hypertrophy of insertions for gastrocnemius, soleus, gluteal, and hamstrings was postulated as due to "jogging while carrying heavy loads." From analysis of such changes in three individuals, Lai and Lovell argued the pattern was, in fact, consistent with a particular occupation: "... the alterations to the skeleton displayed by the Native males are consistent with the habitual lifting, carrying, and rowing performed by *voyageurs*" (p. 230).

Reinhard and colleagues (1994) in the analysis of Native American remains from Nebraska, were also highly interested in activity reconstruction, especially as relating to presumed horseback riding.

> Habitual riding is indicated by the presence of superior elongation of the acetabulum, extension of the articular surface of the femoral head onto the femoral neck, and enlarged muscle attachments for the gluteus medius and gluteus minimis (*sic*), the abductor magnus and brevis, the vastus lateralis, and the medial head of the gastrocnemius muscles (Reinhard et al., 1994: 68).

Reinhard and colleagues argued that the modifications noted above were consistent with horseback riding, as they were found frequently in males. However, similar changes were also observed in females, but, here, attributed to "strenuous labor while planting crops, foraging, burden carrying, and preparing hides."

The assumption, clearly, is that *all* the bone changes were due to activity. Moreover, the patterns can be made to accord with a variety of presumed activities. In this way any skeletal pattern can be explained by *some* combination of behaviors. It, however, is difficult to see how such claims could be falsified. Moreover, it is unclear what can be learned by this approach that is not already "known" from other (generally poor) ethnohistoric or historic accounts.

Lastly, Robb (1994) did a very detailed analysis of 50 different muscle insertion areas in a sample of 56 burials from Italian Iron Age sites. Unlike many similar studies (discussed above), Robb attempted to evaluate a variety of possible influences on the lesions. Most especially, he evaluated age effects and (for all lesions) found a high correlation ($r = 0.69$), agreeing with another controlled study (Shaibani et al., 1992). Moreover, Robb appropriately cautioned that, in addition to chronic stress, surface markings at muscle insertion sites might be influenced by trauma and metabolic conditions.

A problem with much of the work reviewed above is immediately obvious. Trying to relate *all* bone changes to one behavior certainly produces a neat explanation, and one eminently suited to more popular forms of publication. Nevertheless, what seems to be forgotten here is that many of the bone changes probably are produced by etiological factors *other* than activity. Furthermore, even for those skeletal modifications which, at least in part, relate to activity, they are most likely influenced by several different types of behavior. Thus, the willingness to posit overly simplistic explanations and, at the same time, to ignore (or discount) alternative hypotheses, does not serve to advance the field.

One area of enthesopathy research has recently been made more rigorous, and that relates to attempts to more systematically score and record the relevant bone changes. Most notably, the work of Hawkey and Merbs (1995) and Steen et al. (1996) has greatly assisted in this important aspect of investigation. Moreover, Hawkey and Merbs provided a concise overview of much of the contemporary perspective toward the use of enthesopathies to infer behavior. In postulating a relatively simple etiopathogenesis of enthesial involvement, they argued strongly that what they term "musculoskeletal markers (MSM)" are powerful tools to reconstruct behavioral pattern from skeletal data:

> Hypertrophy of bone, in the form of robust muscle attachment, is the *direct result* of this increased stress, and continued stress of a muscle in daily, repetitive tasks creates a well-preserved skeletal record of an individual's habitual activity patterns [emphasis added] (p. 324).
>
> The use of MSM for habitual activity analysis operates under the assumption that *the degree and type of marker are related directly to the amount and duration of habitual stress placed on a specific muscle* [emphasis added] (p. 324).

The stimulus for this approach (and probably also a contributing factor to the manifest confidence expressed in the above quotes) derived directly from Merbs' earlier research on Inuit skeletal remains (Merbs, 1983). As with studies of OA (see Chapter 4), these particular collections (from both Greenland and Alaska) have received considerable attention, especially as they frequently meet criteria presumably facilitating more detailed behavioral interpretation. Most notably, the pre- and proto-historic Inuit are assumed to have engaged in a "limited number of specialized, but *known* activities" (Hawkey and Merbs, 1995: 325) [emphasis added]. As emphasized in Chapter 4, in critiquing the claimed adequacy of these supportive behavioral data, ultimately they do not meet the test of accuracy or precision.

Nevertheless, numerous researchers have focused on Inuit skeletal changes and correlated them with presumed Inuit activity patterns, including for example, kayak paddling, umiak paddling, hide preparation, and harpooning. As noted, this approach was originated and systematically applied by Merbs (1983), although initially the primary focus was on degenerative joint changes.

Several aspects of this research are admirable and provide a good basis for future work. For example, as mentioned above, the identification and scoring criteria have been made much more explicit. Moreover, a *functional* emphasis is employed, focusing on the pattern of involvement of major muscle groups as they theoretically should be affected in various habitual types of motion (and presumably referable to particular activities). Lastly, detailed data were collected relative to side involvement, and thus patterns of asymmetry were analyzed and reported (interestingly, Hawkey and Merbs found a clear right-side predominance among the more affected individuals).

While laudable, there are still obvious gaps in this approach, indeed, the same ones enumerated in the discussion of OA. Without *independent* verification that particular activities (or a range of activities) can and do produce specific lesions, or patterns of lesions, the conclusions from skeletal data *alone* are not convincing. Nor can they be convincing, for the reasoning is obviously circular. Moreover, inherent biases in reconstructing behavior from a skeletal collection (derived from a cultural group) in which behavior is "known," are potentially all the more troubling. Thus, more vigilance must be pursued to ensure independent verification of results. This point is not made to question the integrity of any of the investigators involved in this research. Their integrity is not in question. The objective here is to remind skeletal biologists of the all-too-present pitfalls. Any investigator can easily be seduced by the temptation to "breath life" into our subjects. Indeed, most of us have succumbed at some time. Recognition of the hazards is always the best assurance of avoiding them.

Future Research Directions

As should be now clear, the pressing need is for some type of independent verification, thus providing at least some confidence that enthesopathies, in fact, *do* consistently relate to activity. Controlled clinical approaches, as discussed for OA, trauma, etc., are again the surest, most reliable avenues to attain this goal.

Kennedy defined both some of the limitations as well as opportunities in using such sources:

> Radiographic analysis serves to bring observations of the industrial and sports physician into the purview of the anthropologist, but the fact remains that dried bone is seldom encountered by the clinician in the course of practice, and medical interpretations of bony lesions lack the *sophistication* of observations offered by the trained paleopathologist in human skeletal biology. Given these circumstances, as well as the differences in scientific orientation of medicine and anthropology, it is not surprising that research on markers of occupational stress has developed quite independently (Kennedy, 1989: 133) [emphasis added].

A fundamental problem in much contemporary osteological research is underscored by Kennedy, that is, as overconfidence in the skeletal data. The lack of "sophistication" suggested for our clinical colleagues is a surprising assertion. The rigorous attention to sampling, including requisite stratification, statistical testing, controls for potential confounding variables, and consideration of alternative hypotheses all reflect approaches which medical epidemiologists employ as a matter of course. Clearly, not all clinical research employs these rigorous standards, but in medical epidemiology, especially, such a research model is quite typical; the contrasting situation in skeletal biology need not be further berated here.

Moreover, in declaring "scientific orientations" vary, and thus skeletal biology (*vs.* clinical research) have "developed independently," Kennedy identifies perhaps the most profound difficulty underlying osteological approaches to behavioral reconstruction. I would argue, conversely, that there is only one rigorous form of scientific orientation, and, as just noted, some bio-medical researchers typically employ it. What passes as acceptable science in skeletal biology too often tolerates a lower standard. Ideally, this situation can be soon corrected. And, if research perspectives in the past have developed largely independently, through increased collaboration, this deficiency should also be corrected.

So, what can be done to improve reliability of osteological interpretations of enthesopathies? Admittedly, the immediate clinical literature is not of much direct assistance; as noted, excepting for a few reports dealing with enthesial involvement in DISH or spondyloarthropathy, very little research has dealt directly with such bone changes. There are, however, a number of studies done from an occupational or sports epidemiological perspective which, at minimum, commented on such "periarticular" osseous changes. None of these studies focused primarily on enthesopathy ascertainment, but the observations were made incidentally within a broader investigation of joint involvement (most commonly, of OA).

An immediate limitation with much of these data is they are anecdotal (not surprising, considering the research was oriented toward other questions). Still, the information is helpful, at least in showing how such evidence can be used and pointing the way for more precise and focused research. For example, in three different studies of mechanical drill users, spurring of the olecranon (triceps insertion) has been observed (Rostock, 1936; Barsi and Rossero, 1963; Burke et al., 1977). Some of this research was quite well controlled. In a German study (Rostock, 1936), the prevalence of the lesion was higher than noted in controls, while in an Italian sample, the condition was seen in 21 of 84 miners (25%) (Barsi and Rossero, 1963) and was also reported at a similar frequency in a group of American pneumatic drill-users (Burke et al., 1977). In addition, Hadler (1977) provided a general review of such lesions as they affected primarily the hand, conditions he termed "periarthritis." Most of the involvement reported by Hadler related to tendosynovitis of the wrist and, more recently, such conditions have attracted further attention as part of investigation of "repetitive stress syndrome". Hadler also briefly mentioned possible occupationally related involvement of the rotator cuff. He concluded that evidence was generally lacking but "can be gathered from carefully designed and appropriately chosen and defined populations" (Hadler, 1977: 1019). The urgent, and still unfulfilled, need for such studies is as true today as it was twenty years ago.

The amount of relevant evidence concerning enthesopathy, although still anecdotal, is considerably more for studies of athletes than those from industrial settings. This is perhaps because there have been more studies in sports contexts focused on anatomical areas at greatest risk for such involvement, particularly the elbow and ankle, but perhaps also for more basic etiological reasons. Possibly, enthesopathies (or enthesial-like lesions) develop more so in athletes because the stress is either more repetitive and/or more extreme; this explanation, however, seems improbable. What is more likely is that the lesions *might* develop more so in athletic activities, because often they begin at an early age.

For example, "osteophytes" and "spurs" have been observed in the elbows of young baseball pitchers "commonly at insertion of the triceps muscle" (Adams, 1976: 536). King et al. (1969) also reported "medial traction spurs" in 44% of professional baseball players ($N = 50$), Wilson et al. (1983) mentioned development of large "osteophytes" on the posterior and posteromedial aspect of the olecranon in 5 pitchers (3 college; 2 professional). Bateman (1969) has suggested an etiology for such lesions, arguing they develop as the result of recurrent minor traumata due to stretching forces, and thereby producing a small degree of osseo-tendinous ossification. Bateman further suggested the lesions become enlarged with continued stress.

Ankle involvement ("talo-tibial spurring") has also been recorded as a common occurrence in university athletes (45% prevalence), as compared to 15% in controls (O'Donoghue, 1970) as well as in professional ballet dances (Schneider et al., 1973). Ankle changes have also been noted as "bony outgrowths" in soccer players (Adams, 1979). Vincellete et al. (1972) further noted talar changes ("dorsal exostoses") as well as calcification of the interosseous tibiofibular ligament in 7 (of 23) professional (American) football players. Interestingly, in this small sample, Vincellete and colleagues found some indication of "periarticular new bone" in close to 90% of the athletes (but in only 4% of controls). They attributed most of the modifications to acute trauma, as the vast majority of the football players had a history of such injury.

Alterations corresponding to "tennis elbow" have also attracted considerable interest, although frequently it is unclear which pathological changes are soft tissue and which involve bone. Williams (1979) suggested the condition resulted when "the aponeurotic attachment pulled away from the bone." Priest et al. (1974) further reported "hypertrophic spurs" and "loose bodies" in some advanced tennis players, the latter probably resulting from fracturing of the tip of the olecranon. The distribution of such joint changes, however, might have little to do with activity, at least specific activity. Garden (1961) observed that most patients with "tennis elbow" were not tennis players. Moreover, the condition showed a strong predilection to appear in older individuals (>35 years of age in males; >40 years of age in females).

Javelin throwing has also been found to occasionally produce similar changes to that seen in baseball pitchers, fracturing of the tip of the olecranon (see below), but Miller (1960) reported that in his series of javelin throwers the radiographs were normal (thus bony involvement was unlikely, at least in this group).

Finally, in a recent study of the hand of rock climbers, subchondral cysts and osteophytes were both found in higher prevalence than in controls (Bollen and Wright, 1994) (see Chapter 3). Moreover, and more to the present point, only in climbers "pronounced scalloping" of the necks of proximal phalanges was also seen, and this change was postulated by Bollen and Wright to result from thickening of the distal end of the fibrous portion of the flexor sheath.

It must be emphasized that all these reports are tentative and require further substantiation. As is too often done by osteologists, they cannot be taken collectively to infer more than the evidence permits (that is, as concluded from thoughtful analysis of the individual studies). Clearly, more can be gleaned from further search of the available occupational and especially the sports literature. Moreover, there are other avenues of potentially useful insight. Firstly, the investigations by Stirland (1991; 1997) of the relatively well-controlled *Mary Rose* remains offers considerable promise. Secondly, the even better historically documented sample from Spitalfields provides further potential (i.e., similar to the occupational study of OA done by Waldron and Cox, 1989).

Thirdly, and as noted above, *better* controlled clinical studies are obviously the immediately required avenue of investigation. Most specifically, large studies of individuals with known occupational/sports histories need to be systematically compared with appropriately selected controls. Better radiological imaging of the relevant enthesial sites is also much needed. Moreover, the best data of this sort would ideally come from longitudinal studies of the types already underway relating to OA (e.g., the Framingham study; see Andersen and Felson, 1988; Felson et al., 1991; Hannan et al., 1993). Perhaps, even more appropriate, are the longitudinal data collected on long-distance runners as part of the Stanford study (Lane et al., 1987; Lane et al., 1990; Lane et al., 1993). Indeed, given substantial radiological documentation in this and the Framingham research, perhaps the relevant data already exist to test comprehensively activity influence at least at some enthesial sites. In fact, Lehman (1984) in his discussion of overuse syndromes in runners suggested (in a limited way) a basis for how to approach this potentially rich data source.

As a final option, osteologists working in close cooperation with clinicians could help *initiate* such research. Certainly, the model of such productive (indeed, essential) collaboration already exists for OA research (Rogers et al., 1981; Rogers et al., 1987; Rogers et al., 1990) as well as for spondylolysis investigation (Merbs, 1995; see below). Indeed, as mentioned above, such collaborative work has already begun in a broader, more sophisticated, interpretation of enthesopathy (Rogers et al., 1997). Thus, the future does hold promise in providing a means of testing current assumptions. Osteologists must assist in this venture, and be prepared for the results, whatever they might show!

SCHMORL'S NODES

These lesions, frequently found on vertebral bodies, have attracted even less interest than enthesopathies from both clinicians and osteologists. Moreover, with a few exceptions, they have not generally been used by osteologists as a basis for reconstructing activity/behavior. Given the clinical evidence regarding Schmorl's nodes that does exist, this general hesitancy by osteologists to make open-ended inferences is fortunate.

Schmorl's nodes have been typically described as "cartilaginous nodes" (Schmorl and Junghanns, 1971) which represent "intervertebral herniation of disk material" (Resnick and Niwayama, 1978). The most common sites are lower thoracic/upper lumbar vertebrae, and

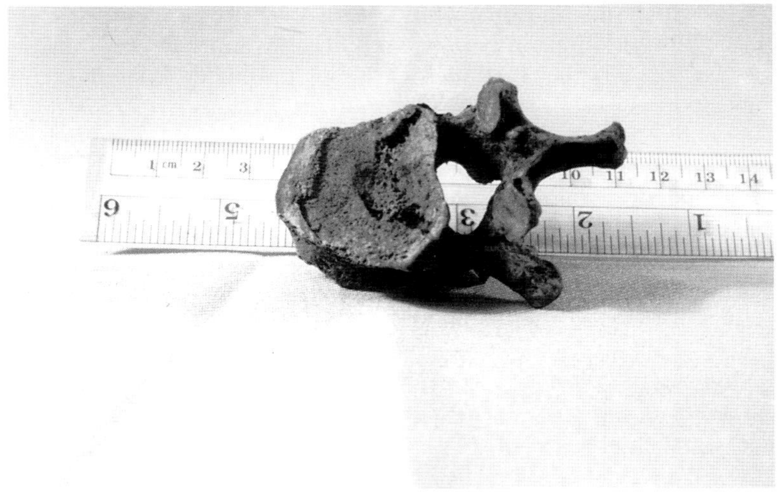

Figure 5-3 Schmorl's node. Thoracic vertebra, Ala-329, Central California. Photograph by the author.

less commonly, mid-lower lumbar segments (Begg, 1954; Hilton et al., 1976; Resnick and Niwayama, 1978). Osteologically, the lesions manifest as shallow depressions with rounded edges, most typically located centrally or slightly posteriorly on the vertebral end plate (Hilton et al., 1976; Kelley, 1982) (Figure 5-3). Moreover, they appear to develop more commonly on inferior end plates as compared to superior surfaces (Hilton et al., 1976).

The condition was first described by Schmorl (1927) and Putschar (1927), independently in the same year. Schmorl's description and more complete interpretation are, thankfully, more commonly available in English as the result of the wide distribution of his co-authored text, *The Spine in Health and Disease* (1971). In this latter work, which is the standard citation in osteological reports, Schmorl and Junghanns provided a detailed description of vertebral "cartilaginous nodes." They attributed the involvement to prolapse (i.e., herniation) of the disk through a gap in areas of weakness where fissures might form. The initial "lesions" (weakness) could develop due to presence of:

(1) "Indentation"—during development along the source of the chorda
(2) A scar—where vessels had been
(3) Necrotic tissue—formed during remodeling (termed "ossification pores"). [Hilton et al., 1976 have argued the condition develops

("invariably") as a result of gaps in the end plate, and they thus prefer to call them "end plate lesions."]

(Schmorl and Junghanns, 1971: 158–159)

It has been suggested that, when the lesions are severe and widespread, they can (rarely) produce an adolescent kyphosis, termed "Scheuermann's disease" (Begg, 1954). However, more likely, Schmorl's nodes are a feature of Scheuermann's disease, not a cause of it, and the condition is thus probably due to a more general weakening of subchondral bone (Resnick and Niwayama, 1978). Moreover, in quite rare circumstances, herniated lumbar disks can occur in adolescents, and when they do occur, they are most commonly the result of acute trauma (Epstein and Lavine, 1964).

Etiology of Schmorl's Nodes

Schmorl and Junghanns discuss a variety of *possible* etiologies producing the lesions, including persistence of growth cartilage, pathological conditions which weaken the plate (infection, neoplastic, metabolic), and acute trauma. Further, Schmorl and Junghanns suggested trauma in childhood may be particularly important, but also remarked that, typically, it is very difficult ever to demonstrate a traumatic cause. The practical applications of the challenges in establishing a *functional* etiology are underscored in Junghanns' discussion of verification of insurance claims.

Resnick and Niwayama (1978) also reviewed a variety of potential etiological influences including, congenital weakening of the end plate, trauma, metabolic and neoplastic conditions, and degenerative disk disease. They concluded that: "These nodes can be produced by any process which weakens the cartilaginous plate—or the subchondral trabeculae of the vertebra"(Resnick and Niwayama, 1978: 57).

From the overall discussion in Schmorl and Junghanns, it is clear that the etiopathogenesis of Schmorl' nodes was uncertain (and still is). However, quite selectively, osteologists have frequently chosen to focus on certain statements. For example, Schmorl and Junghanns did note that "expansive pressure on the nucleus pulposus together with the stress of everyday life produces tears in these specific areas of the plates" (p. 159). However, they also suggested, more generally, that a major influence was elastic pressure of the weight-bearing spine; thus they argued the condition was *not* related to specific activities, but, perhaps, represented a more species-wide phenomenon, i.e., like VOP (see Chapter 4).

Further, Schmorl and Junghanns provided some data showing the lesions are *quite* common (as seen at autopsy), found in 39.9% of males and 34.3% of females. In another comment seized upon by osteologists, Schmorl and Junghanns concluded this sex "difference" in distribution resulted from "the greater stress to which the spine of the male is subject" (p. 164). This claim is, however, problematic, since the condition is so common in *both* males and females; moreover, the difference reported is probably not statistically significant, in any case.

Other similar studies have also consistently reported a very high prevalence of Schmorl's nodes in autopsy materials. In fact, Hilton et al. (1976) found in a small sample ($N=50$) the lesion present in 76% of individuals. Further analysis of this sample suggested the incidence appeared unrelated to either age or bone density.

In the most recent survey, using MRI, 38% of individuals ($N=372$) showed presence of Schmorl's nodes (Stäbler et al., 1997). A point made by all clinical studies is that the lesion most frequently appears in adolescence (Putschar, 1927; Schmorl and Junghanns, 1971; Hilton et al., 1976; Kelley, 1982) and "are common, even in children" (Begg, 1954: 184). Further, there is consensus that the condition is commonly asymptomatic, although it can occasionally produce some pain (Takahashi et al., 1995); interestingly, in these symptomatic cases, MRI revealed inflammation in the vertebral bone marrow. Finally, one recent report, while again finding frequent prevalence of Schmorl's nodes, also mentioned (anecdotally) that in those cases of adolescents with multiple involvement, three (of four) "had a history of hard sports" (Hamanishi et al., 1994).

Anthropological Interpretations of Schmorl's Nodes

As mentioned above, these lesions have not been as widely employed to infer behavior as other osteological manifestations (e.g., OA and enthesopathies). Nevertheless, there are several instances where such inferences have been attempted, and, notably, it appears this trend is increasing. An early innovative use of such data was, characteristically, attempted by Angel (1971), who suggested that Schmorl's nodes were indicative of "sudden pressure from lifting or other movement" (p. 87). Kelly (1982) did a systematic analysis of what he termed "intervertebral osteochondrosis" in the Hamann-Todd collection. He argued these lesions were usually erosive and thus *not* equivalent to Schmorl's nodes. He did, however, suggest they might develop as the result of mechanical stress, although did not provide any independent evidence to support this contention. Additionally, Kelley assessed the prevalence of these lesions in two archaeological samples (Mobridge and Indian Knoll) and attributed the higher frequency in the latter group (14.5% *vs.* 0.6%) to a "harsh lifestyle."

Gejvall (1983) more broadly hinted at a mechanically based etiology, arguing that the lesions resulted from "violent trauma into the vertebral plate" (p. 90). Likewise, in a general osteological assessment, Angel et al. (1987) did not much emphasize these lesions, but did conclude in comparing an historical cemetery sample from Philadelphia with one from Maryland (see above) that: "Arthritis breakdown and Schmorl herniation in vertebrae are slight, compared to Catoctin, indicating less heavy wrenching and lifting, especially at a formative age" (p. 227).

Rathbun (1987) made even more specific claims in attempting to infer activity in a similar (historical) archaeological context. In a quite small sample ($N=36$) derived from a slave cemetery in South Carolina, Rathbun found 35% of skeletal Schmorl's node involvement (i.e., *very* typical; see above). The lesion was seen more commonly in males, which was attributed to "differences in occupation and lifting."

More recently, Reinhard et al. (1994) suggested Schmorl's nodes were due to trauma, and Robb (1994) also concluded the distribution of such lesions were associated with activity "in which tasks dangerous, or 'heavy' or requiring strength were allocated to males" (Robb, 1994: 222). Robb's interpretation was based, again, on a higher prevalence observed in males (69%) compared to that in females (31%) (Total $N=56$). Finally, Larsen et al. (1995) found Schmorl's nodes in 2 of 11 historic burials and argued (as with "other degenerative pathology") the condition resulted from "physical activity," and, more specifically, was "due to high levels of mechanical loading of the back" (p. 148). Given the sample size, and apparent distribution of what is clearly a common condition, this conclusion is surprising. Moreover, its unqualified attribution to "degenerative pathology" is not in accord with most clinical views (where the etiology is seen as complex, but is probably related to some predisposing developmental "abnormality"). In rarer circumstances, however, the condition *might* ensue secondarily from degenerative disk disease (Resnick and Niwayama, 1978).

Schmorl's Nodes: Conclusions

As just noted, the clinical data are not in full agreement regarding the etiology of these common lesions, but the general consensus points to multiple causation (with an underlining bone "weakening" as highly significant). Thus, the condition is most likely primarily a developmental "defect." More appropriately, given its frequent appearance, it would better be termed an "anomaly." Indeed, if the "condition" is as frequent as some data indicate (Hilton et al., 1976), Schmorl's nodes should be regarded as "normal."

Some etiological contribution resulting from eccentric mechanical loading is, of course, possible. However, whatever influence such mechanical factors might have, would, necessarily, have to occur early in life (i.e., adolescence or earlier). As noted, the early onset of the condition is well-recognized clinically, and the likelihood of early onset has also be suggested by osteologists (Kelley, 1982; Angel et al., 1987). One clinical report, in fact, could give some support to this hypothesis of an early onset (partially) stress-related etiology. O'Neil and Micheli (1988) found a rare manifestation of what they termed "atypical Scheuermann's disease" in some young athletes. They speculated the lesions "*may* represent repetitive microtrauma and compressive fractures" (p. 601). Before my colleagues seize too quickly on this intriguing conjecture, it should also be noted that O'Neil and Micheli further argued the condition was probably primarily influenced by predisposing back deformity (thoracic hyperkyphosis and lumbar hypolordosis, i.e., "flatback").

One last epidemiological feature of Schmorl's nodes involvement is worthy of mention. In several clinical studies a higher prevalence of the condition was found in males, although sometimes the differences were slight (Schmorl and Junghanns, 1971; Hilton et al., 1976). A preponderance of male involvement has also been observed in skeletal samples, most systematically by Robb (1994). However, such a population pattern does *not* necessarily argue for a functional explanation, in which males "worked harder." Clearly, systemic influences which predispose adolescent males to developing such vertebral alterations are at least as likely an explanation. Given the common manifestations of these "lesions" in most populations and usually quite frequently expressed in *both* sexes, a more systemic, less localized (i.e., less mechanical) explanation is *more* likely.

Rheumatologists and orthopedists have obviously long known of Schmorl's nodes. Their views on their meaning (or lack thereof) are informative. Hilton et al. (1976: 131) regarded them as a "common but clinically occult spinal defect," and Dieppe (1987: 16.82) again declared them "very common" and "relatively unimportant."

It would be advisable, failing some notable new and reliable clinical support, to defer from making even general behavioral interpretations from these common, relatively innocuous "lesions." Making specific hypotheses such as "lifting" are clearly even less warranted.

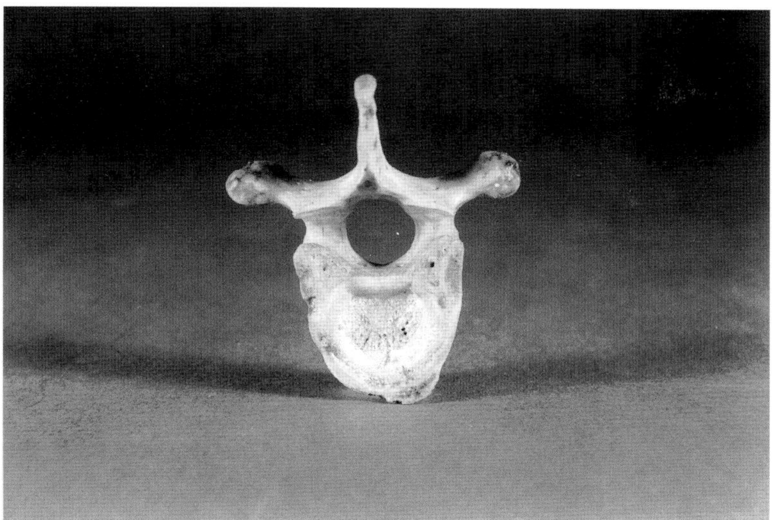

Figure 5-4 Schmorl's node. Thoracic vertebra in a gorilla. (Note: compare with Figure 5-3). Photograph by the author.

In this light, my own observations on quite common appearance of Schmorl's nodes in chimpanzee and gorilla spines call such speculation into further question; unless we build imaginative scenarios of great ape behavior, heretofore undocumented in ethological research (Figure 5-4).

STRESS FRACTURES

The topic of stress fractures could, obviously, also be addressed in the following chapter on trauma. However, these conditions perhaps best belong with the current discussion, as they are, in many cases, documented clinically to occur in contexts of "over-use." Moreover, at least two osteological manifestations, so-called "clay-shoveler's fractures" and, most especially, spondylolysis, are well-established to be associated with activity.

According to O'Neil and Micheli (1988: 598), "A stress fracture is most likely caused by recurrent microtrauma resulting in cortical bone fatigue." They further note their present occurrence is most often associated with "maltraining." These lesions are also called "fatigue fractures" and are not usually caused by acute injury or exotic biomechanical loading. Currey (1984: 279) argued they most often "are caused not by very large stresses, but by stresses merely somewhat larger or more frequent than usual. Nor do these stresses act in unusual directions."

Stress fractures result most commonly from rapid onset of overuse, particularly among individuals who were previously fairly sedentary. The best documented cases relate to military recruits in whom the tibia and metatarsals are most frequently involved (Currey, 1984). Such injury has also been reported more recently among women undergoing military training—and they may, in fact, be at even higher risk than men for such stress fractures (Marti, 1991).

In addition, in recent years, considerable attention has been directed at assessing the risk of stress fractures in young athletes (O'Neil and Micheli, 1988). Such injuries have been well documented in the distal radius of gymnasts (Reed, 1981; Carter et al., 1988; Caine, 1992); the elbows of baseball pitchers (King et al., 1969; Nuber et al., 1992) and javelin throwers (Miller, 1960; Hulkko et al., 1986); and in the calcaneus. These latter lesions, resulting from calcaneal apophysitis (Sever Disease), are, in fact, "the most common cause of heel pain in the young athlete" (Micheli and Ireland, 1987). While most commonly resulting from overuse, the same lesion (avulsion of the posterior calcaneal apophysis) was reported as early as 1917 to also sometimes be due to a single (acute) traumatic episode (Kurtz, 1917). In all these reports (including the latter) the injuries were consistently found in adolescents, that is, before or just after the relevant epiphysis/apophysis was fusing. Indeed, the greater liability of young individuals to such bone injury is probably the most interesting finding, and parallels similar suggestions for some forms of severe OA (see Chapter 2).

Clay Shoveler's Disease

A vertebral stress fracture, occasionally seen in osteological samples, may be quite closely associated with certain types of overuse. Called "clay shoveler's disease," the lesion is usually a fatigue fracture of the spinous process of C7 and/or T1 (Schmorl and Junghanns, 1971) (Figure 5-5). Typically, the lesion is seen in adults (Knüsel et al., 1996), but can occur in adolescents (Weston, 1956), where it occurs as an avulsion injury of the unfused or recently fused apophysis (and is, in this form, termed, "Schmitt's Disease"). In adults, the etiology of the condition has been argued to be due to "direct trauma" resulting from muscle pulls, and Schmorl and Junghanns noted its occurrence

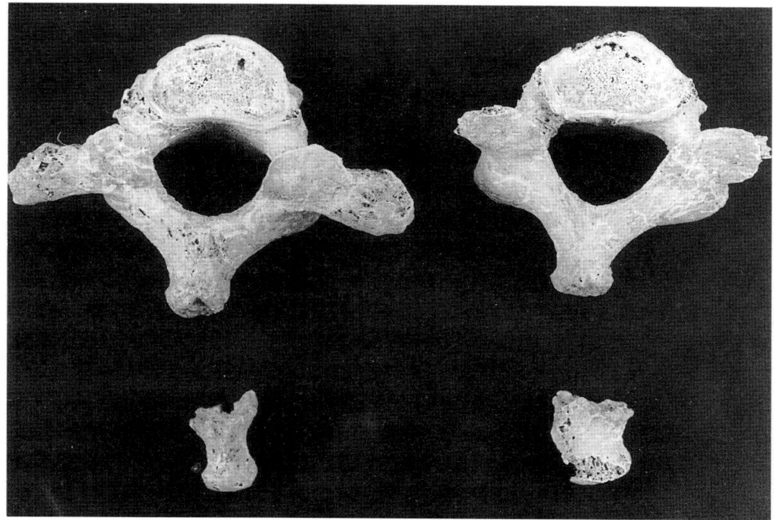

Figure 5-5 "Clay shoveler's disease." Seventh cervical and first thoracic vertebrae, Romano-British (4th C. A.D.), Baldock, Hertfordshire, England. Photograph courtesy, Charlotte Roberts, Calvin Wells Laboratory, University of Bradford (Department of Archaeological Sciences).

in cricket players and, especially in "diggers." In fact, these authors stated: "It can be considered an occupational disease under the proper conditions" (Schmorl and Junghanns, 1971: 261).

Recently, Knüsel and colleagues (1996) have more completely reviewed the clinical literature and list a variety of other occupations/activities which have been clinically reported as sometimes associated with the lesion; these include: weightlifting, metaldipping, root pulling, hay pitching, falling from height, impact on head from falling object, vehicular accident, and some injuries associated with American football. However, in this comprehensive review and discussion, Knüsel et al. argued that the best data associating the lesion with activity were for shoveling behavior, particularly hard clay, as best documented in systematic surveys of German road workers.

Thus, with good reason, Knüsel and colleagues suggested (as did Schmorl and Junghanns) that this is an "activity-induced lesion." Moreover, these authors analyzed the skeletal remains from several British collections and reported on three cases from three different cemeteries (as well as another three cases mentioned by Stroud and Kemp, 1993 from a cemetery in York).

Recognizing some variability in the potential contributing activities, Knüsel and colleagues cautiously concluded that the lesion results from perhaps "several possible manual activities." They went on to suggest that the condition "may be the result of long-standing human subsistence adaptation requiring digging in the soil" (Knüsel et al., 1996: 427).

The conclusion that "clay shoveler's disease," particularly in unhabituated laborers, results from stress fractures due to overuse is now well-substantiated. That such lesions frequently occur in individuals doing heavy digging is also quite well established. What range of *other* activities might also produce the condition is not yet determined, but a variety of sources do suggest that several different types of activity might contribute. Thus, until more complete epidemiological data are available, the first assertion (i.e., activity-induced) made by Knüsel and colleagues appears better substantiated, while the latter (specifically, digging) is an excellent working hypothesis.

Spondylolysis

In osteological contexts, by far the best documented example of such fatigue-related fracture is spondylolysis, another lesion observed in some vertebrae. The osteological manifestation of this lesion, as an incomplete closure of the neural arch through *pars interarticularis*, is most commonly expressed in the lower lumbar vertebrae (Figure 5-6). The condition was noted fairly early by Stewart (1956), but there are clinical (biomechanical) interpretations dating back to 1855 and anthropological reports of the condition as early as 1911 (see Newell, 1995; Merbs, 1996b for reviews). It is, in fact, largely owing to the intensive research by Merbs (1989b; 1995; 1996a,b) that the osteological interpretation of spondylolysis has been put on a sound scientific footing. Beginning with a useful discussion in his 1983 monograph, Merbs has more recently (1995; 1996a,b) produced a series of publications presenting remarkable detail and thorough substantiation. Merbs has considered a variety of etiological agents and cautions to remain prudent in not postulating an overly simplistic single causation. In particular, Merbs has argued that underlying, developmental factors could well play an important role: "What is inherited, however, is not the condition itself, but anatomical features which predispose vertebrae to separation" (Merbs, 1983: 172).

Nevertheless, he has presented a wealth of evidence arguing the lesion *usually* develops as the result of a fatigue fracture, and most commonly, in young individuals. At a more general level, Merbs has also noted the condition is related to bipedal locomotion (especially the lumbar curve), as it is never seen in children prior to walking,

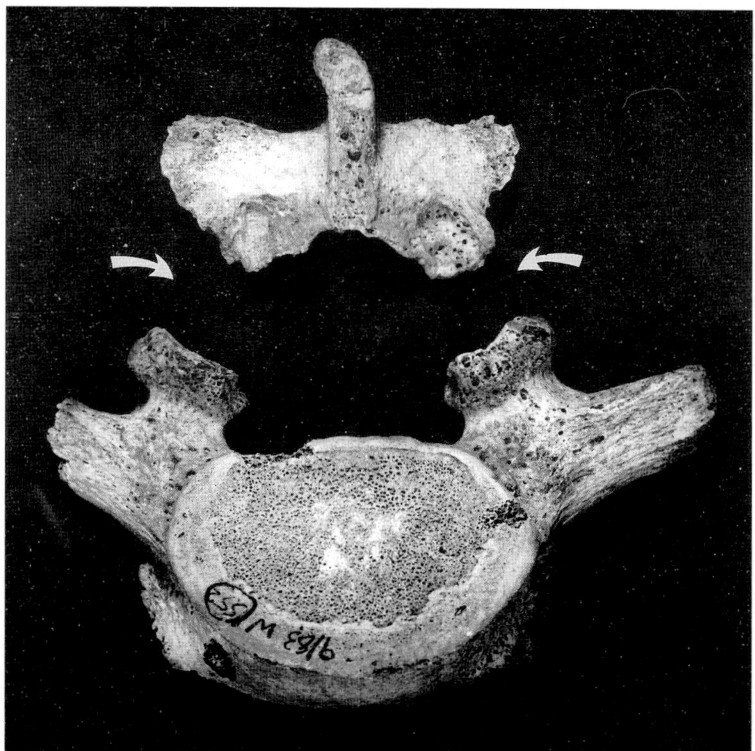

Figure 5-6 Spondylolysis. Fifth lumbar vertebra, Romano-British (4th C. A.D.), Gloucestershire, England. Photograph courtesy, Charlotte Roberts, Calvin Wells Laboratory, University of Bradford (Department of Archaeological Sciences).

nor in older individuals who have never walked. Moreover, it has never been observed in nonhuman primates (Merbs, 1989b) (the latter finding is also further confirmed by own work on gorilla and chimpanzee spines).

As noted, most commonly, spondylolysis is seen in L5 (or L4), but has also been reported by Merbs (1996a) to occasionally be found in the sacrum (S1). Here, as in the lumbar spine, the lesion is thought, most typically, to result from a stress fracture. Merbs has more recently concluded: "The picture that emerges is one of mechanical stress in the lumbosacral region, created in a vertebra by movement to that below, producing a stress fracture" (Merbs, 1996b: 210)

From a systematic analysis of all types of lesions (partial and more complete), Merbs (1996b) suggested the situation is quite dynamic, leading to considerable healing in older individuals. Bridges (1989c) found an almost opposite pattern in her skeletal sample from northwest Alabama, with a higher incidence in older individuals. She thus suggested that, in females, age-related bone loss may be an important contributing factor to onset of spondylolysis late in life.

Of course, osteological studies, constrained to a cross-sectional perspective, cannot fully resolve these more precise details of pathogenesis. Micheli and Wood (1995) have recently observed a possible contributing factor that might partially explain Bridges' finding. In a comparison of young athletes with older individuals (all individuals presenting with spondylolysis and low back pain), 48% of the adults *also* had disk abnormalities, compared to 11% in the adolescents.

In order to support his functional hypothesis, Merbs reviewed a considerable clinical/sports literature in which spondylolysis has been observed among individuals engaged in (with numbers of citations from Merbs, 1996a,b): gymnastics (8 sources); contortionism (1 source); dancing (2 sources); diving (1 source); hockey (1 source); javelin throwing (3 sources); rowing (1 source); canoeing (1 source); American football (3 sources); handball (1 source); weight lifting (3 sources); and cricket (1 source). Further, Merbs contended from his own skeletal research that there may also be some association with kayaking, wrestling, paddling, and harpooning.

Clearly, there is substantial evidence relating a variety of activities, but also this range argues that many different types of motion might be significant, a point also made by Bridges (1989c). In fact, Merbs (1996b) from the diverse occupational and sports literature, which he has so completely mastered, further commented the condition has been "associated with certain activities, but at a disappointingly low predictive level" (Merbs, 1996b: 365). He thus concluded that attribution to specific behaviors "will probably prove elusive."

The osteological materials which Merbs has most intensively investigated, and which first alerted him to the condition, are those of Inuit people. In his earlier report of the Sadlermuit, he found a very high prevalence of 44 involved cases in 70 spinal columns (i.e., 63%). In a later analysis (N = approximately 400 individuals) from a number of Inuit sites, Merbs also included cases of partial separation (lysis), and for the full sample, he found a prevalence of approximately 12.5% (51 involved cases; total of 110 affected sites) (Merbs, 1995). For sacral involvement only, a lower prevalence was found among the Inuit, with 16 cases observed in 373 individuals (4% prevalence).

Isolated cases of spondylolysis have been reported fairly widely in the literature, but most descriptions are anecdotal, and thus lack the paleoepidemiological controls permitting determination of prevalence. Bridges (1989c), however, did consistently control her sample, and found a 17% prevalence of spondylolysis in males and a 20% prevalence in females (Total $N = 157$). Likewise, Stirland (1996) also presented some prevalence data from an English Medieval cemetery sample (8.5% of males involved) and compared these with prevalence of the lesion among the *Mary Rose* individuals (11.1%).

Thus, it can be seen that the prevalence of the lesion varies widely among populations, but is consistently highest among the Inuit (than for other groups thus far reported). Whether varying patterns of activity alone account for these population differences, or whether genetic/developmental predisposing features also contribute, is not yet fully established. To this point, however, the systematic, long-term research by Merbs provides an exemplary model of how such skeletal/behavioral research should be conducted.

SCARS OF PARTURITION

Although not an activity in quite the sense as those imputed in the etiology of enthesopathy, Schmorl's nodes, or stress fractures, parturition is an event which produces mechanical stresses on portions of the skeleton. Given the potential for useful information, it is not surprising that osteologists have sought to find specific markers of this activity—what have been termed "scars of parturition." Indeed, if such markers could be reliably established, the data would be of great utility in helping reconstruct a central paleodemographic aspect (i.e., fertility) of ancient groups. Moreover, such information might prove useful in forensic applications, especially those relating to personal identification. Owing to this potential forensic utility, "scars of parturition" have especially attracted the interest of forensic anthropologists.

Fairly early on, there were several mentions in the clinical literature of potential effects of parturition on the skeleton, including those by Lynch (1920), Putschar (1931), and Heyman and Lunquist (1932). None of these was initially much noticed by osteologists (although Putschar did republish his findings in 1976 in *The American Journal of Physical Anthropology*). The first mention of the relevant skeletal changes that came to the attention of osteologists was probably that by Stewart in 1957. He later described his "discovery:"

> In retrospect, I look upon the circumstances that led to my 1957 publication, which was probably the first to bring the features in question to the attention of physical anthropologists, as serendipitous in the true sense

> of the word: that is, they made possible in the words of one definition, 'a discovery of valuable or agreeable things not sought for.'
>
> ... my surprise was great, therefore, when I noted among the Eskimo very different shaped symphyses [from those in male Korean war dead] including, in some instances, depressions and/or cavities on the adjacent dorsal surfaces. Immediately, I realized that the specimens that deviated from my concept of normality were of the female sex. Thus, although I had looked at Eskimo skeletons many times before, for the first time I was then 'seeing' these particular features as abnormal and limited to the female sex. The final step, of accounting for the abnormality, was easy, of course (Stewart, 1970: 128).

J. Lawrence Angel, a colleague of Stewart's at the Smithsonian, was also making similar inferences at about the same time, and Stewart (in the publication of a 1970 symposium) referred to Angel's suggestion the previous year: "Since adequate documentation on parity is so limited in this assemblage of material, I am sure Dr. Angel regards his deductions as tentative, especially as regards their use in forensic interpretations" (Stewart, 1970: 127).

In these provisional suggestions by osteologists, the lesions themselves were not fully described, nor was there much in the way of documentation from well-controlled samples (i.e., those with *known* parity), a limitation readily admitted by Stewart. "Greater accuracy in interpreting abnormal female pubic bones, in terms of parity, can best be achieved, of course, through the assembly of an *adequate series of documented specimens*" (Stewart, 1970: 128) [emphasis added].

In the following year Angel enthusiastically suggested that parturition markers could provide valuable information: "Thus, anterior exostoses, posterior pits, and erosion of different degrees of the pubic symphysis give a good clue to the number of children a given women has produced—that is, too her fecundity [fertility]" (Angel, 1971: 72).

Over the years, three separate regions of the os coxa have been postulated to show identifiable changes which result from parturition. Angel et al. (1987) nicely summarized these:

> There are two specific sets of changes at the pubic symphysis: (1) From each birth an increasing irregularity of the posterior margin of the pubis at the joint, forming a sharp projection, plus development of pits which eventually join into a deep fossa; *number of pits if they are present, is not equal to number of births*. (2) On the external surface of the pubis bone roughening of the spiral fossa below the pubic tubercle produced by stress from the rectus and other muscles of the abdominal wall. The first of these results from actual births including the effects of the hormone relaxin, plus tearing of ligaments and hemorrhaging at time of delivery, and the latter directly from pregnancies (independent of full-term births). There is also deepening and widening of the preauricular sulcus [the third area of change] and

often some changes in the sacroiliac joint from stress of carrying fetuses and producing newborns, plus aging and outside hard work (Angel et al., 1987: 214–215).

From the above quote, it is clear that Angel and colleagues recognized some of the limitations in the interpretation of parturition markers. They further stated the "number of births is hard to read from the pelvis." Nevertheless, they went on to argue that "individual variation tend to cancel each other out and allow a relatively mean estimate [of fertility]." This last assertion is somewhat surprising, especially in light of a number of increasingly well-controlled studies during the late 1970s and early 1980s that indicated such interpretations were not supportable. Gilbert and McKern (1973), for example, could not establish any clear correlation of pubic changes with parity (in a sample of 140 individuals with known parity); accordingly, they suggested a variety of possible explanations for the wide variation in expression of pelvic indicators, including, fetal size, pelvic size/shape, and obstetrical practices.

Hermann and Bergfelder (1978), from a sample of 49 pubic bones (obtained at autopsy from women with known parity and number of abortions) also could find no osteological indicators directly associated with parturition; indeed, they saw distinct changes in nulliparous women, and in some elderly women with many childbirths, no changes were seen. In a less well-controlled sample (the Hamann-Todd collection), but one still providing some decent documentation regarding parity, Holt (1978) reported similar negative findings as those of Hermann and Bergfelder, and thus concluded: "It is suggested that bony changes previously used as an indicator of child bearing must be reexamined in light of the fact that 'birth scarring' has been found in women known not to have had children" (Holt, 1978: 91).

The largest, most systematically controlled study was reported the following year by Suchey et al. (1979). In this investigation, 486 paired pubic bones were obtained at autopsy (Los Angeles County Coroner) from an ethnically mixed population. The authors established *some* association and noted: "A statistical association was found between the number of full term pregnancies and the degree of dorsal pitting. However, *the correlation is not strong*" (Suchey et al., 1979: 517) [emphasis added]. Moreover, Suchey and colleagues pointed out a number of confounding factors including especially the age at which the pregnancies occurred and interval from last pregnancy until death.

While most of the emphasis has been directed at establishing criteria from modifications to the dorsal aspect of the pubic symphysis (so-called "pits" and "depressions"), some investigation has also been aimed at the other areas described above by Angel et al. (1987). Firstly, the preauricular sulcus area, especially changes within the sulcus, was suggested as a likely location for parturition modification by Houghton (1974) as well as Kelley (1979). However, no consistent association of changes to this region with parity could be established by Andersen (1986), Tague (1988) or Cox and Scott (1992).

Secondly, the pubic tubercle has also been postulated (initially by Angel in 1969) to reflect parturition changes. Kelley (1979) also argued that parturition markers could be analyzed in this region, and, more recently, Cox and Scott (1992) provided some systematic data to support this claim. In this latter study, from the well-documented Spitalfields sample, 94 female skeletons were analyzed (with corresponding historically reconstructed obstetrical histories). Indeed, Cox and Scott (1992) noted that, in most cases, there were good data relating to number of children, birth spacing, and age at first and last births (quite remarkable, and pointing again to the great utility of this sample). Nevertheless, repeated studies from precisely documented contemporary contexts have not yet provided the required verification for these potential parturition markers of the pubic tubercle. Moreover, Galloway (1995) has noted the changes seen here are mostly of a qualitative nature, and are, thus, difficult to standardize.

Another recent attempt (Gilmor, 1990) to isolate more specific dorsal pubic alterations ("circles") also has proven, on reexamination (Suchey and Pierce, 1992), not to be replicable. Some attempts have been made by other researchers to utilize multiple indicators (dorsal symphysis, preauricular sulcus, pubic tubercle) to assess parturition (Kelley, 1979), but further (more detailed) attempts to utilize such a methodology (Andersen, 1986) failed to confirm the utility of using multiple indicators.

In fact, the studies noted above (by Kelley, 1979 and that by Andersen, 1986) were both conducted on the *same* collection (Hamann-Todd) (as was that by Holt, 1978), and all found different results. Galloway (1995) has regarded this inconsistency as quite disturbing and indicative of a wider problem in assessing markers of parturition,

> A broader question which might be asked with regard to this material is: Why have contradictory results been found, and, furthermore, why *despite the lack of congruence*, have the three markers been adopted, at least in part, within the forensic anthropological community (Galloway, 1995: 83).

Galloway (1995) further concluded that the ability to distinguish (skeletally) parous from nulliparous women is questionable; assessment of the number of births is even more questionable.

What can one conclude from these often-conflicting findings relating to "parturition scars?" Clearly, the mechanical/physiological stresses relating to birth *could* alter the proximate osteological structures, perhaps even in a consistent fashion. Tague suggested (from animal studies) that increased secretion of estrogen might promote bone resorption, but had to conclude (in light of still incomplete findings): "... further study of nonhuman mammals is required to delineate the proximate and ultimate causation of bone resorption of the innominate" (Tague, 1988: 265).

In her excellent recent review of the value of osteological markers of parturition, Galloway (1995) cautioned that there are several factors influencing pelvic remodeling:

> Joint instability, regardless of reproductive status, appears higher in women and, consequently, appears to result in higher instances of pelvic joint change. Secular change in reproduction, obstetrical practices, physical activity, and hormonal levels also interfere with an interpretation of these indicators (Galloway, 1995: 83).

That some of the pelvic osteological changes called "parturition scars" result from birth trauma is thus probable. But they do so at a *low* level of predictability. In analysis of skeletal samples, what can be confidently concluded is that adult females were more often parous than adult males, obviously, an interpretation offering minimal insight. With very well preserved samples, it could also be concluded that many adult females were parous, and a good proportion of these were likely multiparous. However, again in most human societies, this finding is completely expected. Without considerably better resolution, parturition markers, therefore, provide little, if any, additional information in reconstructing past behavior (and demographic profiles) of past populations. At present, they are not sufficiently predictive to allow even rough estimates of fertility (not in individuals, nor, obviously, in population samples). Thus, it could be asked: Of what real use are they?

AUDITORY EXOSTOSES

A condition with apparently a relatively straight-forward etiology is the development of auditory exostoses, that is, "bony growths in the auditory canal" (Hrdlicka, 1935: 1) (see Figure 5-7). Kennedy (1986) noted that a more general term for such lesions is "auditory hyperostoses" (referring to "any bony auditory lesions"); she suggested the

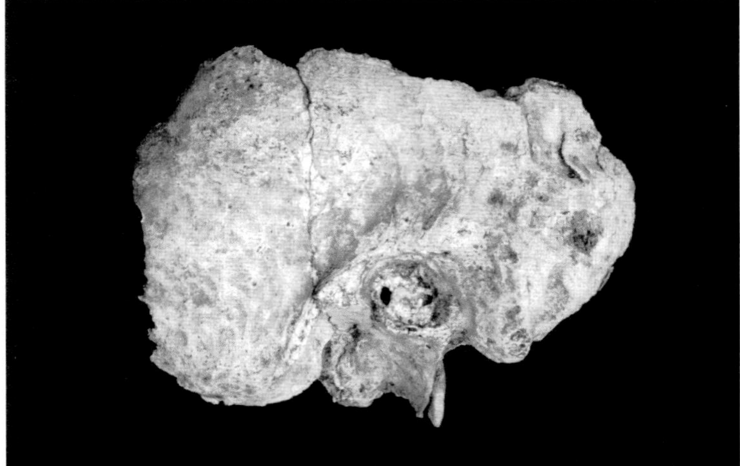

Figure 5-7 Auditory exostosis. Beckford, Worcestershire, England. Photograph courtesy, Charlotte Roberts, Calvin Wells Laboratory, University of Bradford (Department of Archaeological Sciences).

term "exostoses" be used to "denote all discrete bony auditory lesions" (p. 404). Hutchinson et al. (1997) further emphasized the distinction between osteomas and exostoses; they noted the latter are "often multiple in number, bilateral and symmetric" (p. 418).

Recognition of such hyperostotic bone lesions has been documented in the clinical literature as early as 1809 (for excellent reviews of this clinical literature, see Di Bartolomo, 1979; Kennedy, 1986). In the 19th and early 20th centuries a variety of etiologies had been suggested, including, genetic/congenital, traumatic, and masticatory stress. Hrdlicka (1935) in a monographic treatment of the subject concluded ear exostoses were due to chronic "irritation." The source of such irritation had already been suggested by Wyman (1874) and Field (1878) as likely associated with swimming. Indeed, Wyman made this argument (apparently, the first to do so) on the basis of skeletal remains (from Peru), once again displaying the highly fruitful results of anthropological/clinical interaction. Further research in the 20th century has gone far to further substantiate a clear association of auditory exostoses with exposure to water, most especially cold water (van Gilse, 1938; Fowler and Osmun, 1942; Adams, 1951; Harrison, 1951; 1962). Clinicians working with surfers on the west coast of the US have also reported the same condition, here termed "surfer's ear" (Seftel, 1977). In an unpublished study, Scott (n.d.) found high prevalence of ear obstruction in competitive surfers, with 84% showing at least a 30% blockage ($N=275$ cold water surfers). More recently, similar results have also been found in cold-water surfers in Oregon (Deleyiannis et al., 1996) and in a sample of 87 Navy divers (Karegeannes, 1995).

From a superb review of the available literature, Kennedy concluded,

> The clinical assumption that auditory exostoses are caused by chronic, long-term exposure to cold water, recently termed the "thermal aquatic hypothesis" (see Di Bartolomo, 1979) is currently widely, if not universally accepted by clinicians (Kennedy, 1986: 404).

In order to test further the thermal aquatic hypothesis in skeletal samples, Kennedy collated a huge data base from 90 different reported osteological samples (including a total of thousands of individuals). Moreover, she rigorously was able to provide prevalence rates for each sample (as she excluded all samples of <30 individuals). The results were interpreted specifically from a geographic (latitudinal) perspective. In Kennedy's tabulation, "with only few exceptions," the conditions of the hypothesis were met. She thus concluded:

> It seems clear from extensive clinical data, experimental replication, ethnographic observation, archaeological reconstruction, and latitudinal analysis that there is a causative relationship between diving for cold water resources and moderate to high frequencies of auditory exostoses (Kennedy, 1986: 412).

In the samples collated by Kennedy, as well as other osteological and clinical sources, the onset of the condition is usually in adulthood (after age 20) and more common in males than in females; the latter finding is generally attributed to a sexual division of labor. Kennedy, however, noted and explained one particularly intriguing exception:

> The Tasmanian sample provides a particularly elegant test of the hypothesis since here, unlike almost everywhere else where males were the exploiters of aquatic resources, the females were major providers of such foods. Here, unlike almost everywhere else, it is the females who had the exostoses, while the males had none (Kennedy, 1986: 412).

In a few more recent osteological studies some further evidence supporting the aquatic hypothesis has also been reported. In two Roman samples, Manzi et al. (1991) found a higher prevalence of auditory exostoses in a group (mostly males), and they suggested the ancient Romans were more frequently using baths than another group with less access to such facilities. In the Ala-329 collection from the coastal/bayshore region of Central California, Douglas (1993) found indication of moderate-large exostoses in 11% of adult males ($N=61$) and 1% of adult females ($N=76$). Sakalinskas and Jankauskas (1993) also reported moderately high prevalence (especially in males) in a large archaeological sample from Lithuania. Additionally, in a recent systematic study of Chilean skeletal materials from 43 different sites (Total $N=1149$), Standen and colleagues (Standen and Arriaza, 1996; Standen et al., 1997) found the highest incidence of auditory exostoses among coastal residents (31%), low incidence among residents of lowland valleys (2%), and no involvement whatsoever was observed among highland people. Interestingly, the high coastal incidence persisted in both pre-agricultural and agricultural groups, suggesting to the authors that marine resources were exploited fairly intensively within both subsistence regimes.

Finally, reminding us again of the potential complexities even in ostensibly simple conditions, Hutchinson et al. (1997) suggested a variety of potential etiological factors influencing auditory exostoses. In addition to cold water immersion, they noted possible effects of otitis externa, eczema, trauma, infection, and mechanical and chemical irritation. Given such possible complexities in etiology, Hutchinson and colleagues concluded: "Our research indicates that it is unlikely, given the anatomical and chemical complexity of the external ear canal, that cold water can serve as an *exclusive* etiology for auditory exostoses" (p. 421) [emphasis added].

CONCLUSION

In this chapter a variety of skeletal conditions have been reviewed, each advanced in recent years to be related to some form of activity. The verifiability of these different "markers" vary greatly, at least as far as available data can substantiate them. The interpretations of two spinal fatigue fractures, spondylolysis and the initial results concerning "clay shoveler's fracture," are well-grounded and provide good models for how substantiation in paleopathology should be attained. Moreover, the investigations of auditory exostoses have provided good insight and considerable clinical corroboration. It should be noted that these conditions all have relatively straight-forward etiologies, and ones that are well-recognized by clinicians from studies of clearly documented (at risk) population samples.

The investigations of parturition markers over the last three decades also offer instructive insight. As noted, most of the research has been conducted by osteologists working within the forensic sciences;

accordingly, there has been considerable effort directed at establishing well-documented series for validation as well as establishing repeatability of observations. The results of these various attempts have generally been disappointing, at least relative to the proximate goal (of establishing reliable indicators of parturition). However, in terms of a more general goal of applying rigorous scientific standards to such behavioral interpretations, the research, in its total, provides another excellent model for skeletal biologists.

The establishment of reliability for enthesopathies and Schmorl's nodes as markers of activity, however, has proven even more disappointing. In both cases, the amount of independent substantiation is very weak, leading researchers to draw behavioral conclusions directly from the skeletal material. Molleson, for example, concluded from the distribution of some modest bone modifications in a fragmented Near Eastern sample:

> It is the repeated occurrence of the specific changes, in the absence of other anomalies or pathologies, that suggest they are part of an acquired adaptation to a particular activity; an activity that is carried out frequently and for long hours (Molleson, 1989: 357).

This type of approach, of course, quickly becomes teological, a point raised by Galloway in referring to research on parturition. "A frequent problem has been the lack of documented cases in which parity can be established. *This has resulted in the development of circular arguments equaling the presence of pitting with parturition*" (Galloway, 1995: 85) [emphasis added].

What is especially sobering in applying Galloway's critique more widely is that parturition data are (and have been for some time) *far* superior and far better documented than any comparable research relating to enthesopathy. As in research relating to other skeletal conditions, the best hope again lies in mutual interaction between osteologists and clinicians. As related to enthesopathy, the first fruits of such collaboration are now becoming apparent (e.g., Shaibani et al., 1993; Rogers et al., 1997). While the findings are not necessarily what was expected (or hoped for; i.e., they do not support a primarily activity-based etiology), they are in many ways *even* more interesting. Indeed, similar to the recent research on OA and trauma, such substantial collaborative contributions by osteologists have yielded unexpected insights. In many ways, thus, this research provides us increasing illumination of etiology, epidemiology and more: Data which can, obviously, increase understanding, if osteologists can escape the myopia of activity-only scenarios.

Chapter 6

Trauma

> Serious problems may arise in distinguishing between intentional and accidental bone damage, and even when damage is convincingly intentional, the actual intent may not be obvious (Merbs, 1989a: 187).
>
> On this fragile foundation we shall not make bold to theorize on who fell most or who suffered violence (Bennike, 1985: 58)

INTRODUCTION

Traumatic lesions are certainly among the most easily observed and occasionally among the most dramatic types of changes found in osteological material. Moreover, from contemporary experience, we know that severe trauma can, and often does, leave distinctive markers in the skeleton—sometime relating to severe, even life-threatening injury. Thus, it is no surprise that paleopathologists have long been interested in describing and oftentimes *explaining* traumatic involvement in human skeletal materials.

"Trauma" is defined by Roberts (1991: 226) as "any bodily injury or wound, and [it] may affect the bone and/or soft tissues of the body." Lovell (1997: 139) includes within trauma "injury to living tissue that is caused by a force or mechanism extrinsic to the body." Under the overall umbrella of trauma, are a variety of different types of injuries, including, most distinctively, fractures. As defined by Roberts (1991: 226), a fracture is "the result of any traumatic event which leads to a complete or partial break in the continuity of bone." Fractures may be partially or completely healed, or they may have been perimortem, i.e., at or near the time of death. In cases of excellent preservation, and certainly in contemporary forensic contexts, it may be possible from the form of the lesion(s) to distinguish such perimortem damage from that occurring postmortem. However, in most archaeological contexts, such a distinction is frequently impossible. Accordingly, most of the osteological data relating to fracture focus on healed lesions. Of course, at the other end of the spectrum, when a fracture has healed quite completely with no obvious macroscopic deformity, it may also be impossible (at least impractical) to recognize the lesion. It should be emphasized that systematic radiological assessment is also highly useful (Lovejoy and Heiple, 1981; Roberts, 1991; Lovell, 1997; Lambert, 1997) and may prove invaluable in distinguishing long-standing healed lesions, as well as assessing adequacy of healing (and possible treatments which may have been attempted). Roberts (1991), particularly, made the point that evaluation of the amount and nature of the fracture callus, as well as the amount of displacement, are better assessed on X-ray than on gross examination. Nevertheless, even with good supporting radiological evaluation, diagnosis of long-standing, well-healed lesions presents a major challenge to osteologists.

In addition to typical fractures, Steinbock (1976), within the topic of trauma, included crushing injuries, wounds (due to sharp instruments, including mutilation, scalping, and trephination) as well as dislocations and growth-arrest lines (the latter, not normally classified as trauma, but more commonly, viewed as a growth/metabolic/nutritional insult). To the above sub-categories, Ortner and Putschar (1981) add deformation, traumatic lesions due to pregnancy (discussed in Chapter 5), and, as also treated by Steinbock, all forms of ancient surgery. Finally, Merbs (1989) in an excellent overview of the entire topic of skeletal trauma, further enumerated the classification to include fatigue (stress) fractures, pathological fractures, and tooth loss. The latter has often been overlooked by osteologists, but is a major aspect of contemporary facial/dental injury, especially as the result of interpersonal aggression. Moreover, such dental trauma is recognized as a significant component of traumatic injury in nonhuman primates (Kilgore, 1989).

Since fractures are fairly easily recognized, and are more common than exotic types of injuries (discussed below), most of the emphasis in paleopathology has addressed patterning of fractures in a variety of skeletal collections. This same emphasis will be followed here for the majority of the discussion of skeletal trauma.

However, occasionally more unusual and highly dramatic types of injuries have been observed by skeletal biologists. These traumatic injuries have been interpreted as resulting from scalping (Milner et al., 1991; Willey, 1990; Bridges, 1996a—discussed further below); decapitation (Harman et al., 1981; Bennike, 1985; Willey, 1990; McKinley, 1993; Liston and Baker, 1996); hanging/strangling (Bennike, 1988; Waldron, 1996); and amputation (Morel and Demetz, 1961; Webb, 1995;

Mays, 1996). Finally, there has also been considerable interest in osteological patterns relating to possible cannibalism (Turner, 1993; White, 1992) (see below for further discussion).

Much of the paleopathological literature concerning trauma has been in the form of case reports. However, in recent years more controlled paleoepidemiological approaches have been pursued, for example those by Lovejoy and Heiple (1981), Bennike (1985), Jurmain (1991a), Grauer and Roberts (1996), Kilgore et al. (1997), and Jurmain and Bellifemine (1997). In this more general type of analysis, population *patterns* are investigated, and usually the prevalence rates are reported by element (thus facilitating comparisons; see below and Tables 6-3 and 6-4) .

The case reporting format does, of course, provide a ready means for descriptive details, and thus can be useful in clarifying diagnosis. Nevertheless, occasionally it can lead readily to embellished behavioral interpretations. Perhaps, in fact, traumatic lesions are particularly suited to such "storytelling," providing as they do an inherent plot line: Victim is injured, victim goes through arduous recuperative phase, victim recovers (or not).

As in other types of skeletal conditions, discussed throughout this book, behavioral interpretations are understandably a major focus of skeletal trauma research. And, indeed, many of the behavioral interpretations that have been made in recent years are both provocative and oftentimes informative. Kennedy (1989) in his overall review of skeletal "markers of occupational stress" listed several types of fractures, although he primarily concentrated on osteoarthritic changes and enthesophytes (discussed, respectively, in Chapters 4 and 5). As noted above, Merbs (1989) focused his review more clearly on trauma and provided a very useful overview of classification, healing processes, and an in depth discussion of a wide variety of traumatic conditions. Merbs also augmented his discussion with several examples from his own research.

As the opening quote to this chapter reveals, Merbs takes a cautious approach to behavioral interpretation of skeletal trauma. He does list, however, without comment, a variety of different types of fractures with names ostensibly implying behavioral associations: boxer's fracture (metacarpal); grenade thrower's (humerus); hangman's (C1); parry (ulna); and sprinter's (anterior superior and inferior iliac spines). Of these, only so-called "parry fractures" have any adequate clinical documentation (and this topic also is still far from resolved; see below for detailed discussion). The remainder of these imaginative proposals should be treated with appropriate skepticism.

One of the central issues relating to interpretation of traumatic lesions, and one emphasized in this chapter, concerns identification of likely causation, that is, whether the injury was the result of accident or was deliberate (the result of interpersonal aggression). One attempt at general behavioral correlation was that by Cohen and Armelagos (1984), in which a group of researchers was asked to identify osteological aspects of the transition to agriculture. In this effort, each participant was requested to evaluate patterns of trauma (as well as OA—discussed in Chapter 4). Rathbun, in his discussion of southwest Asian materials, emphasized one major limitation of interpretation:

> Expression of trauma in the skeleton appears to decrease through time. However, I hasten to mention that much of the trauma seen in the latter phases appears to be from human violence. It is impossible to differentiate at other times among human violence, trauma through accident in rugged terrain, and possible injury from work with animals (Rathbun, 1984: 159).

Other investigators participating in the symposium had similar difficulties; of the 19 reports included in the symposium publication, only two (using original data) suggested any link between presence of trauma and interpersonal aggression. A problem seen in many of the reports as well as much of the paleopathological literature, was that traumatic involvement, including fractures, was not recorded in a systematic fashion. Cohen and Armelagos concluded from the generally inconclusive findings of all the reports: "For the most part, however, the reporting of trauma at the resolution of accidental and violence-related trauma is not sufficient to permit statements about trends in the two classes" (Cohen and Armelagos, 1984: 591).

Another symposium focusing on trauma was published in 1996. The editors stated that the various contributors generally consider "... the wider behavioral application of trauma, using the presence, patterning and distribution of traumatic lesions as powerful tools for interpreting lifestyle from the skeleton" (Eisenberg and Hutchinson, 1996: 1). However, despite these optimistic assertions, many of the contributors (particularly Bridges, 1996a; Smith, 1996; Stirland, 1996; and Wakely, 1996) remained quite cautious in drawing behavioral conclusions. Stirland, for example, noted: "It is not possible to specify which injuries were associated with specific activities (and this emphasizes one of the major difficulties for paleopathologists)" (Stirland, 1996: 100).

STUDIES OF TRAUMA IN FOSSIL HOMINIDS AND NONHUMAN PRIMATES

As with more recent archaeological contexts, most of the investigations of trauma in more ancient hominid remains have been limited to case reports (e.g., myositis ossificans in the Trinil femur and the perforation in the temporal bone of the Broken Hill cranium—Wells, 1964). However, systematic research on this topic has more recently been pursued, especially in work by Trinkaus. In his comprehensive analysis of the Shanidar Neandertals, Trinkaus (1983) described traumatic injuries in four (of the six reasonably complete) adults. In a comparative overview Trinkaus also briefly described other healed injuries (fractures) in remains from La Chapelle (rib), La Ferrassie I (proximal femur), La Quina 5 (humerus), Krapina 180 (ulna), and Sâla 1 (supra-orbital torus).

In a highly innovative approach, Berger and Trinkaus (1995) extended this research and attempted further to interpret patterns of traumatic involvement (and their possible behavioral correlates). They firstly provided a comprehensive tabulation of traumatic lesions found in Neandertal skeletal remains, including primarily fractures (involving 14 such lesions in 11 individuals).

Indeed, with rigorous controls relating to sample size, statistical tests, and, most especially, substantial clinical support, Berger and Trinkaus exemplify in this study the direction that behaviorally oriented research should pursue. The use of large comparative clinical (hospital) samples, using data first collated by Wood Jones in 1910, together included reports on more than 13,000 traumatic injuries. To this, Berger and Trinkaus added another clinical sample from New Mexico (workers' compensation reports of injuries). In addition, three further archaeological samples were utilized for further comparison (see discussion below). Lastly, and most creatively, the authors also drew from injury reports of professional rodeo performers collated over a nine-year span ($N = 2593$ reported injuries). All these data, both contemporary samples and archaeological ones, were then compared with Neandertals for patterns of injury (specifically, healed fractures).

The closest parallels (quite striking, in fact) are between Neandertals and rodeo participants, leading Berger and Trinkaus (1995: 841) to conclude regarding Neandertal behavior: "The similarity to the rodeo lesion distribution suggests frequent close encounters with large ungulates unkindly disposed to the humans involved."

The most significant aspect of the fracture patterns related to the relatively high involvement of the head and neck, seen both in rodeo performers and also among the Neandertals. Of course, human activities proximate to large ungulates could be one etiological factor accounting for such a pattern of involvement. Nevertheless, as discussed below, there may be other possibilities as well.

A less systematic investigation, but still yielding intriguing findings, was done by Wendorf (1968) on late Paleolithic Nubian skeletal material. From the site of Jebel Sahaba, Wendorf observed a very high prevalence of projectile wounds, leading him to conclude "almost half the population probably died violently" (Wendorf, 1968: 993). This assertion, however, is likely overstated. From the data presented, it appears that there were six confirmed embedded projectiles in four individuals. The remaining 20 involved cases were interpreted on the basis of archaeological context, that is, the association and location of certain artifacts with the burials. As discussed further below, more conservative interpretation should rely on the *direct* evidence, i.e., embedded weapons; accordingly, the frequency of projectile involvement in this Nubian sample (in relatively complete adults) translates to 9.8% (4/41). This revised prevalence is more in keeping with comparable data discussed below, especially those from California.

Lastly, Frayer (1997) has recently argued that the Mesolithic site of Ofnet (Bavaria) provided evidence of "a massacre in which a large number of men, women, and children were slaughtered with selected parts buried in two mass graves." While not all sites provide such dramatic evidence, nevertheless, such findings in Nubia, Bavaria, and elsewhere certainly help substantiate the general claim made recently by Keeley (1996) of the widespread presence of some warfare, both through time and across cultures.

Trauma in Nonhuman Primates

In addition to evaluation of osteoarthritis (discussed in Chapter 4), skeletal biologists have investigated patterns of trauma in various nonhuman primate species. In fact, most of this recent research, especially that dealing with great apes, has been done by the same investigators (Lovell, 1990; Jurmain, 1989; 1997). The systematic work on trauma done by Adolph Schultz, especially his pioneering study of healed fractures in gibbons (1939), is well known. Schultz also reported fracture data for orangutans (1941) and gorillas (1950). In addition, there has been some systematic investigation of patterns of trauma in Old World monkeys, most notably Buikstra's research on rhesus monkeys (1965) and that done by Bramblett on savanna baboons (1967). In Buikstra's sample of 43 adult rhesus monkeys, 11 (26.6%) showed some fracture. Of the total number of fractured elements, 17 were of long bones (including the clavicle), and these data are thus roughly comparable to those shown in Table 6-1 for great apes. Bramblett found an even higher prevalence of fractures (all elements) with 30 of 37 animals (81%) showing some involvement.

More recently, the research focus has again concentrated on patterns in great apes, most notably in the systematic research by Lovell (1990). In this very detailed report, Lovell presented data on quite large samples of chimpanzees ($N=49$), gorillas ($N=85$), and orangutans ($N=46$). Most of the traumatic lesions were healed fractures, with a total of 81 such lesions found in 44 individuals (24% of the total sample). Between-species comparisons showed orangutans with the highest prevalence of trauma (46% of individuals), followed by chimpanzees (33%), with gorillas (combined mountain and lowland samples) having 15% of individuals affected. Prevalence of healed fractures in long bones in gorillas and chimpanzees is shown in Table 6-1—where the data are compared with my own results.

Lovell also reported on cranial involvement, finding a total of 29 cranial lesions. Of these, however, only four were of the face, with the remainder of the vault. Given the apparent lack of facial involvement, it is not surprising that Lovell found no statistically significant sex differences (see below).

My own research has also recently focused on traumatic lesions in African great apes (Jurmain, 1989; 1997). This work initially surveyed the skeletal remains from the Gombe chimpanzee population (Jurmain, 1989) and has more recently been expanded to include larger comparative museum samples. For long bone involvement (including the clavicle), results showed the Gombe chimpanzees had a high prevalence of healed fractures (4/13=31% of individuals).

Table 6-1. Long bone fracture incidence by element, African great ape samples

	Powell-Cotton samples		Samples analyzed by Lovell[1]	
	Chimpanzees	Gorillas	Chimpanzees	Gorillas[2]
Clavicle	6/172=3.5%	1/116=0.9%	1/34=2.9%	0/47=0%
Humerus	2/184=1.1%	1/124=0.8%	0/34=0%	0/57=0%
Ulna	7/184=3.8%	1/123=0.8%	2/34=5.9%	1/49=2.0%
Radius	3/184=1.6%	1/123=0.8%	1/34=2.9%	2/48=4.2%
Femur	1/182=0.5%	7/121=5.8%	1/31=3.2%	1/53=1.9%
Tibia	1/182=0.5%	1/123=0.8%	0/32=0%	1/46=2.2%
Fibula	1/178=0.6%	1/121=0.8%	0/32=0%	1/42=2.4%

[1]Lovell, 1990; [2]Combined lowland and mountain gorilla samples

Likewise, in the comparative materials, chimpanzees also had the highest prevalence (20/92 = 22%), followed closely by gorillas (11/62 = 18%) and bonobos (2/15 = 13%). Prevalence rates by element (with comparisons with Lovell's results) are shown in Table 6-1.

Lovell (1990) postulated possible behavioral influences on the fracture pattern she observed, most particularly the highest involvement being seen in the more arboreal orangutan. In the comparative data presented here, for African great apes, the highest incidence for most elements is found among chimpanzees. The only exception (and the only difference that is statistically significant) is the markedly higher frequency of femoral involvement in gorillas. Of the seven femoral fractures observed in gorillas, five were of the diaphysis and a sixth was of the neck; the remaining lesion was an apophyseal fracture of the greater trochanter. Interestingly, of the six diaphyseal or neck fractures in gorillas, five were in females. It is not obvious why gorillas (overall) should have more femoral injuries than chimpanzees, but it is perhaps noteworthy that female gorillas, who are more arboreal than males, show the preponderance of these lesions. Moreover, it is unclear why bonobos, which are typically the most arboreal of the African apes, should have the least appendicular fracture involvement (although sample size here is limited). Finally, it is of note how well healed were these potentially highly serious injuries (Figure 6-1). In human archaeological samples, such femoral injuries are only infrequently seen (with but few exceptions, e.g., Kilgore et al., 1997). Thus, in support of Schultz's (1939) conclusions regarding the frequency and healing of fractures in free-ranging nonhuman primates, these data further show how even life-threatening injuries can be accommodated.

Patterns of cranial injuries are perhaps even more informative than those seen in the postcranial skeleton. In the African ape samples (Jurmain, 1997), again, the highest prevalence was found among the Gombe chimpanzees (*P. troglodytes schweinfurtheii*) (see Table 6-2). However, for the larger comparative samples, lowland gorillas were more affected that the chimpanzee sample from West Africa (*P. troglodytes troglodytes*). Bonobos were clearly the least affected of the African apes.

When facial injuries (including bite wounds) are considered separately, a significant sex difference ($p < 0.001$) emerges. Of the total of 20 facial injuries in all samples, 18 were found in males. It is probable that this consistent sex difference is informative and might well relate to gender-based behavioral differences, most especially the frequency and severity of inter-individual aggression (Jurmain and Kilgore, n.d.).

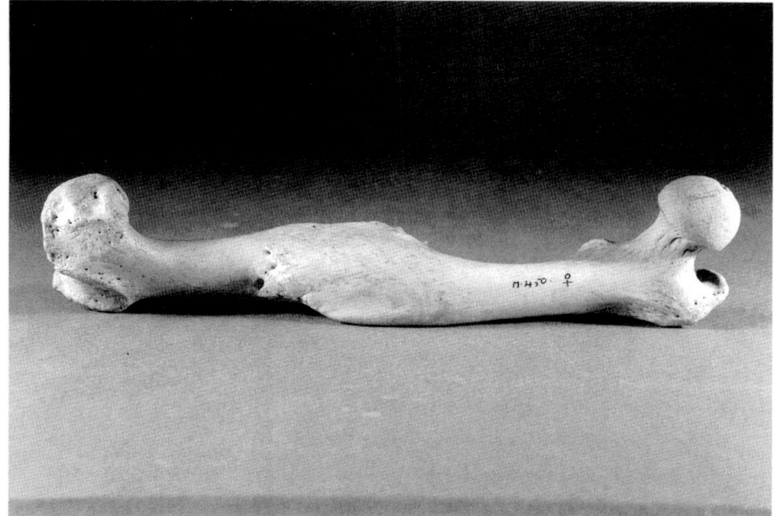

(a)

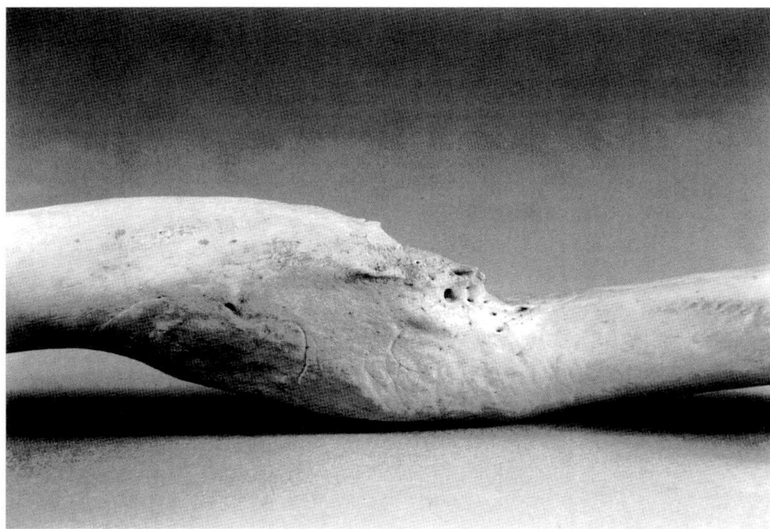

(b)

Figure 6-1 **(a)** Femoral midshaft fracture; adult female gorilla; **(b)** Closeup (ventral aspect). Photographs by the author.

Table 6-2. Cranial trauma in African great apes

Sample	N	Vault fractures	Facial fractures	Bite wounds
Gombe Chimpanzees	14	0	1	2
Powell-Cotton Chimpanzees	116	1	3	2
Gorillas	135	5	5	5
Bonobos	56	0	2	0

Moreover, some human data on cranial injuries (Walker, 1989; Jurmain and Bellifemine, 1997; Lambert, 1997) show similar gender-related patterns, which in humans as well might be partially explained by behavioral influences (see below).

Along these same lines, the nonhuman primate data are also informative concerning some postcranial lesions. In particular, "parry" fractures have been interpreted in humans as clear indicators of interpersonal aggression. However, as discussed below, the etiology of such lesions is not as unambiguous as has been commonly assumed. Nevertheless, in one example from Gombe National Park, the not completely healed, still-active, ulnar fracture in an adult female was almost certainly caused by a blow delivered by another chimpanzee during an attack. This individual, known as "Madam Bee," was a member of the Kahama group which had split from the main Kasakela community. During the early 1970s all the adult members of this Kahama group were known to have been killed by Kasakela individuals (or they disappeared) (Goodall, 1986). Madam Bee was one of the individuals who died from multiple injuries, suffered during at least three observed highly aggressive attacks (and a fourth unobserved incident was likely). Both the timing as well as the location of the forearm injury (as well as observed incapacitation) are strongly suggestive that the fracture did indeed occur while parrying a blow. It is perhaps ironic that it is from some nonhuman animals that the quality of documentation and recognition of individuals allows such a specific behavioral interpretation. For humans, in skeletal data derived from archaeological contexts, only very rarely, if ever, can such resolution be achieved.

Additionally, two further examples of skeletal injuries related to reasonably well-documented incidents have been described in the Gombe materials (Jurmain, 1989; 1997). These include a long-standing traumatic hip dislocation in the adolescent Michaelmas (which occurred when he was four years old), and a healed crushed ankle in the adult male, Hugo.

CRANIAL TRAUMA IN HUMANS

Similar to the nonhuman primate data, cranial injuries in humans are highly intriguing and potentially informative. Not many systematic studies of cranial trauma have been done from which data on prevalence of specific traumatic lesions can be gleaned. Numerous studies of cranial trauma have been done on Old World skeletal materials, although work prior to 1980 rarely presented consistent data allowing assessment of prevalence. For example, Brothwell (1961) presented data from Britain, including samples spanning from Neolithic to Anglo-Saxon periods. From the combined samples, thousands of individuals were likely represented, from which Brothwell identified 48 cranial lesions (mostly vault fractures).

In perhaps the most detailed study done on cranial trauma in the Old World, Bennike (1985) detailed the prevalence of various traumatic lesions as seen in Danish skeletal remains. Her findings from different time periods showed rates of fractures varying from 0% up to 8.3% in the Mesolithic (although here, $N = 12$). Otherwise, the highest prevalence was found for the Middle Ages (4.6%, $N = 65$).

Walker (1997) has recently assembled highly systematic data on craniofacial trauma from a total of 2280 individuals derived from samples from Siberia, Spain, the United Kingdom, and the U.S. Walker found the highest prevalence for nasal fractures (7.0%), followed by the frontal (4.6%), and the parietal (3.9%). Walker's discussion, as well as that in the contribution by Lambert (1997), provide an excellent overview of cranial trauma, including problems of diagnosis and excellent support from clinical analogs.

Roberts (1991) was more focused on healing patterns of fractures, but she did note that in a very large sample from Britain, a total of 56 cranial fractures was found (in a sample of >6,000 individuals; the cranium was, in fact, the most frequently involved region of the body). The recent study by Kilgore et al. (1997) concentrated more on postcranial injury, principally because only one cranial lesion was identified (in an adult sample, $N = 146$). Robb (1997) also recently reported five involved individuals, in a sample of 56, from an Iron Age Italian site.

From the Pacific and Australia two other notable studies reporting cranial trauma have recently been published. Owsley et al. (1994), in a large sample comprising about 500 skeletons from Easter Island, found the prevalence of cranio-facial trauma to be 2.5%. The authors further commented that such lesions were more common in males, a pattern also seen commonly in other samples (see below) as well, as already noted, in African apes. Moreover, from "sporadic historical documentation" it was suggested that warfare was quite common and lethal in late prehistory; however, from the skeletal evidence, Owsley and colleagues concluded such historical accounts most probably were "exaggerated" (a useful reminder regarding the precision and reliability of many historical sources).

Webb's comprehensive review of paleopathology in Australian prehistoric remains included a detailed analysis of cranial trauma (most typically, fractures). Webb's results showed a high prevalence of cranial trauma in populations from most of the continent; indeed, at almost all sites *both* males and females had frequencies of cranial traumatic involvement exceeding 15% (the only exceptions from the tropical region, where prevalence was 6.5%). In one area (Rufus River) the rates of cranial involvement for males and females respectively were a remarkable 26.2% and 27.7%, exceeding even the very high frequencies reported from California (see below). Further, the equally high (or even higher) prevalence among women, as compared to men, is also unusual.

Webb, as discussed in Chapter 4, has been quite circumspect in drawing behavioral interpretations from osteoarthritic involvement. However, for traumatic lesions he more willingly interpreted the patterns as largely behavioral in origin, arguing, "the study of trauma can tell us about the behavior and lifestyle of Aboriginal groups from different parts of the continent" (Webb, 1995: 188).

Webb postulated, for example, that the greater number of left parietal lesions in males suggests they were likely struck from the front; conversely, the greater number of right parietal injuries in females argues they were struck from the rear, quite possibly during "domestic squabbles." It should be emphasized, however, that such speculation leaves little room for adequate testing. With what appears to be but minor adjustments in inferred behavioral scenarios, the behavioral explanation can be made to fit almost *any* pattern of skeletal involvement (points well-emphasized relative to OA; see Chapter 4). Webb does not, however, become too enamored of trying to infer particular weapon use from skeletal lesions, and thus cautiously concluded:

> Rather than describe the weapons used to produce the injury, it seems to be more appropriate to determine where the injury occurs on the head and from there the possible reasons for variation which occurs across the continent (Webb, 1995: 205).

Compared to research done in the Old World, there have been a greater number of systematic analyses of cranial trauma done on New

World archaeological samples. For example, several reports which provided data on cranial fracture incidence include: Hooton (1930) on Pecos Pueblo; Snow (1948) on Indian Knoll; Newman (1957) on California Central valley sites; Stewart and Quade (1969) on Pueblo Bonito and Hawikuh; Morse (1969) on Illinois' sites; Miles (1975) on Wetherill Mesa; and Ferguson (1980) on two New Mexico Pueblo sites. In all these investigations, prevalence of cranial fractures was fairly low, ranging from about 2% to 5% of individuals. Further reports (following 1980) of cranial injury which also indicated low prevalence rates of cranial trauma were those by Lovejoy and Heiple (1981—who focused more on postcranial patterns) and Merbs' (1989) mention of isolated injuries from Arizona (Nuvakwewtaqa) and Peru (Cinco Cerros).

Other reports have indicated considerably higher rates of cranial injury, including several contexts suggestive of systematic violence (see below). In addition, and most especially from sites in California (and northwestern Mexico), further evidence has indicated high levels of cranial injury. The most meticulous of these investigations was that done by Walker (1989), who analyzed a total of 744 crania from several sites in the Santa Barbara Channel region of southern California. From this work, Walker ascertained very high levels of cranial vault fractures—with a combined frequency (all sites) of 19.4% of individuals, Indeed, on one channel island (San Miguel), the incidence of cranial fractures was an exceptional 24.4%. Walker's analysis was quite detailed and included descriptive information regarding size, depth, and location of lesions—the majority of which were restricted to the frontal and parietals. From essentially the same sample, Lambert (1994; 1997) reported a 17% prevalence in a sample of 753 crania.

Walker carefully considered various etiological factors, but, given the standardized shape and location of lesions, concluded the likely cause was deliberate (i.e., interpersonal violence). Moreover, as the cranial fracture lesions were so frequent, Walker suggested the *intent* of the perpetrators may have been non-lethal, ritualized aggression (drawing an analogy with the ax fights among the Yanomama described by Chagnon, 1983). While non-lethal regular aggression certainly may be a possible explanation for the pattern of cranial lesions seen in the Santa Barbara Native Californians, other evidence from here (and elsewhere in California) argues the intent may indeed have been to inflict serious, even terminal injury (see below).

An even higher prevalence of cranial vault injuries was reported by Tyson (1977) for a site in Baja California, where an astonishing 30.6% of individuals ($N=49$) were involved. Tyson also inferred a behavioral interpretation, but unlike Walker, suggested, instead, the individuals were inflicting wounds via some form of self-mutilation (an intriguing suggestion, but one as yet without substantive contemporary parallels, except, perhaps, with the exception of Australia; see Webb, 1995).

More recently, in some of my own data from Central California (Jurmain and Bellifemine, 1997) similar (but considerably less frequent) types of injuries to those reported by Tyson and Walker were found. For the skeletal materials from the Ala-329 site on the east side of San Francisco Bay, seven cranial lesions were observed (in a sample of 365 crania). Five of these injuries were of the vault, and the remaining two of the face (Figure 6-2). The incidence was (for vault lesions only) 2.3%; in other words, the overall prevalence was only about one-eighth that observed by Walker (1989) and less than one-tenth that found by Tyson (1977) (the differences between Ala-329 and these other areas in California were significant, $p < 0.001$). Thus, while cranial injury was not uncommon at Ala-329, it was apparently markedly lower than in neighboring regions located to the south. The incidence, however, is similar to that found by Newman (1957) for a Central Valley site (2.3%, $N = 140$), in the region immediately adjacent to the east of San Francisco Bay. In addition, similar frequencies of cranial involvement to that seen at Ala-329, have been reported from

Figure 6-2 Cranial depression fracture. Frontal bone, Ala-329, Central California; male, 13–15 years. Photograph by the author.

the American Southwest (Hooton, 1930; Ferguson, 1980), the Midwest (Wilkinson, n.d.), and the southeastern U.S. (Smith, 1996).

Central California thus looks fairly typical for areas in prehistoric North America where periodic conflict was likely persistent. Southern California, however, appears as more anomalous, with dramatically frequent prevalence of cranial trauma (equaled, in fact, only in Australia).

Another interesting aspect of the population patterns, in both central and southern California, may also pertain to understanding the cultural dynamics involved. In Walker's samples, as well as that from Ala-329, males were more frequently involved than females (Santa Barbara Channel, males = 24%, females = 10%; $p < 0.001$; Ala-329 males = 5.2%, females = 0%; $p < 0.05$). The higher prevalence of cranial injury among males is a quite consistent pattern among human samples, although at least two well-documented exceptions have been reported, from the Riviere aux Vase site in Michigan (Wilkinson and Van Wagenen, 1993; Wilkinson, 1997), and from the La Plata River valley in Colorado (Martin, 1997). In the former case, the sex difference for a quite large sample ($N = 212$ crania) was statistically significant, but in the latter ($N = 23$ crania), the difference was not statistically significant. In addition, as noted above, in Australia males and females are about equally involved for cranial trauma (Webb, 1995). The generally greater degree of male cranial involvement is probably not irrelevant—and may reflect culturally patterned behavior relating to interpersonal aggression. In fact, as discussed above, a similar pattern of apparent greater level of male cranial injury has also been reported in African apes (Jurmain, 1997).

Archaeological Contexts Suggesting Patterned Violence

Cranio-facial injuries certainly provide highly suggestive evidence relating to interpersonal aggression. Moreover, there have been a few well-documented examples of more systematic aggression that have left telltale signatures on skeletal remains (especially affecting the cranium).

The earliest well-documented occurrence was at the Mesolithic Ofnet site in Bavaria (Frayer, 1997). In addition, there are numerous other more recent illustrations, most especially in North America. For example, Pfeiffer (1985) reported for an Archaic sample from Frontenac Island in New York State that, of 100 partial skeletons, seven individuals showed cranial trauma (two of which also had embedded projectile points). Other lesions included both depression fractures and cut marks. While prevalence rates cannot be confidently assessed in this sample (since the number of crania was not specified), it is clear that quite serious cranial trauma was common. Even more clear examples of patterned violence, including mutilation/scalping, have been reported from South Dakota at the Larson site (Owsley et al., 1977) and from Crow Creek (Willey, 1990; Willey and Emerson, 1993); from Illinois at the Norris farms site (Milner et al., 1991); and from Alabama at the Koger's Island site (Bridges, 1996a).

Crow Creek is the largest and most dramatic of these assemblages, testifying to at least one example of extreme interpersonal aggression in prehistoric North America. In his meticulous monograph treatment of the Crow Creek massacre, Willey (1990) described a variety of traumatic lesions in the remains of a minimum of 486 individuals (representing, most probably, the inhabitants of an entire village).

Among the Crow Creek remains, in the cranium especially, the lesions revealed the fate of the villagers. Of the preserved frontal bones, a minimum of 90% showed cut marks (and perhaps as many as 99% were involved). 42% of crania showed evidence of depressed fractures (mostly localized to the frontal and parietal regions—i.e., similar to the pattern documented by Walker, 1989). Moreover, approximately 25% of first cervical vertebrae showed cut marks, which Willey interpreted as indicative of decapitation. A curious negative finding was the lack of projectile wounds; Willey does not specify exactly how many such wounds were found, but did suggest some of the points may have been removed during post-mortem mutilation.

Even given the dramatic circumstances involved, and what must have been considerable inclination to speculate regarding behavior, Willey recognized a variety of limiting factors constraining behavioral interpretation, and thus cautiously concluded:

> Another problem is that identifying perimortem trauma from bones is difficult in the first place, and attributing trauma to interpersonal aggression even more tenuous. And even when interpersonal aggression is indicated, determining whether the trauma occurred before and perhaps caused or contributed to death or was a mere mutilation following death is usually impossible (Willey, 1990: 93).

Although not resulting from as massive a violent episode, the remains from the Norris Farms #36 site in Illinois also strongly indicate intensive aggression. From a cemetery sample consisting of 264 burials, Milner et al. (1991) suggested that 43 of these "died violently." Assessment of such a cause of death was made from perimortem (unhealed) injuries, mutilation, and carnivore chew marks. In addition, some individuals showed embedded chert fragments (the number and location were not indicated).

Moreover, 14 crania displayed cut marks interpreted as signs of scalping, and 11 individuals were apparently decapitated. Of the 43 individuals who were judged to have died from traumatic injuries, a similar proportion were male (35%) and female (29%). Again, as at Crow Creek, the demographics suggested residents of a village were attacked (not a specialized raiding party). Still, the absence of children (41 of the 43 victims were >15 years old), suggested not all members of the village were equally represented among those who were slain.

Another interesting feature of the mutilated individuals suggested a possible risk factor (influencing who was attacked). An unusually high proportion (42%) of the victims showed some other pre-existing debilitating condition (healed rib fractures, shoulder and hip dislocations, and infections). Milner and colleagues thus suggested that sporadic raids by enemies may have been directed at vulnerable individuals—especially those at exposed localities, away from the village.

More recently, Bridges (1996a) has interpreted similar patterns of cranial trauma in northwestern Alabama (at the Koger's Island site) as also indicative of periodic warfare. From several mass graves consisting of 108 individuals, excavated in the 1930s, but recently reevaluated, Bridges found four individuals with cranial fractures (three were healed, one perimortem). In addition, Bridges interpreted cut marks in seven individuals as evidence of scalping (all lesions were carefully recorded for location). From all the evidence of cranial trauma, Bridges concluded 23 individuals at Koger's Island experienced "violent death" (translating to 21% of all individuals; 29% of adults; 37% of adult males).

Beyond the recognition of deliberately patterned aggression, Bridges draws only a few cautious behavioral interpretations. For example, the pattern of cut (scalping) marks was highly variable individual to individual, prompting Bridges to suggest the aggressors either rushed or were inexperienced. Overall, the pattern of violent activity at Koger's Island is similar to that seen at Crow Creek and Norris Farms, but also with some interesting differences. Most especially, the midwestern sites show definite evidence of post-mortem mutilation of the bodies, while in Alabama, such behavior was not indicated.

Cannibalism

Another topic which has generated considerable interest and controversy relates to interpretation of possible cannibalism. Claims of specialized treatment of human remains, potentially including exploitation of flesh, have been made even for quite ancient contexts. Among the earliest such findings, with some substantiation, is that of the Middle Pleistocene Bodo archaic *H. sapiens* cranium from the Middle Awash region of Ethiopia. White (1986) very carefully described a variety of telltale lesions on this cranium, using S.E.M. also to characterize the bone changes. In what is probably the most convincing paleoanthropological demonstration of such treatment of the dead, White concluded the cranium displayed "patterned intentional defleshing of this specimen by a hominids(s) with a stone tool(s)" (White, 1986: 508). White, however, remained highly conservative in implying specific behavioral intent/motivation, appropriately cautioning: "It is impossible to falsify hypotheses of cannibalism, cannibalism combined with curation, simple curation, mutilation or decoration with the evidence at hand" (White, 1986: 508–509).

Most of the claims relating to cannibalism during the Pleistocene concern interpretation of bone changes seem among the Neandertals. Russell (1987a), for example, carefully assessed the pattern of breakage in the Krapina remains from a variety of perspectives (and testing various alternative explanations). She concluded that "the morphology of the prehistoric breakage is inconsistent with the cannibalism hypothesis" (Russell, 1987a: 373). Russell also examined such changes as microscopic "incised striations" and compared these lesions with those seen in more recent (human osteological) ossuary remains from Michigan (Russell, 1987b). Although the more recent contexts here are reasonably well-controlled (quite clearly demonstrating secondary burial), obviously more specific behavioral resolution is not feasible. Russell, nevertheless, still felt confident in asserting "the striations on the Krapina Neandertal remains are consistent with postmortem processing of corpses with stone tools, probably in preparation for burial of cleaned bones" (Russell, 1987b: 381).

Employing an even more explicit, rigorous approach, White and Toth (1991) compared purported osteological indications of cannibalism found in the Grotte Guatteri Neandertal cranium (Circeo 1) with that observed in a series of 100 trophy skulls from Melanesia. White and Toth isolated several specific markers of cannibalism in these better documented contemporary remains, including: polish, cuts/scraping marks, peeling, and adhering flakes. The Circeo Neandertal, in fact, displayed *none* of these changes, but did show some other features apparently not typically associated with defleshing. "Not one of the 100 Melanesian crania shares this combination of features with Circeo I" (White and Toth, 1991: 123). Using highly explicit (and documented) criteria to test such behavioral claims is clearly the most scientific approach, and Toth in his reply to comments aptly reminds us of the wider implications: "The crucial message here is that all such claims of early hominid cannibalistic, ritualistic, or symbolic behavior must be scrutinized with similar analytical detachment and rigor to ascertain whether such interpretations have any validity" (White and Toth, 1991: 135).

For modern osteological contexts, the region that has yielded the most consistently patterned bone modifications widely interpreted as signs of cannibalism has been the Southwest United States. Especially from Anasazi archaeological contexts, the work of Turner and colleagues has again approached the topic using explicit criteria. To the bone changes suggested by White and Toth above, Turner (1993) adds: extensive perimortem cranial and postcranial bone breakage, anvil-hammerstone abrasions, burning, and many missing vertebrae. Turner's explicit methodology also quite usefully included enumeration of what postmortem alterations should, alternatively, have produced and argued that: "Perimortem breakage is differentiated from dry bone breakage by spiral fracturing, smooth unstepped fracture surfaces, and embedded bone chips"(Turner, 1993: 426). He was thus able to conclude for the sample of six individuals excavated from a charnel pit at Chaco Canyon, New Mexico that,

> Because there are no sure signs of sun bleaching or ground surface weathering cracks, and a complete absence of scavenger damage, that is, carnivore puncture marks, gnawed ends of long bones, and the distinctive gnawing incisions of porcupines, wood rats, and other rodent teeth, it is likely the remains were not left exposed on the ground surface for an extended period of time before deposition (Turner, 1993: 428).

These observations led Turner further to suggest that trauma, butchering, and burial,

> were part of an episodic event that took place over a relatively short period of time (Turner, 1993: 428).

In what must be the most detailed evaluation of the entire topic of prehistoric cannibalism, White (1992) also focused on Anasazi remains from the Southwest (specifically, the Mancos 5MTUMR-2346 site in southwest Colorado). Similar to the methodology employed by Turner (as well as White and Toth in their comprehensive review of earlier hominid cannibalism), White here further elaborated highly specific osteological signatures of such bone treatment. To those listed above, White further suggested: pot polish, percussion pits, interconchoidal scars, chipmarks, and scrapemarks.

White, with participation from a team of able laboratory assistants, also attempted the demanding task of conjoining as many of the 2,106 initial fragments (representing a minimum of 29 individuals) as was possible. This intensive approach is derived directly from archaeological studies of butchering techniques (e.g., Binford, 1981; 1984; Villa, 1990). The scientific rigor of such a methodology is rarely matched in contemporary human osteological research and is most typically absent in most of paleopathology. The lessons to be learned from the demonstrated rigor of our archaeological colleagues, and which is further so well exemplified by White in his investigation, should be quite obvious.

Even given the enormous detail of accumulated data (achieved over what must have been thousands of hours of laboratory preparation and analysis), White still remained cautious in drawing overly specific behavioral interpretations. For example, he chose not to conclude what the evidence from Mancos inferred relating to "violence" and "warfare." Rather than being drawn into such an interpretive abyss, White instead reminds us,

> The terms "violence" and "mutilation" have been based, in the assessment of Southwestern human bone assemblages, on evidence of percussion-driven fracture and scarring of the bones. Unfortunately, these terms carry with them implications that the processor(s) of these remains carried out their activities with an intent beyond mere economic processing. These terms, for example, are never applied to artiodactyl remains showing fragmentation patterns (White, 1992: 360).

What is so noteworthy regarding these investigations of cannibalism, most especially as illustrated in the research of Turner and White, is that they explicitly state the hypotheses, clearly specify alternative explanations, and thus, ultimately provide testable interpretations (i.e., they are falsifiable). Unfortunately, to this point, very little else in the realm of osteological reconstructions of behavior meets these rigorous standards.

While not all skeletal assemblages provide as dramatic or as clear manifestations of human behavior, further evidence of trauma in the postcranial skeleton sometimes affords useful insights. Still, as will be shown, interpretation of postcranial trauma is rarely as unambiguous as lesions found in the cranium—and certainly not as obvious as at the sites yielding evidence of massacre, defleshing, and general mayhem.

EVIDENCE OF TRAUMA IN THE POSTCRANIAL SKELETON

Osteologists have investigated postcranial traumatic lesions much in the same manner as they have those in the cranium. Indeed, the number of systematic paleoepidemiological analyses of traumatic lesions in the postcranium exceeds that for the cranium, and such studies have become especially more frequent since 1980.

Some earlier reports, while generally not as controlled, are still worthy of note. As early as 1910, Elliot Smith and Wood Jones reported detailed data on fracture prevalence from various sites in Lower Nubia (i.e., southern Egypt). They quite methodically tabulated prevalence data by element and also provided excellent comparable clinical data on fracture prevalence. Indeed, these same data, both archaeological *and* clinical were creatively employed by Berger and Trinkaus (1995) in their study of Neandertal fracture patterns (see below for further discussion). This research on trauma, most notably that by Wood Jones (1910), thus was exemplary in several respects, and today still provides greater data resolution than many contemporary reports.

Bourke (1969) reworked the Nubian fracture data first accumulated by Elliot Smith and Wood Jones and drew several intriguing conclusions. Firstly, as in most other archaeological samples (see Table 6-3), the most commonly fractured region was the forearm (accounting in the Nubian materials for 31% of all postcranial lesions). A second finding was that 11.4% of fractures occurred below the knee, but this region was considerably more involved in a contemporary hospital sample (with a 25.7% prevalence of such lesions). Elliot Smith and Wood Jones (1910) argued that this difference reflected a better locomotory accommodation in barefooted Nubians as compared to shoe-clad modern Londoners, an explanation restated by Wells (1964), in his comparison of modern patterns with Anglo-Saxon remains.

Other investigations which provide some useful data on postcranial fracture pattern include that by Merbs (1983) on the Sadlermiut Inuit, in which he found a surprisingly few 10 such lesions in 94 adults (and of these, only five were of long bones). Despite such a low prevalence of fractures, and those found could easily be manifested in *any* group, Merbs suggested rather specific behavioral etiologies in the Sadlermiut. For example, the two fractures of vertebral spinous processes and three clavicular lesions seen in males were attributed to "violent wrestling matches," while the lone finger lesion was suggested to have occurred during a "wrestling match or on a hunting trip." Merbs also collected data on vertebral compression fractures, which were considerably more frequent in his study population than other postcranial traumatic injuries (found in 36 of 80 adults with preserved vertebral columns). Again, Merbs inferred a behavioral causation, in

Table 6-3. Prevalence of long bone fractures (population comparisons)

Element	Kulubnarti[1] *n* (%)	Ala-329 (Calif. Indians)[2] *n* (%)	Great Lakes[3] *n* (%)	Libben[4] *n* (%)	Danish[5] *n* (%)	British Medieval[6] *n* (%)
Clavicle	262 (0.4)	291 (0.7)	66 (3.0)	260 (5.7)	386 (1.3)	—
Humerus	276 (3.6)	300 (0.3)	140 (2.1)	450 (0.7)	703 (0.1)	891 (0.8)
Radius	259 (6.2)	301 (4.3)	103 (3.9)	369 (5.4)	608 (1.5)	770 (1.3)
Ulna	260 (13.1)	290 (5.2)	124 (2.4)	351 (3.1)	607 (2.1)	752 (1.5)
Femur	281 (1.1)	313 (0.0)	112 (0.9)	347 (2.6)	998 (0.0)	937 (0.1)
Tibia	232 (0.0)	315 (1.6)	82 (2.4)	349 (1.4)	852 (0.7)	864 (0.7)
Fibula	218 (1.4)	237 (0.0)	—	257 (3.5)	364 (0.6)	725 (0.8)
Totals	1788 (3.7)	2047 (1.8)	727 (2.1)	2388 (3.0)	4518 (0.8)	4938 (0.8)

Sources:
[1]Kilgore et al., 1997; [2]Jurmain, 1991; [3]Pfeiffer, personal communication; [4]Lovejoy and Heiple, 1981; [5]Bennike, 1985; [6]Grauer and Roberts, 1996

this case as a result of riding on sleds. For support, Merbs referred to the highly anecdotal evidence of "snowmobiler's back." Further, a slight difference in pattern of involvement of these vertebral compression fractures between males and females (mid-thoracic peak in females, lower thoracic/lumbar peak in males) led Merbs to an even more specific behavioral inference: "The occurrence of the condition higher in the vertebral column of the Sadlermiut women is attributed to their carrying heavy burdens on the back, particularly the carrying of infants in this position while sledding" (Merbs, 1983: 168).

In a review of large British collections, Roberts (1991) did not report specific element prevalence, but noted that (after the cranium) the most involved elements were the fibula, followed by the radius. Cohen et al. (1994) reported some interesting data from the Maya site at Tipu. In a skeletal collection numbering 457 individuals, Cohen and colleagues found the overall fracture involvement by element at less than 2%. Moreover, they noted that 8.7% of adults displayed some fracture involvement and provided more specific data for three elements:

Tibia	13/710 = 1.8%
Fibula	7/610 = 1.1%
Humerus	1/649 = 0.2%

In all respects, the data provided by Cohen et al. are quite typical for archaeological samples (see Table 6-3), and their interpretations of causation were appropriately careful as well, suggesting, for example, that a facial fracture "may reflect violence."

Pietrusewski and Douglas (1994), in a very large Hawaiian collection (>6,500 elements), found a prevalence in the prehistoric materials of 1.6%. The prevalence rates of fractured elements were as follows:

Ulna	3.6%	Humerus	1.2 %
Radius	2.0%	Clavicle	1.4%
Fibula	1.8%	Femur	0.8%
		Tibia	0.8%

As with the Mayan sample, these Hawaiian data again fit very closely with those obtained elsewhere (see Tables 6-3 and 6-4).

Stirland (1996) did not provide prevalence rates by element, but she did indicate individual incidence of non-vertebral postcranial trauma for three different English Medieval samples: 7.9% for Norwich; 12.4% for Backfriars (Ipswitch); and 9.5% for the *Mary Rose*. Stirland also approached interpretation carefully, and noted that, compared to the *Mary Rose*, a sample of sailors and soldiers, all of whom died in a shipwreck in 1545, the higher prevalence of trauma in the Blackfriars cemetery (a sample of friars and lay benefactors) was "hard to explain."

The most rigorous contemporary paleoepidemiological approach to the investigation was exemplified in the research by Lovejoy and Heiple (1981) on the large Libben skeletal collection. Following this lead, several other well-controlled studies (Bennike, 1985; Jurmain, 1991a; Grauer and Roberts, 1996; Kilgore et al., 1997) have also been reported in the last decade (see Table 6-3). Most importantly, in these more systematic studies, data were controlled as follows:

1. Prevalence rates were reported *by element*.
2. Elements were included in calculations, only when complete (following Lovejoy and Heiple, 1981, when "intact").
3. Where appropriate, diagnoses were confirmed by radiological assessment.
4. Age was rigorously controlled and evaluated in some tabulations.
5. Where appropriate, (e.g., male/female, between-population comparisons), statistical testing was employed.
6. Given the above, relatively large initial samples were required to carry out such systematic paleoepidemiological investigations.

Another innovative approach used by Lovejoy and Heiple was the assessment of fracture prevalence as a function of age—that is, computing fracture rates in terms of "years at risk." This same methodology was also followed by Burrell et al. (1986) in their investigation of a Sudanese Nubian sample (a collection recently reassessed by Kilgore et al., 1997; see below).

Bennike (1985), in her general analysis of Danish remains, also included detailed data on fracture prevalence. As shown in Table 6-3, as with the other archaeological samples, the forearm is the most common site of fractures (here followed by the clavicle). It should be noted that the Libben sample is the only one in which the clavicle was the *most* commonly fractured element; in all other populations, it was the forearm which consistently had the highest incidence.

In my analysis of the Ala-329 collection from Central California (Jurmain, 1991a), very similar patterns of long-bone fracture incidence were found to that reported by Bennike. In this sample, including 2047 long-bone elements (including the clavicle), the ulna and radius accounted for 78% (28/36) of all fractured elements. Likewise, forearm fracture was the dominant pattern in the remains from the Kulubnarti site in Sudanese Nubia (Kilgore et al., 1997). Here, of the total of 1788 elements (from 146 adult skeletons), a total of 67 fractures was observed. Of these, 50 (75%) were lesions of the forearm. As shown in Table 6-3, at Kulubnarti, the prevalence of forearm fractures was higher than in any other sample—in most cases, dramatically so. Further, an interesting aspect of the Kulubnarti fracture pattern was the unusually common occurrence of multiple lesions. Of the 48 individuals who displayed some healed fracture, 13 (27%) had more than one lesion. Indeed, the proportion of individuals, within the entire sample of 146 burials, with *multiple* lesions at Kulubnarti (8.9%) equals that of many other archaeological samples for simple (single element) involvement.

In the Kulubnarti material the proportion of individuals with at least one fractured long-bone element was 32.9%. Kilgore and colleagues cautiously made a behavioral interpretation of this pattern, arguing it related to falls in a harsh, steep, boulder-strewn terrain. The lack of cranial injuries and a careful analysis of the forearm lesions (see discussion below) supported this conclusion: "Without corroborating evidence of interpersonal aggression [cranial injury/projectile wounds], the most parsimonious explanation for the high prevalence of fractures and multiple lesions at Kulubnarti, is accidental falls" (Kilgore et al., 1997: 112).

Recently, for Old World collections, in addition to the pioneering work of Bennike (1985), further systematic analyses of postcranial fracture patterns have been completed—particularly for materials from the United Kingdom. Grauer and Roberts (1996) carefully recorded fracture prevalence at St. Helens-on-the-Walls (in the City of York) in 1,014 individuals represented by almost 5,000 long-bone elements (not including the clavicle). Of these, a total of 41 elements was involved, giving a very low overall incidence of fracture of just 0.8%. Indeed, the only other contexts where such low long-bone fracture prevalence was also observed were also from the Old World—from Denmark (Bennike, 1985) and from other (Medieval) sites in the United Kingdom, listed by Roberts and Manchester (1995) (see Table 6-4).

These five other U.K. sites come from Gloucester, Whithorn (Scotland), Fishergate (York), Chichester (Sussex), and St. Nicholas Shambles (London). Excepting the restricted London materials ($N = 296$ elements), all the other skeletal samples were substantial—with the number of elements ranging from 1,861 to more than 9,000 (at Whithorn). Overall prevalence of long-bone fractures varied from 0.8% up to 6.1% (in the small sample from London; see Table 6-4). Indeed, including the sample from York reported by Grauer and Roberts (1996), of the six Medieval British samples, four show combined

Table 6-4. Prevalence (%) of long bone fractures, British Medieval samples*

Sample	N (elements)	Hum	Rad	Uln	Fem	Tib	Fib
London St. Nicholas Shambles[1]	9563	5.3	8.8	8.2	3.8	6.0	4.1
Blackfriars[2]	1861	0.3	1.4	0.5	0.5	0.5	0.3
Fishergate[3]	3235	0.4	0.8	0.8	0.2	0.5	1.7
Whithorn[4]	9563	0.0	0.5	0.1	0.2	0.4	0.8
Chichester[5]	1554	4.2	3.2	2.8	0.4	2.3	7.2

*Data compiled in Roberts and Manchester (1995: 76)
Original Sources
[1]White, 1988; [2]Wiggens et al., 1993; [3]Stroud and Kemp, 1993; [4]Cardy, 1994; [5]Judd, 1994

fracture incidence of less than 1%, one (Chichester) had a 2.6% overall incidence, and thus only one sample (from St. Nicholas in London) exceeded 3%.

In the last few years, for the first time, comprehensive fracture data from varied archaeological contexts allow for systematic analysis and meaningful comparisons. In most instances, fracture prevalence is low. Moreover, usually the forearm bones are the most frequently involved elements (There are some exceptions in the British Medieval samples, where fracture incidence was consistently very low for *all* elements). From a behavioral perspective, interpretation is, however, not unambiguous. Many researchers would interpret a considerable proportion of forearm fractures (especially those of the ulna) as indicators of interpersonal violence. However, this behavioral inference is far from convincingly established (see below). Of course, interpretation of fractures alone and in one portion of the body is not usually meaningful. What is required is an evaluation of the pattern of traumatic involvement—including other types of potentially informative lesions as well as consideration of craniofacial involvement.

OTHER TRAUMATIC LESIONS

In addition to fracture, a variety of other types of traumatic lesions have been evaluated by skeletal biologists. Some of the more exotic forms of trauma (e.g., decapitation, hanging) were mentioned briefly earlier, and consideration of injury related to scalping was also discussed above in more detail as related to special contexts of organized aggression.

Dislocation

Another type of lesion occasionally seen in skeletal remains is dislocation. However, such injuries are probably more rare than fractures (in actual occurrence); moreover, they leave less clear skeletal evidence. As a result, as compared to fracture, dislocations are much less frequently diagnosed and discussed in the osteological (paleopathological) literature. There are, however, a few exceptions. Wood Jones (1910) reported just one dislocation of the hip, and Ortner and Putschar (1981) described two examples of shoulder dislocation and one case of traumatic hip involvement. Particularly for hip dislocations, it is important to differentiate congenital causation from a traumatic etiology—a point well emphasized by Ortner and Putschar (1981) as well as Roberts and Manchester (1995). In my research of the Ala-329 site in central California, three hip dislocations were identified, two of which were diagnosed as traumatic and one congenital (Jurmain, 1991a).

Kilgore et al. (1997) reported one further case of traumatic hip dislocation (from Kulubnarti, in an individual with multiple other injuries), and Milner et al. (1991) alluded to cases of both shoulder and hip involvement (but provide no details). Lastly, I also have discussed an interesting case of traumatic hip dislocation in a chimpanzee from Gombe National Park (Jurmain, 1989; 1997).

Thus, at present, the paleopathological data relating to dislocation are simply a collection of case reports. Since the prevalence of such injuries (at least skeletally diagnostic ones) is apparently quite low, perhaps not much more can be attained. It is possible, however, that osteologists frequently have failed to recognize the subtle diagnostic osseous markers of such injuries, and, thus, greater attention to more precise diagnosis is encouraged.

Wounds/Projectile Injuries

A number of researchers have commented on the presence of traumatic lesions resulting from weapon use. Especially from Europe, many examples of sharp weapon use, oftentimes with dramatic injuries produced, have been reported (e.g., Brothwell, 1961; Wells, 1964; Bennike, 1985; Hawkes, 1989; Wenham, 1989; Waldron, 1994b; Anderson, 1996) (Figure 6-3). Fewer cases of sharp weapon use have been reported in the New World, but several examples of apparent metal weapon use (from a site in post-contact Florida) have recently been reported by Hutchinson (1996).

More commonly, from prehistoric contexts in the New World, direct evidence of weapon use comes primarily in the form of embedded

Figure 6-3 Probable sword wound, frontal, Iron Age, Beckford Worcestershire, England (Note the injury shows some healing). Photograph courtesy, Charlotte Roberts, Calvin Wells Laboratory, University of Bradford (Department of Archaeological Sciences).

projectile points. Indications of projectile injury were mentioned by Hooton (1920) at an Ohio site as well as by Black (1979) for a Mississippian site in Missouri. Pfeiffer (1985) also noted two cases from New York with pieces of projectiles embedded in crania. More detailed evidence was presented by Merbs (1989) of foot involvement from a Hohokam site in Arizona, and seven other cases of cranial involvement in Tennessee, mostly from the Cherry Site, were also noted by Smith (1996; 1997).

The most frequent findings of projectile injury in North America (and, perhaps, from anywhere) have come from California. Lambert and Walker have reported very high incidence of projectile lesions in California Indians from the Santa Barbara Channel region (Lambert and Walker, 1991; Lambert, 1994; Lambert, 1995). Two of the lesions were found in crania (Walker, 1989).

Lambert (1994; 1997), in a total sample of 1744 individuals, found a total of 38 individuals with projectile points embedded in bone (2.2% incidence). She further noted that, also counting suggestive bone nicks and suspicious archaeological contexts, an additional 20 individuals also were potentially involved. Lambert argued that to ignore those contexts not actually showing embedded projectiles would be "likely to result in a serious underrepresentation of the actual number of victims" (Lambert, n.d.). However, this view misses another obvious point, i.e., most estimates of pathological prevalence, from bone involvement solely, underestimate actual morbid conditions. The object is not entirely to reconstruct morbidity from skeletal materials, but to record data which can be systematically compared among different samples. As in Wendorf's interpretation (1968) mentioned above, many of the cases of 'suspicious' context alluded to by Lambert derive from interpretations made at the time of archaeological recovery. However, since such collection techniques (and quality of preservation) vary so widely, it is difficult to see how such data could be utilized in rigorous comparative studies. I would argue, to the contrary, that the best data here, as in many aspects of paleopathology, are those that can be most systematically recorded and thus most easily compared among various samples.

Additionally, from a large sample (>2,000 individuals), mostly deriving from Central California, Tenney (1986) reported on 18 embedded projectiles found in 13 individuals. In my analysis of the Ala-329 collection (also from Central California), 12 embedded projectiles (or fragments of projectiles, all obsidian) were found in 10 individuals (Jurmain 1991a; Jurmain and Bellifemine, 1997) (Figure 6-4). Similar to the two cases mentioned by Walker (1989), one of these

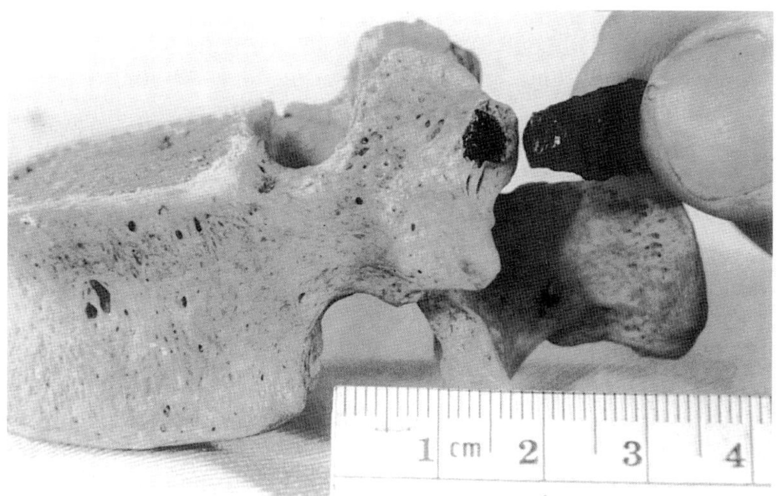

Figure 6-4(a) Obsidian projectile point embedded in transverse process of first lumbar vertebra. Ala-329,Central California, male, 25–35 years. The broken portion of the obisian point was found with the burial.

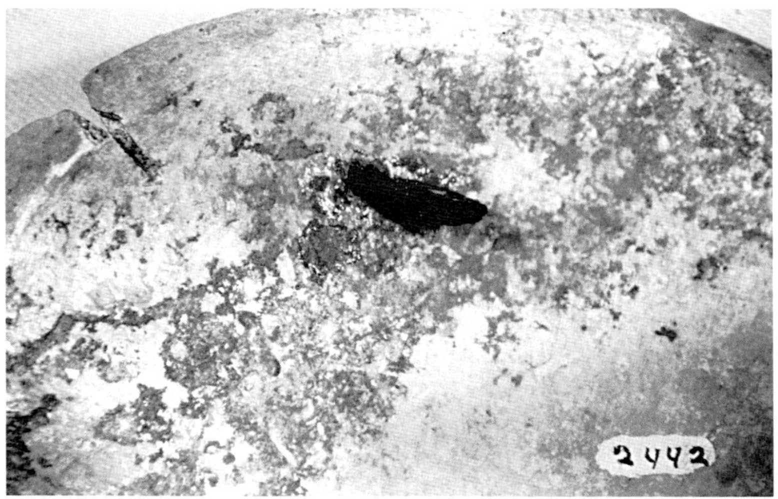

Figure 6-4(b) Large obsidian point embedded in left ilium. Ala-329, Central California, male, 18–25 years. In this specimen no healing is evident. Photographs by the author.

lesions also was of the cranium. Finally, Murad and Mertz (1982) remarked on one individual from the Sierra region of northeastern California who had three projectiles, all embedded within vertebrae.

It should be emphasized that these same groups from southern and central California that have the highest frequency of projectile injury are *also* the same groups showing high (or very high) rates of cranial injury (Walker, 1989; Jurmain and Bellifemine, 1997). It is from such a more consistent pattern of overall traumatic involvement that behavioral reconstruction is most justifiably attempted. Isolated lesions offer little hope for systematic interpretation (see below for further discussion).

BEHAVIORAL INTERPRETATIONS OF TRAUMATIC SKELETAL INVOLVEMENT

Some traumatic lesions provide unambiguous evidence of behavior, especially as relating to interpersonal aggression. Interpretation of embedded projectile points is usually obvious (although Merbs, 1989 does mention that the foot injury he described could have been accidental). Modern examples of gunshot wounds, containing fragments of bullets or shot, are analogous.

Lesions with a slightly less secure interpretive basis include those relating to likely sword or ax wounds as well as "blunt-force" trauma producing depressed cranial fractures (Courville, 1962; Lambert, 1997). Similarly, many facial fractures (e.g., of nasals or zygomatics) are also frequently, although not always, caused by blows.

Conversely, some traumatic involvement is clearly more suggestive of accident, particularly trauma to the distal forearm. Most especially, fractures of the distal radius, particularly those of the styloid, that is, Colles' fractures, are almost universally the result of a fall onto an outstretched (pronated) hand. Inferring that Colles' fractures are usually due to falls does not, however, permit precise resolution of the particular activities leading to such accidents. An example of such an unrestrained inference was that by Wells concerning the higher male Colles' fracture involvement in a group of coastal California Indians:

> This probably reflects a sexual division of labour in which men scrambled over slippery rocks amid the surf—in search of molluscs and small fish—whilst the women restricted their food gathering to more accessible pools left by the ebbing tide (Wells, 1964: 55).

In most circumstance Colles' fractures convey generally useful, if not precise, information. There are, however, other traumatically induced lesions, of the forearm especially, which do not yield such a clear behavioral etiology. These fractures of ulnar shafts are most commonly termed "parry fractures," implying the pronated forearm is struck while fending off a blow (oftentimes, suggested as caused by a club) (Figure 6-5). It has been known for some time that, in contemporary groups, a blow to the forearm can produce an isolated fracture of the ulnar shaft, the so-called "nightstick fracture" (Dymond, 1984; Rogers, 1992). However, while a deliberate (i.e., aggressive) explanation is clearly a *possibility*, that does not argue that the etiology of such lesions is *always* explained simply in this way (see below).

The first osteologists to infer interpersonal aggression as the cause of ulnar shaft fractures were Elliot Smith and Wood Jones (1910), who coined the term "parry fracture" to describe their apparent causation. Like other simple behavioral explanations in paleopathology with readily remembered names, the concept of "parry fracture" has been both popular and enduring.

Thus, over the past century the persistent consensus has been that ulnar fractures *are* parry fractures. Indeed, the name itself has come to describe *both* the nature and cause of the lesion. This type of thinking has thus led to many an uncritical evaluation. Not long ago, in the interpretation of ulnar shaft fractures in Central California Indians,

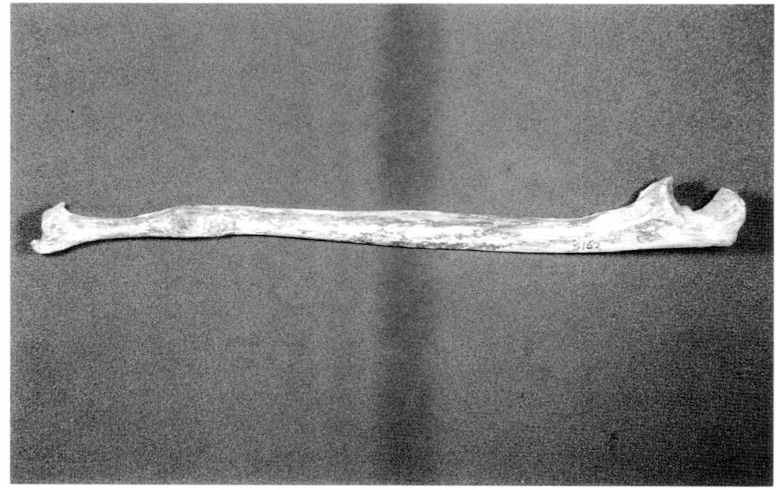

Figure 6-5(a) Diaphyseal fracture of right ulna, frequently termed as a "parry fracture." Kulubnarti, Sudanese Nubia, male, 41–50 years. Photograph by the author.

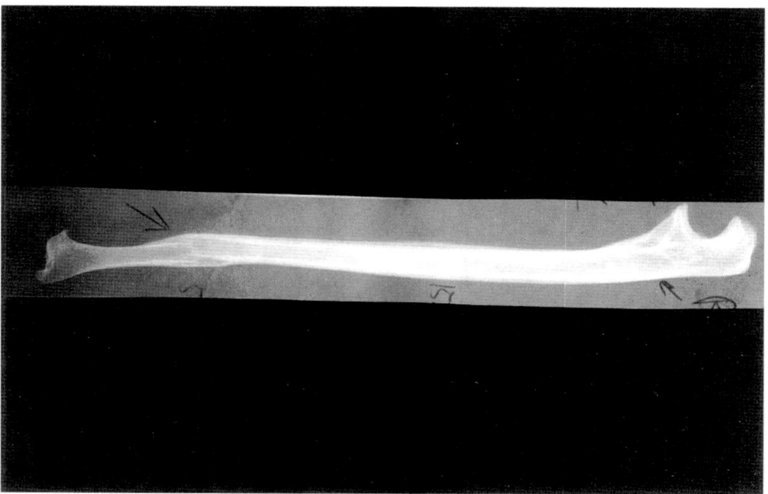

Figure 6-5(b) Radiograph. Note the original cortex is still clearly visible, indicating that, at the time of death, the injury was still healing. Photograph courtesy, Lynn Kilgore.

I made as a simplistic and uncritical assessment as one is likely to find in the literature:

> The frequency of forearm midshaft injuries suggests interpersonal violence (i.e., parry fractures), which is further indicated by the predilection for left side involvement. Arguing against intergroup fighting is the similar incidence of female involvement of the forearm as that seen among males (13 male, 14 female). This observation does not of course preclude violence directed at women within the group (Jurmain, 1991a: 244).

Three decades prior, as relating to his comprehensive review of British skeletal materials, Brothwell expressed essentially the identical view:

> There is no doubt that this high incidence results from the forearm being used to ward off blows to the head region, but the fact that at least seven of the twenty-five British cases were female, suggests that the violence may have been as much the result of localized (even domestic) disputes as an indication of inter-group warfare (Brothwell, 1961: 333).

Writing at about the same time, Calvin Wells characteristically added his own imaginative twist to interpretation of parry fractures, suggesting such injuries "... point to short tempers and aggressive conduct being a common feature of the society and, as many of these occur in females, wife-beating or general low status of women may be implied" (Wells, 1964: 53).

Such explanations are obviously straightforward as well as neat; they *might* also even be correct. What these explanations do not provide, however, is any means of demonstrating falsifiability. In paleopathology usually it is initially assumed that the lesions are always caused by aggression. In males it must be one kind of aggression; if that does not match the population pattern, then it *must* be another type of aggression in females. Any skeletal pattern can thus be made to fit the hypothesis. The meaning of variation within populations can be obscured by such simplistic reasoning. Moreover, the basic assumption inherent at the start—i.e., that such lesions result from "parrying" should also not go unchallenged.

Still, most osteologists have at least tacitly continued to accept this conclusion, perhaps as most enthusiastically supported by Webb in arguing that in Australia "parrying" fractures were:

> the most common fractures among east coast females and desert males, which naturally leads to the conclusion that women on the east coast fought either with males or with each other and that fighting between desert men might be more common than it is in groups from elsewhere (Webb, 1995: 200).

Webb, in fact, readily interprets *all* distal ulnar fractures as "parrying" lesions, even when accompanied by an ipsolateral radial fracture. Clinical data (discussed below), however, do not support such a simple explanation, and in the case of paired forearm involvement, actually indicate a quite different etiology.

As might be expected, not all paleopathologists have been as completely accepting of the "parry fracture" concept. Angel (1971), in his analysis of the Lerna skeletal materials from Greece stated that the ulnar fractures might have resulted "plausibly from direct violence." Angel, however, did not remain as cautious in assessing two fractures of fifth metacarpals—which he argued were due to "striking a misdirected blow."

Bennike (1985) was characteristically careful and balanced in her approach to interpreting what she referred to as "so-called parry fractures," which she further noted "can also result from a fall." Given such ambiguity of causation (as quoted at the beginning of this chapter), Bennike refrained from speculating further on such a "fragile foundation."

More recently, Smith (1996) has more comprehensively reviewed the etiology of forearm fractures. Like Bennike, Smith concluded that ulnar shaft fractures might result as much (or even more) from accidents as they do from interpersonal aggression. Smith assessed the gender-related patterns of forearm involvement in several samples from Tennessee ($N = 809$ individuals). The results led her to be suspicious of simplistic aggression-based interpretations. Moreover, she made the crucial point that cranial trauma did not usually accompany these forearm lesions, and thus concluded: "More importantly, craniofacial trauma data do not support an aetiology that would explain mid-shaft forearm fractures as a result of interpersonal violence" (Smith, 1996: 84).

Smith further cautioned that other researchers have not always paid appropriate attention to the full pattern of traumatic involvement. "However, uncritical interpretation of interpersonal violence is often made in the absence of evidence that any (not necessarily co-occurring) pattern of blunt trauma cranial injury occurs" (Smith, 1996: 84).

Finally, Kilgore et al. (1997) and Lovell (1997) have similarly argued that ulnar fractures might result from a variety of causes—and thus not solely from aggression. Further, as Smith found in her Tennessee data, the forearm lesions at Kulubnarti (Kilgore et al., 1997) are present in high frequency in the *absence* of craniofacial injuries.

Forearm Fractures: Clinical Perspectives

As suggested above in some of the comments made by osteologists, there are obviously a variety of causes that can produce forearm fractures, and clinical perspectives can help identify many of these. Indeed, as for OA and other skeletal manifestations, the *only* way to achieve any degree of precision relating to etiology of such conditions is from well-documented clinical contexts.

Contemporary data accord well with osteological interpretation of distal radial lesions (i.e., Colles' fractures). There is not much ambiguity or debate relating to the cause of such injuries: They result almost always from falls (Dobyns and Linscheid, 1984). However, there is some debate relating to isolated fractures of the ulnar shaft.

Good clinical documentation has shown that both ulnar shaft and radial shaft injuries can result from a fall and can produce lesions almost anywhere along the shaft of either forearm element (Evans, 1949; Rogers, 1992). However, since the two bones are tightly joined to each other, proximally by the annular ligament and along the diaphysis by the interosseous ligament (Wiley et al., 1974), most forces strong enough to fracture one element will usually also affect the other forearm bone in some way as well (although not necessarily producing bone changes). A brief review of forearm fracture nomenclature (after Schultz, 1990; Rogers, 1992) would be useful:

Colles'—Fracture of the distal end of the radius, often accompanied by dislocation of the proximal ulna.

Moore's—fracture of distal radius (Colles') associated with ulnar styloid fracture and dorsal subluxation of distal ulna.

Monteggia—Fracture of ulnar shaft with associated radial head displacement, if not actual dislocation. The direction of displacement is either anterior or antero-lateral (Evans, 1949). If the radial head is not displaced, the alternative result is the radial diaphysis being fractured simultaneously with the ulna.

Galeazzi—Isolated fracture of the radial shaft (usually at the junction of the middle and distal thirds) and associated with dislocation of the distal radio-ulnar joint.

Nightstick (also called parry)—Isolated fracture of the ulnar shaft—usually in the middle or distal third, produced by a blow and *not* involving the radius.

Essex-Lopresti—Fracture of the radial head or neck, occasionally associated with dislocation of the distal radio-ulnar joint.

In most contemporary contexts, ulnar shaft fractures are caused by falls (with some exceptions discussed below). The trauma produces a Monteggia fracture, and the mechanism is nicely explained by Evans:

> When a patient falls forward on the outstretched hand the forearm is already pronated and at the moment of impact the hand becomes relatively fixed to the ground. To the downward momentum of the falling body a rotation force is added when twisting of the trunk causes external rotation of the humerus and ulna. If this force continues until the normal range of pronation at the radio-ulnar joints is expended, something must give. The ulna cannot rotate, because it is fixed below by the ulnar carpal ligament and above by its articulation with the humerus. The ulna is therefore liable to fracture, and the combination of rotation force and the bending force set up by longitudinal compression may produce an oblique transverse or butterfly fracture. At the same time the radius is forced into extreme pronation and lies across the ulna, at the junction of the upper and middle thirds. As the ulna fractures, the two bones come into contact and the point of contact forms a fulcrum over which the upper end of the radius is forced forward. As the pronation force continues, the radius is either levered forward out of the superior radio-ulnar joint or is fractured in its upper third (Evans, 1949: 578–579).

As Evans suggested, some additional data indicate most Monteggia fractures occur in the upper third of the ulna, and Rogers (1992) noted the following distribution of such lesions:

proximal third:	89%
middle third:	10%
distal third:	1%

An almost identical distribution of Monteggia fractures was also reported by Boyd and Boals (1969). Rogers further commented that nightstick (parry) fractures usually occurred in the distal third of the ulna, "less commonly in the middle third, and rarely, if ever [were] encountered in the proximal third" (Rogers, 1992: 816). If, indeed, the contrasting locations of nightstick *vs.* Monteggia fractures proved consistent, location of the lesion along the ulnar shaft might be helpful in assessing probable causation in osteological specimens. Moreover, ulnar shaft fractures appear to be influenced variably, depending on population studied. Evans, working in an English urban clinical setting, stated that all the ulnar shaft lesions he saw were the result of falls. Conversely, Du Toit and Gräbe (1979), working with South African patients, argued that all ulnar fractures they observed in adults were due to "direct trauma." Their patient sample was clearly one involved in regular violence, as 69% of the patients had a history of assault. In this group, which is perhaps a good model for individuals engaged in systematic aggression, 67% of lesions were to the left side.

As also noted by Dymond (1984), a frequent weapon used in South Africa (by police as well as assailants) is a baton, some of which were a particularly heavy-ended version called "knobkerries." Another interesting aspect of the pattern of involvement among the South Africans was that cranial trauma was often associated with forearm injury; among those individuals with ulnar fractures, 27% also had injuries to the head or face (Du Toit and Gräbe, 1979).

Another recent study of trauma victims in Papua New Guinea showed the greatest proportion of injuries (55%) resulted from interpersonal aggression (Matthew et al., 1996). Moreover, of a total of 93 total fractures in the sample, 34 (37%) were of the elbow or forearm. Interesting, however, a surprisingly high number of femoral fractures (13) were also observed in this rural New Guinean sample. The severe nature of many of the injuries might relate to the pattern of violence: 24% of all traumatic injuries were attributed to "tribal fights."

Another point, also of potential interest to osteologists, is that nightstick fractures usually affect *just* the ulna, while diaphyseal injuries due to falls (Monteggia fractures) will normally cause a displacement of the radial head. It would appear, however, that the radial displacement does not usually produce permanent osteological remodeling; nevertheless, when ulnar shaft fractures are identified, it would be profitable to always closely inspect the radial head and proximal ulna to assess any possible involvement.

Monteggia fractures can also occur in children, and here, as with adults, the lesions are usually of the proximal or middle third of the ulnar shaft. In one study, conducted in Tunis, several lesions were not treated immediately (the delay following injury lasting up to two years) (Fowles et al., 1983). Some of the long-standing injuries had persistent radial head dislocation, but with no apparent loss of function. In fact, some clinicians think it better not to attempt to surgically correct long-standing, previously untreated Monteggia fractures in children (Fowles et al., 1983).

Some researchers have also claimed the radius can be fractured by a direct blow, producing a Galeazzi fracture of the radial shaft. Houghston (1957), in fact, claims the "usual" cause of such lesions is a "direct blow to the dorsoradial aspect of the forearm," a view, challenged by Rogers (1992), who thinks such an etiology unlikely. Mikic (1975), however, did find in a clinical series from Yugoslavia that about 6% of the Galeazzi lesions treated there were attributable to a direct blow (most of the lesions resulted from traffic accidents or in accidents associated with heavy machinery). In 25% of cases in this Yugoslavian group with Galeazzi lesions of the radius, the ulnar shaft was *also* fractured.

In fact, an important consideration of forearm fracture etiology is whether both elements are or are not involved. Evans (1949) noted that a likely outcome of Monteggia-type injury was also simultaneous fracture of the radius, and Smith and Sage (1957) found from analysis of 338 patients with forearm *shaft* fractures that a majority showed multiple lesions:

61% of forearm fractures—both elements involved
25% of forearm fractures—only ulna involved (with and without displacement of the radial head)
15% of forearm fractures—only radius involved (with and without dislocation of the distal ulna)

Moreover, in this series, Smith and Sage (1957) noted that 60% of all lesions which involved both elements were in the middle third of the shaft. Unfortunately, for these data, there was no patient history information presented as to causation. However, according to Rogers (1992), the majority of the lesions almost certainly resulted from falls—as he argues that both ulna and radius are rarely both fractured as the result of a blow (but are frequently both affected following a fall). In circumstances of extreme force, according to L.D. Anderson (1984), both elements can be fractured by a "direct blow," most commonly seen today in automobile and motorcycle accidents. Clearly an important point is raised here: "Blows" can result from sharp forceful impacts in mechanical vehicles or in severe falls (e.g., against a boulder), and not *just* from deliberate attacks.

OVERALL PATTERNS OF TRAUMATIC INVOLVEMENT

Recently, some clinicians have initiated research into understanding patterns of deliberate violent injuries, as reflected in involvement throughout the body. This work has particularly been prompted by Breiting, Aalund, Danielsen, and colleagues, who have begun an international comparative study of such patterns of deliberate violent trauma in Denmark (Breiting et al., 1989), Argentina (Danielsen et al., 1989), and Chile (Aalund et al., 1990). Moreover, similar investigations have been carried out in England (Shepherd et al., 1990), Sweden (Ström et al., 1992; Boström, 1997), and South Africa (Butchart and Brown, 1991). All these studies reported on soft-tissue injuries as well as fractures (and most lesions were of the soft tissues).

The study conducted recently in Sweden (Ström et al., 1992) assessed facial injuries in assault victims. Of 222 individuals with facial injuries, 138 had fractures. The most common site was the nasal bone, followed by the jaws (mandible and maxilla, respectively). Interestingly, the zygomatics, orbits, and calvarium were considerably less frequently affected. An important issue for osteologists is that nasal bone involvement is a likely outcome of deliberate aggression, a point also mentioned by Merbs (1989). However, osteologists have rarely presented consistent data on facial lesions, and nasal bone involvement is almost never addressed. One factor conditioning this perspective is, of course, the influence of poor preservation. Nevertheless, more careful evaluation, of the nasal region in particular, should be considered.

Similar results of more specific investigations of assault victims have also shown the craniofacial region to be most frequently involved. In one systematic English investigation, 86% of all fractures in assault victims ($N = 139$) were of the head, and most especially, the face (Shepherd et al., 1990). As in the Swedish study mentioned above, the nasal bones were most frequently affected, but zygomatic and mandibular lesions were also quite common. In another recent English study (Asadi and Asadi, 1996) most mandibular fractures resulted from interpersonal assault, while in an urban Maryland group of baseball bat assault victims, severe injuries were concentrated in the mid-facial area (Ord and Benian, 1995).

Ström et al. (1992) also commented briefly on asymmetry of facial injuries, but no side predilection was found. For example, of the 12 zygomatic fractures, five were of the left side, and seven of the right. Ström and colleagues suggested this lack of consistent asymmetry might be caused by the fact many of the victims were struck from behind (while fleeing). However, in a study of facial fractures in Scotland, a greater preponderance of left-sided injuries was observed (Brook and Wood, 1983). This greater left-sided involvement was found in cases of assault as well as injuries due to industrial accidental falls, but not in those resulting from sports (equestrian) injuries. Again, as noted above, osteologists should recognize such complexities, and their implied constraints: that is, as asymmetries might imply being hit from the front, from the rear, or neither, the inferences possible in archaeological contexts are far from secure.

The data from Argentina (Danielsen et al., 1989) are intriguing. As in other contemporary studies, most injuries were of soft tissues (e.g., wounds, contusions, concussions, internal injuries). However, systematic data on fractures were also presented. Of the total of 45 fractures observed in this study, the relative proportions of involvement by body segment were as follows: head/neck (33%); upper extremity (all elements, 31%); trunk (27%); lower extremity (9%) (see Table 6-5).

Table 6-5. Patterns of fracture involvement—all body segments[1]

	Head/neck	Trunk/pelvis	Up. append.*	Low append.**	Total
Neandertal[1]	7 (30%)	4 (17%)	8 (35%)	4 (17%)	23
Libben[2]	6 (6%)	20 (21%)	28 (30%)	40 (43%)	94
Nubia[3] (Lower)	17 (11%)	17 (11%)	88 (55%)	38 (24%)	160
London[3] (Hospital)	108 (6%)	124 (7%)	968 (56%)	530 (31%)	1730
New York[3] (Hospital)	1640 (14%)	1534 (13%)	5649 (47%)	3136 (26%)	11,959
New Mexico[1] (Worker's Compensation)	13 (2%)	116 (15%)	370 (47%)	293 (37%)	792
Rodeo[1]	71 (39%)	24 (13%)	58 (32%)	28 (16%)	181
Denmark[4] (Emerg. ward)	174 (61%)	17 (6%)	76 (27%)	19 (7%)	286

Argentina[5] (Emerg. ward)	174 (61%)	17 (6%)	76 (27%)	19 (7%)	286
Torture[6] victims	5 (16%)	12 (39%)	5 (16%)	9 (29%)	31

Statistical Comparisons:	*chi-square*	*p*
Nean *vs.* Rodeo	0.762	0.859
Nean *vs.* Denmark	11.30	0.01
Nean *vs.* Argentina	1.59	0.661
Nean *vs.* Torture victims	5.89	0.117

* Includes shoulder and hand
** Includes foot
[1]From Berger and Trinkaus, 1995; [2]From Lovejoy and Heiple, 1981; [3]From Elliot Smith and Wood Jones, 1910 (as tabulated in Berger and Trinkaus, 1995); [4]From Breiting et al., 1989; [5]From Danielsen et al., 1989; [6]From Rasmussen, 1990 (as tabulated in Aalund et al., 1990)

Further, an interesting finding was that, of all the upper extremity injuries, 14 were without any weapon, 17 were with use of a sharp instrument, and only three were with use of a stick or baton (four others were with a firearm). It would have been useful to have had precise data on which upper appendage elements were involved and in what manner they had been fractured. Still, two points are worthy of note. First, bare hands (and feet) alone can inflict serious damage, and, secondly, "sticks," "batons," or "clubs" are used more in some groups (recently, particularly in South Africa). The types and location of injuries, however, do not seem to vary with the type of weapon employed (a cautionary tale for osteologists). A similar study from Chile (Aalund et al., 1990) found many fewer fractures (only 9 in a total of 236 victims of violence).

The data from South Africa, collected in Soweto in the 1980s, and representing 1,282 victims of violence, are also of considerable interest. Here, the great majority of injuries (84%) was found in males, compared to 16% in females. Similarly, approximately 80% of injuries recorded in the Danish study (Breiting et al., 1989) and 84% in a Swedish emergency room sample (Boström, 1997) were among males; even more gender-biased was the huge discrepancy in perpetrators, where males in Denmark accounted for 95% (745/784) of the attackers. (Note: sex bias in injury ratio has also been reported in California archaeological samples and among Africa great apes). Moreover, as in Argentina, a fair proportion of injuries was inflicted with just hands and/or feet (13.7% of injuries in females and 8.7% in males). Lastly, the most commonly affected areas of the body, combined, for all types of lesions, was again the head/neck (which included 99 diagnosed fractures), followed by the trunk. Similarly, in the most recent study of Swedish assault victims (Boström, 1997), the great majority of injuries (82%) were of the head, and a large proportion of these (74%) were inflicted by fists or kicks.

The contemporary information relating to body segment distribution of injury is useful in comparison to that presented by Berger and Trinkaus (1995). As discussed briefly above, these authors systematically compared data relating to fracture patterning, both from archaeological samples as well as from contemporary contexts. The fracture data from contemporary groups at significant risk of violent injury are shown in Table 6-5. Berger and Trinkaus included in their tabulations information specific to a wide variety of body segments, including: head/neck; trunk; shoulder/arm; hand; pelvis; leg; and foot. It is, however, difficult to judge how comparable such information is (between samples), especially as the major comparisons are done using *proportionate* levels of involvement. Even in contemporary samples, for example, rib injuries are not always consistently reported, nor are those of the hands and feet. Serious problems of preservation affecting archaeological materials could confound such comparisons even further. In archaeological contexts, representation of ribs, vertebrae, the pelvis, as well as hand and foot elements, will generally be much less adequate than for the cranium and major long bones. Moreover, given variable disturbance factors and quality of archaeological recovery, the degree of such non-representativeness will differ, and will do so in a mostly unpredictable fashion. Thus, it would seem that injuries to the less well-preserved osteological elements will be under-represented in the archaeological record, and this bias will from vary site to site. Accordingly, in comparisons of archaeological samples made below (and shown in Table 6-6), Berger and Trinkaus' data on proportionate anatomical involvement are recalculated to emphasize the better preserved osteological elements.

The data collected from contemporary hospital emergency room contexts (shown with some of Berger and Trinkaus' data in Table 6-5) are only minimally retabulated, and is done so to make the categories compatible across the different studies. Two of these series, from hospital emergency wards in Denmark (Breiting et al., 1989) and in Argentina (Danielsen et al., 1989) are discussed above; the third

Table 6-6. Patterns of fracture involvement (using reduced segment categories—see text)

	Head/neck	Up. append	Low. append	Total
Neandertals[1]	7 (41%)	7 (41%)	3 (18%)	17
Rodeo[1]	71 (55%)	47 (36%)	11 (8.5%)	129
Nubia, Lower[2]	17 (12%)	85 (62%)	36 (26%)	138
Libben[3]	6 (8.5%)	28 (39%)	37 (52%)	71
Ala-329[4]	8 (18%)	31 (71%)	5 (11%)	44
Kulubnarti[5] (Sudanese Nubia)	1 (1%)	61 (90%)	6 (9%)	68

Statistical Comparisons:

	chi-square	*p*
Nean *vs.* Rodeo	1.93	0.38
Nean *vs.* Ala-329	6.43	0.04
Ala-329 *vs.* Kulubnarti	10.66	0.005
Ala-329 *vs.* Libben	19.56	0.0001
Ala-329 *vs.* Nubia (Lower)	4.46	0.108

[1]From Berger and Trinkaus, 1995; [2]From Elliot Smith and Wood Jones, 1910 (as tabulated in Berger and Trinkaus, 1995); [3]From Lovejoy and Heiple, 1981; [4]From Jurmain, 1991; Jurmain and Bellifemine, 1997; [5]From Kilgore et al., 1997

sample is that of 200 victims of torture, as originally presented by Rasmussen (1990), are collated in Aalund et al. (1990: 198).

Berger and Trinkaus make the clear argument that the Neandertal pattern of fracture involvement more closely parallels that of rodeo performers, as compared to that of modern clinical samples or materials from archaeological sites (see Table 6-5). The most obvious similarity is in the relatively high proportion of head/neck injuries in both Neandertals and rodeo performers. However, the victims of violence also show similar patterns of intensified head/neck involvement (in the Danish sample, dramatically so). The clearest parallel with the Neandertal fracture distribution is among the Argentine patients presenting at a hospital emergency clinic. In fact, statistical testing (Chi-square) failed to distinguish between the Neandertals and this sample of aggression victims ($p = 0.661$) to almost the same degree as between the Neandertals and rodeo performers ($p = 0.859$) (see Table 6-5). The difference between Neandertals and the Danish victims was significant ($p < 0.01$), but, interestingly, was not in comparison of Neandertals with the torture victims ($p = 0.117$).

A possible conclusion, then, is that the heightened Neandertal craniofacial involvement might have resulted from interpersonal aggression rather than "close encounters with large ungulates." It seems more probable, however, that the rodeo analogy is a better one than the various samples of assault victims. Nevertheless, one should consider a *range* of possible explanations and utilize a wide diversity of contemporary and archaeological data to test alternative hypotheses. Certainly, Berger and Trinkaus' innovative approach and highly accessible presentation of raw data allow (indeed, invite) such further exploration. As such, it is one of the best current examples of rigorous behaviorally oriented osteology.

Applying such a methodology could also be useful in behavioral interpretations of other archaeological samples (that is, more recent ones than Neandertals). To this end, in Table 6-6, fracture distribution data comparing Neandertals, rodeo performers, and archaeological samples are tabulated (here, the categories are restricted to eliminate the archaeologically more poorly represented elements, i.e., trunk, pelvis, hands, and feet). In these comparisons, the relatively high proportion of head/neck injury among Neandertals is not matched in any of the more recent archaeological samples (and, again, matches most closely that of rodeo performers, the only sample from which the Neandertal pattern is not statistically significantly different).

However, within the archaeological samples themselves, the Central California site (Ala-329) considerably exceeds the other three in the proportion of head/neck fractures (and the difference is significant in comparison to Libben, $p<0.0001$, as well as to Kulubnarti, $p<0.005$). Clearly, this approach of comparing the anatomical regional patterning of fractures could prove most informative and of considerable utility. Of all the archaeological sites shown in Table 6-6, Ala-329 has the most conclusive corroboratory evidence (see below) relating to fairly systematic aggression. And, not incidentally, its pattern of fracture distribution stands out from the others.

Possibly also highly pertinent to interpreting the meaning of fracture distribution is the *wider* pattern of traumatic involvement. In California, there is other unambiguous evidence of interpersonal aggression, as evidenced by embedded projectile points (at Ala-329 and elsewhere). In Lower Nubia, Elliot Smith and Wood Jones (1910) alluded to some weapon use. Apparently, at Libben such corroboratory evidence is less obvious, and is least so at Kulubnarti, where no signs of projectile injury or other direct weapon use were found.

This is not to argue that the picture makes for a perfect fit. The fact that such disparate activities as riding bulls and encountering aggressive, poorly socialized humans can produce similar fracture patterns should remind us of the underlying complexities. Nevertheless, the more complete and coherent are the patterns in a variety of samples, the more likely they can be systematically compared; furthermore, such analysis facilitates falsification, or, alternatively, at least tentatively, provisional acceptance of an hypothesis.

CONCLUSIONS/RECOMMENDATIONS

The most important consideration regarding comprehensive interpretation of trauma in skeletal remains was just discussed, that is, assessment of the overall pattern of involvement of relevant lesions. Interpretation of behavioral influences, specifically an accidental *vs.* intentional etiology, requires some demonstration of craniofacial involvement in addition to postcranial patterns (a point well-emphasized for osteological studies by Smith, 1996; Kilgore et al., 1997; and Jurmain and Bellifemine, 1997, as well as noted in clinical contexts by Du Toit and Gräbe, 1979). Moreover, more standardized methods of recording traumatic skeletal involvement would be most beneficial (see Lovell, 1997). More generally, methodological approaches to identification and reporting of fractures (and other traumatic lesions) have not been particularly precise, and thus the following recommendations are offered:

1. Fractures (particularly of long bones and the cranium) should be reported *by element* (and prevalence rates calculated on the basis of

complete elements—following the approach employed by Lovejoy and Heiple, 1981; Bennike, 1985; Jurmain, 1991a; Grauer and Roberts, 1996).

2. Lesions should be recorded quite precisely as to location (in the cranium as by Walker, 1989; and in the postcranium as by Kilgore et al., 1997; also see further recommendations in Lovell, 1997). Using such a methodology, for example, in the assessment of long bone diaphyseal fracture involvement, would be especially helpful in interpretation of forearm fractures (i.e., Monteggia fractures *vs.* parry fractures of the ulna).
3. Multiple element involvement should also be carefully noted, and, again, would be important in interpretation of forearm injury.
4. In addition to healed fractures themselves, other types of traumatic lesions, including wounds, projectiles, and cuts marks, should be recorded. Further, more subtle lesions, such as remodeling of ligamentous insertions and dislocations need to be more carefully assessed.
5. Fractures which fail to join (i.e., ununited fractures) also require more detailed evaluation. While such lesions (in long bones) have been thought rare (Stewart, 1974), more recently, they have been more regularly recognized and described (Merbs, 1989; Jurmain, 1991a; Webb, 1995).
6. Facial fractures also are probably more common than are typically reported osteologically (as suggested in clinical contexts, e.g., Ström et al., 1992; Boström, 1997). More attentive evaluation of such lesions (particularly affecting the nasal bones) thus is also needed. Indeed, a more rigorous application of this approach has recently been completed by Walker (1997).
7. More general assessment of the proportional distribution of lesions (after Berger and Trinkaus, 1995) would also be informative; moreover, as in their study, detailed enumeration of lesions (including raw frequencies) allows data to be compiled (and compared) in a variety of ways.
8. As in other aspects of behavioral interpretation from osseous involvement, healthy skepticism is always encouraged. Highly specific assertions linking behavior with particular traumatic lesions are rarely justified. Even in recent contexts (e.g., Hershkowitz et al., 1996) where temptation is perhaps greatest to make such specific interpretations, exercise of caution is all the more necessary (as so well exemplified by Wakely, 1996).

Chapter 7

Studies of Bone Geometry: The Shape of Things to Come?

> Every change in the ... function of bone ... is followed by certain definite changes in ... internal architecture and external confirmation in accordance with mathematical laws (Wolff, 1892; quoted in Carter, 1984: S19).

INTRODUCTION

The most recent innovative approach to reconstructing behavior (specifically, activity) from skeletal remains derives from investigations of bone diaphyseal geometry. Moreover, some clinical and osteological research concerning bone mass and density relate to similar, although less precise, aspects of bone remodeling. In recent years, most especially in the research of Ruff and colleagues, this approach has been formulated and a variety of intriguing osteological applications have been suggested (Ruff and Jones, 1981; Ruff and Hayes, 1983a,b; Ruff, 1987; 1995; Brock and Ruff, 1988; Ruff and Larsen, 1990; Ruff and Runestad, 1992; Ruff et al., 1993; Ruff et al., 1994; Larsen and Ruff, 1994; Trinkaus et al., 1994).

The main goal of this research is to determine from cross-sectional geometric properties of long bone diaphyses indications of the loading histories that ultimately shaped these dimensions. Reviews of the theoretical bases of this approach can be found in several recent publications (Ruff and Hayes, 1983a; Larsen and Ruff, 1994; Trinkaus et al., 1994; Bridges, 1996b). Essentially, engineering principles are applied to understanding "global geometry" of cortical bone from a biomechanical perspective (Ruff and Hayes, 1984) to facilitate calculation of geometric strength properties as they relate to shape (Ruff and

Hayes, 1983a). Long bone structure is viewed most crucially in terms of its internal architecture, that is, in the amount and distribution of cortical bone as seen in cross section (Bridges, 1996b). Most especially relating to these principles, it is assumed, rigidity, rather than compliance, is crucial to the appendicular skeleton, particularly as it is adapted to bending and torsional loads. "There are few physiological constraints on diaphyseal structure; thus dramatic alterations in diaphyseal cross-sectional size may occur with changes in mechanical loadings" (Ruff and Runestad, 1992: 409).

Earlier studies had attempted more roughly to assess bone size (Frayer, 1980; Hamilton, 1982), mass, volume, or density, (Nilsson and Westlin, 1971; Dequeker, 1975; Smith et al., 1984; Kennedy, 1985) but:

> Changes in bone geometry, or shape, may be more informative with respect to *specific* environmental adaptations, in particular adaptation to *specific* mechanical forces, which are indicative of functional use and thus behavioral differences (Ruff, 1987: 392) [emphasis added].

Similarly, Bridges concluded: "An explanation of diaphyseal structure can then yield insights about the usual forces placed on the bone during life and, indirectly, about the activities of earlier humans" (Bridges, 1996b: 112).

FACTORS INFLUENCING BONE GEOMETRY

As revealed by the above quotes, and much emphasized below, the greatest emphasis by osteologists has been directed at linking activity influences to changes in bone architecture. Nevertheless, there has been at least some recognition of the likely complexity of the processes underlying remodeling of long bone diaphyseal shape.

Bridges (1989a), for example, noted that nutritional effects, including malnutrition, age-related osteoporosis, genetic differences, and certain diseases (especially infectious ones) all might play some role. Ruff et al. (1994) further suggested the possible role of hormones acting differentially with age, i.e., producing different results in childhood *vs.* later adolescence and adulthood (see below for discussion of age effects). Potential effects of hormonal control were also postulated by Kennedy (1985) and Trinkaus et al. (1994). Potential genetic influences were strongly suggested some time ago by Garn and colleagues (1964) who found significant differences in amount of compact bone (as assessed from hand radiographs) in Americans of European descent as compared to Americans of Asian descent. These authors argued that some aspects of bone turnover might well be under control of relatively simple, i.e., Mendelian, genetic mechanisms (a view also suggested here). More recently, differences in amounts of both cortical and cancellous bone were observed in white as compared to black samples (Han et al., 1996).

As discussed in Chapter 5, general rates of bone turnover might well affect development of both osteophytes and enthesophytes (Rogers et al., 1997). Such *systemic* influences might also differentially affect diaphyseal geometry, and could relate to underlying genetic mechanisms. Indeed, recent well-controlled data strongly suggest that genetic factors can influence geometric patterning of long bones. In a study of monozygotic and dizygotic twins, high heritability estimates were found, both for bone mineral density (BMD) as well as geometric variation of the femoral neck (Slemenda et al., 1996). Moreover, there was no evidence that length of the femoral neck was under such apparent genetic control. The authors thus argued:

> It is concluded that there are significant familial influences on the distribution of femoral bone mass and on the calculated structural strength of the proximal femur, but not on femoral neck length. If the assumptions of the twin model are correct, this is evidence for genetic factors influencing these traits (Slemenda et al., 1996: 178).

Age and Sex Effects

Among the obvious, and best documented, influences on changes in bone geometry (as well as density) are those relating to age and sex. Garn (1970), Martin and Atkinson (1977), and Burr (1980) emphasized differential age effects on both bone density and geometry in males compared to females. While males show continued apposition throughout life, females more often exhibit gradual decline in such bone apposition after age 35. This same point was further reinforced by Ruff and Hayes (1983b; 1984) as well as by Bridges (1996b), who further noted that periosteal remodeling appears most responsive in preadolescence, while later in life the endosteal surface becomes more responsive. Goulding et al. (1996) found a very high correlation of femur axis length with age ($r=0.92$) in girls aged 3–16 years. Further complicating the picture, however, is the observation from another study (Bouxsein et al., 1994) that, in aging women, changes in bone geometry in the ulna differed substantially from those seen in the radius.

The best-controlled clinical/epidemiological data allowing evaluation of age effects are those relating to humeral asymmetry in professional tennis players (Jones et al., 1977; Priest et al., 1977; see below).

To specifically assess the influence of age, these same data were recently reanalyzed by Ruff et al. (1994) from an explicitly ontogenetic perspective, in order to evaluate "possible differential response of cortical bone dimensions to mechanical stimuli during periods of growth." In this recent contribution, Ruff and colleagues considered a variety of factors potentially influencing the degree of bone remodeling, including age at which play began, number of years played, and age at ascertainment. Interestingly, the only significant correlate affecting humeral asymmetry was *age at which play started*. Ruff and colleagues suggested that after about 8 years of age the maximum degree of asymmetry was attained, and no further notable changes in robusticity of the playing arm were achieved. Agreeing with the views mentioned above, Ruff et al. further argued that most of the hypertrophy resulted from periosteal deposition quite early in life (childhood and early adolescence), as there was much less asymmetry and more endosteal remodeling in individuals who began the activity later. "... the results ... strongly suggest a change during development in sensitivity of bone-forming surfaces to increased mechanical loading" (Ruff et al., 1994: 39).

Ruff and colleagues also postulated possible differential age influences of hormones. In agreement with Garn's earlier suggestions (1970), levels of growth hormone in children may have greater influence, but, following adolescence, testosterone and estrogen might play the more significant role. The dynamics relating to endocrine control of bone turnover in early adolescence may be especially complex, since the crucial variable may not be chronological age, but stage of puberty (Grimston et al., 1993). This variable will obviously differ individual to individual, but more disturbingly (to those trying to interpret earlier skeletal groups), it may also differ population to population. The degree of skeletal maturation might provide some control in evaluating this possible confounder, but the challenges inherent in controlling adequately for such potentially conflicting variables are considerable.

Further, and related to such endocrine influences in later life, females, on average, show considerably more endosteal resorption than males (Noble et al., 1995; Feik et al., 1996; Han et al., 1996). This process is especially marked in post-menopausal women, leading in many to osteoporosis. A large clinical literature exists detailing the dynamics of this process. For osteologists the clear message is that analyses of either bone density and/or geometry must be rigorously controlled for sex as well as age (essential methodological requirements for evaluating all the other major osteological "markers" discussed in prior chapters).

Mechanical Factors: Theoretical Considerations

It is quite certain that mechanical factors considerably influence bone remodeling, as reflected in bone volume, bone density, and cross-sectional geometry. Indeed, mechanical (especially weight-bearing) loading is *required* for maintenance of healthy bone, as demonstrated in studies of weightlessness in astronauts (Mack et al., 1967; Morey and Baylink, 1978) or in patients following enforced bedrest (Donaldson et al., 1970; Krølner and Toft, 1983). Weightlessness or inactivity "reduces strength, stiffness and energy absorbing ability of bones" (Currey, 1984: 245).

These circumstances are, of course, extreme. What is of more interest in osteological reconstruction, but clearly more difficult to ascertain is: What effects does mechanical loading have on bone within the *normal range of activities*?

At present, there are only minimal data to answer adequately this question (see below). Moreover, at a more theoretical level, there are further reasons to recognize complexities in the etiology of bone remodeling (see also discussion in Chapter 5). First, even though the crucial contribution of mechanical influences is recognized, the precise nature of the mechanism is not well understood. Evidence from animal experimentation (discussed below) has shown that, in addition to *type* of activity, duration (repetivity), amplitude, and sense (i.e., torsion *vs.* compression) are all significant. Furthermore, these factors may vary somewhat independently and produce variable effects (when considering age, sex, and especially, skeletal site). Lanyon (1984: S58) noted that strain magnitude, strain rate, and strain distribution were "only three of the variables so far demonstrated to be of importance." The differential effects of the amplitude of strain as opposed to duration have also elicited comment by clinical researchers. Lanyon and colleagues concluded:

> Unfortunately, little is known at present of the mechanisms by which continued intermittent deformation influences bone structure and so the relative importance of large or small deformation cycles and the significance of their alignment is at present speculative. *Except for knowing that mechanical conditions are 'important' we know nothing of which aspect of them is relevant and how it acts* (Lanyon et al., 1975: 267).

Frost (1991) further noted that most likely it was amplitude of loading (as opposed to duration) which most stimulates remodeling. He thus concluded that the most significant consideration was a "time-averaged value of the mechanical forces on bone, (which) assigns disproportionately more weight to larger rather than to small loads, no matter how frequent the latter" (Frost, 1991: 430). Silman and Hochberg (1993: 446–447) concurred in noting: "The normal skeleton appears to adapt rapidly to the stimuli associated with physiological loading; thus repetitions of habitual activities are unlikely to confer additional benefits. In fact, short periods of loading in nonphysiological directions appear to be most osteogenic."

It may, in fact, require *unusually* severe mechanical loading, and one occurring early in life as well, to provoke a significant response. "Under normal conditions, however, there appears to be a broad range of physical activity in which bone is relatively unresponsive to changes in loading history. With *some* repeated loading, bone hypertrophy can be pronounced" (Carter, 1984: S19).

A further, and likely crucial complication in evaluating the role of activity on bone geometry is, not just understanding the nature of the activity, but the *specific* bone site affected. Lanyon et al. (1979) observed from strain measurements of bone cortex that responses varied animal to animal, bone to bone, and even site to site within the same element. They thus suggested that "variable but crucially site-specific strain thresholds may exist throughout the skeleton" (Lanyon et al., 1979: 546).

This marked regional variability in bone response has also been observed by Carter (1984), Silman and Hochberg (1993), and Kuiper et al. (1997). Carter concluded from mixed results of experimental studies that "The strain remodeling response of bone must be, therefore, either (1) very imprecise, (2) precise but site specific, or (3) both site specific and imprecise" (Carter, 1984: S20). Carter favored the last alternative, a conclusion not very encouraging to osteologists seeking to make *specific* behavioral inferences from bone geometric variability (see below).

Moreover, there could well be metabolic influences, especially as differentially affected by diet. While considerable attention has been directed at understanding basic bone physiology/metabolism in both clinical (Dempster, 1992; Marcus, 1992) and archaeological contexts (Sandford, 1993), not much work has been done regarding potential effects of diet specifically on bone geometry. Some recent work (e.g., Daly et al., 1997) has shown that dietary calcium intake is correlated with bone mass as well as bone architecture. Since consideration of diet has obvious import for understanding bone changes linked achaeologically to subsistence changes (see below), Bridges (1989b) as well as others have suggested more research is urgently needed to address the potential role of diet.

Quite clearly, then, in addition to activity, there are a variety of factors influencing bone remodeling and, consequently, bone geometry.

Currey (1984: 245) succinctly observed that "It is inconceivable, on the other hand, that the shapes of bones could result *merely* from mechanical adaptive loads placed on them."

Clinical investigators have just begun to actively assess the array of potential etiological factors influencing bone geometry. From provisional results a multifactorial model is evident, emphasizing, at minimum, both mechanical and endocrine factors. Such considerations prompted Lanyon to conclude:

> However, it is only by considering the interactions of these influences on the behavior of the relevant bone cell populations, that the course, pattern, balance, and objective of remodeling throughout the skeleton can be understood (Lanyon, 1984: S61).

Carter similarly argued:

> These observations open the possibility that bone atrophy and hypertrophy are controlled by different mechanisms. Therefore, two (or more) complementary control systems may be involved in the regulation of bone mass by bone cyclic strain histories (Carter, 1984: S19).

ACTIVITY AND BONE GEOMETRY

While acknowledging potential complexity in the etiology of bone remodeling, those osteologists who have pioneered this approach nevertheless prefer to emphasize mechanical explanations. A representative view was expressed by Bridges: "Variation in cross-sectional dimensions and strength are most likely to be related to habitual activity patterns" (Bridges, 1989a: 388).

This orientation is understandable, given very similar motivations (and biases) in the assessment of OA. Moreover, there are some data, both from experimental work with animals and specific epidemiological surveys of human samples which provide *at least partial substantiation* for a mechanical interpretation of patterning of bone geometry. Such a perspective, however, should obviously not ignore the manifest complexities raised in the above discussion.

Experimental Animal Studies

Extensive research has been done to evaluate the effects of exercise on bone density and remodeling in experimental animals. Most of this work has focused on responses of bone quality/density rather than explicitly addressing bone geometry. A notable exception, however, is the work of Woo et al. (1981) in which bone "strength" in pigs was assessed through computation of 1st and 2nd moments of inertia. The results showed that, while bone density and biochemical content did not change, geometric properties responded clearly to long-term exercise, increasing cross-sectional area by 23% (as compared to controls). Further, the response was one that obviously increased the structural strength of the bone.

Lanyon and colleagues have also evaluated the effects of exercise through use of strain gauges in sheep (Lanyon and Baggott, 1979) and pigs (Lanyon et al., 1979) as well as by ostectomy of a portion of the ulna in roosters (Rubin and Lanyon, 1984b). Matsuda et al. (1986) also performed experiments with (immature) roosters, and in this latter study actually found exercise to reduce bone strength, as measured by 2nd moments of inertia, by up to 50–70% after eight weeks. After 12 weeks there were no substantial differences from controls, but the more exercised animals' bone was still more subject to fracture (than in controls). These results conflict sharply with those obtained by Woo et al. (1981), prompting Meade (1989) to suggest the findings by Matsuda et al. indicated, "Both geometric and material properties seem to be affected negatively by high levels of exercise during this period of *rapid growth*" (p. 230) [emphasis added]. In studies of human response to exercise, as well, this initial loss of bone volume and bone strength has also been observed (see below).

Lanyon and Baggott (1976) argued that mechanical factors obviously played a role, if not a clearly deterministic one:

> It is reasonable to suppose, however, that whatever the relevant mechanical influence on the bone is, it will be related to, if not consequent upon, the changes in surface bone strain which result from an animal's normal activities (p. 441).

Given the state of knowledge at the time, they went on to conclude:

> It would be optimistic to anticipate that a limited series of experiments would reveal much of the nature of the relationship between bone's mechanical input and its response in terms of modeling and remodeling. What we believe it has done is to indicate some of the questions which require answering and assumptions which must be challenged, if this relationship is to better understood (p. 442).

Twenty years later many of the same assumptions still persist and the same questions remain largely unanswered.

General comparisons across a variety of species have also not provided much illumination (at least, as relating to clarifying the utility of simple functional explanations). In a review of patterns of bone geometry in numerous, differently sized, animals, Rubin and Lanyon (1984a) found that in small *and* large animals there was a fairly uniform functional strain on the skeleton, regardless of size. They referred to this overall property as the "dynamic strain similarity" and suggested it manifested independently of body size or locomotory function. Rubin and Lanyon surmised that strain capacities, as well as bone geometry, might be most under control of *species-wide* characteristics.

Ruff and Runestad (1992), in analyzing their own data comparing structural adaptations in three primate forms, found only minimal variation across species in bone geometric properties: "However, in general, locomotor adaptations in the limb bones of the primates examined here (macaques, great apes, humans) are more apparent as proportional changes in bone length to body mass than as cross-sectional proportions to body mass" (Ruff and Runestad, 1992: 414).

Neither of these more general comparative studies addressed specifically individual plasticity (the *main* issue of the other studies reviewed here). However, the overall trends indicate that perhaps most of the geometric properties inferred from skeletal structure are under broad, species-wide, adaptive control. Individuals might then deviate clearly from this underlying modal adaptation only in *extreme* circumstances.

There are a number of constraints relating to the animal experimental data, most notably limitations imposed by studies restricted to captive animals. Both experimental samples and controls, of course, are confined to largely artificial environments; thus, the range of "activities" assessed is quite limited. Moreover, many of the studies either implanted strain gauges or performed partial ostectomies, both requiring surgery. The consequent effects of such intervention on bone remodeling and/or exercise patterns in the recovering animals are difficult to assess. Extrapolation of this research to humans expressing "normal" ranges of activity is thus problematic.

Effects of Activity: Human Clinical Investigations

Certainly, the most appropriate data from which to extrapolate to human osteological contexts are those directly related to contemporary human samples (with *known* activity patterns). Not a great deal of focused research has been done specifically evaluating the relationship of bone geometry and activity. As noted above, considerable work has been carried out on older adults, especially focusing on bone loss in women. Such investigations, however, have only limited value for archaeological samples in which most adults usually did not reach an age where post-menopausal osteoporosis became a significant factor. Moreover, osteologists have concentrated their attention, not so much on endocrine-related factors, but more specifically on ones presumably related to *activity*. And, as seen above, this aspect acts most dramatically quite early in life.

Thus, there is a notable "gap" between those data clinicians have traditionally gathered and what is required to help establish a precise activity-mediated mechanism. It is not surprising, however, that such information has, to date, been scarce. As with enthesial reactions, the immediate implications of "normal" bone remodeling in children and younger adults are not considered a serious health problem. Moreover, data collection involves usually some (slight) risk to participants, in the form of radiography.

Nevertheless, a few systematic studies have been completed and thus provide some useful provisional insights. Nilsson and Westlin (1971) compared 64 fairly advanced athletes with controls (although the latter were not generally sedentary). Bone density of the distal femur was assessed through photon-absorption, and results showed increased density among athletes. Moreover, the differences were most apparent among those engaged in activities requiring higher active weight bearing, as swimmers did not show greater density in comparison to controls. Nilsson and Westin concluded that "This finding indicates that even moderate physical activity may increase the bone mass" (p. 182).

In another study, of middle aged women (aged 35–65 years), Smith et al. (1984) found that in those individuals who exercised regularly (three times/week) mineral mass eventually increased, as compared to sedentary controls. However, reminiscent of results seen in some animal studies, the bone content (as measured by photon-absorbtiometry) actually *decreased* in the exercising women during the first year. Indeed, in these more active subjects, it took almost three additional years to make up this initial loss.

A recent intriguing investigation analyzed bone density in children (aged 10–16 years) who were engaged in different types of sports (Grimston et al., 1993). The researchers divided the sample respectively into what they termed "impact loading" activities (running gymnastics, tumbling, dancing) *vs.* "active loading" (i.e., less weight-bearing, as in swimmers). Bone density was ascertained in L2–L4 as well as for the femoral neck. In carefully matched samples (for age, sex, stage of puberty, body weight, and race), results showed significantly more bone mineral density in the femoral neck for both sexes combined (and in the lumbar spine when contrasting males separately) for participants in impact sports. From these data, the authors concluded:

> The results of this study demonstrated significant differences in bone mineral density measures between children matched for race, gender, stage of puberty, and body weight as a function of mechanical loading regime. Specifically, children competitively involved in sports producing impact loads to the skeleton of greater than or equal to 3 times body weight had significantly greater femoral neck BMD than children involved in the non-weight-bearing (active load) sport of swimming (Grimston et al., 1993: 1207).

Three recent well-controlled studies evaluating bone mineral as related to exercise in female samples provide further useful insights. Lee et al. (1995) assessed 62 female college athletes (engaged in volleyball, basketball, swimming, and soccer) as compared to two groups of controls (moderately active and sedentary). From bone scans of total body, femur, and lumbar (L2–L4) BMD, higher bone mineral density was found in athletes, and, most especially, in more "active" impact-loading sports (i.e., volleyball and basketball). Moreover, and particularly applicable to ongoing research by anthropologists (Bridges, 1989a,b; Ruff et al., 1992; Trinkaus et al., 1994), Lee and colleagues focused especially on patterns of *bilateral asymmetry*. Here, results indicated more asymmetry in the upper limbs, and the only group not showing significant upper limb asymmetry was swimmers. Moreover, consistently the right arm had higher bone density, except, again in swimmers (where the upper limbs were equal).

A potential underlying complicating factor, however, was also observed by Lee and colleagues. In many of the controls, both those engaged in moderate levels of activity *and* the more sedentary individuals, there were also significant asymmetries, in fact, almost to the same degree as observed in active sports participants. Perhaps, thus, degree of asymmetry might reflect more hand dominance and *not* activity level. Of course, in extreme cases, the situation may be less ambiguous (see below).

In another study, women (ages 25–41) engaged in exercise (walkers and aerobic dancers) were compared to more sedentary controls (total $N = 93$) (Alekel et al., 1995). Lumbar spine, proximal femur, and total femur BMD were compared (controlling for a variety of confounders including height, age, body weight, and calcium intake). Results showed that both walking and aerobic dancing contributed to higher BMD, but as both forms of activity produced similar results, *type of exercise* was not a contributing factor.

Two other recent investigations have further assessed the influence of activity in growing males. Daly et al. (1997) compared bone mass at several skeletal sites utilizing ultrasound. Comparisons were made between elite gymnasts and "normoactive controls," and results showed greater bone mass in the gymnasts for the calcaneus, distal radius, and proximal phalanx. Moreover, the use of broadband ultrasound attenuation "appears to provide a safe, noninvasive means of comparing the skeletal status of exercising and normal children" (Daly et al., 1997: 401).

Nordström and colleagues (1997) also evaluated the effects of exercise in young adult men (average age = 25 years), but concentrated not on elite athletes, but on individuals engaged in low and moderate levels of physical exercise. Nevertheless, even in such relatively inactive individuals, "at the time of peak bone mass attainment, physical activity is an important predictor of the clinically relevant proximal femur in young men with a low or moderate level of physical activity" (Nordström et al., 1997: 332).

The findings of the above four studies, however, are at variance with results reported by other investigators. For example, two studies of the effects of exercise in post-menopausal women (Aiola et al., 1978; Krølner et al., 1983) failed to find consistent differences among the individuals who exercised as compared to controls. The latter study, however, while not finding a significant difference in mineral content of the radius, did find such a difference in the BMC of lumbar vertebrae in exercising women. Further, it also could be argued that the studies were short-term (the former just lasting one year) and included quite small samples. Moreover, the effects of exercise in *older* individuals is, in reality, not the main focus of osteologists trying to reconstruct prehistoric behavioral patterns.

Given these limitations, the most significant contradictory results. perhaps, are those reported by Myberg and colleagues (1991). In their study of bone bending stiffness, elastic modulus, cross-sectional moment of inertia, and mineral content as well as external metrical determinations of the ulna, "activity did not correlate with any variable." The sample, however, was again quite small ($N = 48$) and the assessment of activity patterns, as defined by questionnaire, could also be critiqued.

Finally, Lindholm et al. (1995) evaluated bone mineral content in young (former) gymnasts compared to controls. However, no difference was found between the former athletes and the controls. The authors were also concerned with the potential long-term (negative) effects of intense activity and reduced dietary intake in young women.

In this light, it is important to note that such levels of activity in young females *might* produce mixed effects relating to the balance of bone in the skeleton, for example, by delaying puberty.

Perhaps, what is most sobering, is that for the bone quality measurements (BMD) assessed here, just a few years following intense training by these young women, the athletes were *not* different from sedentary controls. Whether their skeletons showed other features that might indicate the gymnasts' radically different loading histories, is not ascertainable from the information reported. However, it is not at all certain that there are unambiguous permanent skeletal manifestations produced in young females engaged in such levels of activity. Thus, skeletal biologists searching for such skeletal markers might be frustrated, not for lack of determination or enthusiasm, but for lack of specificity in the underlying etiological mechanisms; additionally, these mechanisms may be more sex-specific than previously recognized. It could also, however, be reasonably argued that some 'markers' might be more useful than others. In this light, assessment of cross-sectional geometry is quite likely considerably better than measurements of bone mineral, at least as reflecting *general* activity levels.

There have also been a few recent investigations explicitly addressing the influence of handedness on bone dimensions. Schell et al. (1985) collected anthropometric data (arm and calf circumference) on 135 adolescents and found upper limb asymmetry associated with handedness, especially in right handed boys. Left handers, however, did not show a significant difference, nor did girls (right or left handers). Schell and colleagues suggested the sex difference in degree of asymmetry might be explained genetically, i.e., "females are genetically better canalized and thus show less asymmetry in body dimensions" (p. 321). They also noted, however, that the pattern was consistent with muscle use hypertrophy, "if one further assumes that these adolescent females are not as involved as their male counterparts in activities that stress the side of the body that is preferred for use" (ibid.). Schell and colleagues thus cautiously concluded: "At this time, the explanation of asymmetries in terms of muscle use hypertrophy is advanced, primarily for the purpose of discussion and to suggest further research" (ibid.).

A more recent, larger, and better controlled study of the influence of handedness on upper limb asymmetry was reported by Roy et al. (1994). From hand radiographs of more than 1000 adults (of known hand preference), significant differences were found in external dimensions (and cross-sectional geometry) between the dominant and non-dominant side. These differences were again most marked among male right handed individuals (although sample sizes for females and left handers were small).

It could also be noted that a third study (Allander et al., 1974), while not evaluating bone dimensions or quality, also found gender effects in range of motion associated with handedness (again, the most significant differences were in right handed males).

Correlations of Bone Remodeling Effects with Specific Activities

The available data relating specific activity to aspects of bone remodeling, especially as influencing cross-sectional geometry, are extremely sparse. Most certainly, at present, such specific information pales in comparison with that collected for OA (see Chapter 3). Given the probable strong age-related dynamics of bone remodeling, not surprisingly, most of those data that do exist come from investigations of young athletes, especially tennis players and runners.

Two groups of tennis players have been investigated and, as noted above, one of these groups has yielded particularly useful information (Jones et al., 1977; Priest et al., 1977; Ruff et al., 1994). The other report by Montoye et al. (1980), however, is also illuminating. Montoye and colleagues in their analysis evaluated older tennis players (mean age = 64 years) and found an increase of 13% in bone mineral content in the playing arm as compared to the contralateral limb. These results are especially interesting, for while not indicating the extreme degree of asymmetry found in tennis players who compete at an advanced level, and who began at a very *young* age, the level of asymmetry is still surprising in a sample of this age.

The more systematic set of studies of a group of elite tennis players (Chinn et al., 1974; Priest et al., 1974; Jones et al., 1977; Priest et al., 1977) has, without question, been the most influential clinical data used by osteologists to support a functional hypothesis concerning aspects of bone remodeling. As good radiographic evidence was collected initially (with biplanar views), further analyses of cross-sectional geometric features could be accomplished (see particularly, Ruff et al., 1994).

Several aspects of this sample of elite athletes as well as the patterns of bone remodeling they show are highly interesting. In the initial analysis of asymmetry of the upper limbs (Jones et al., 1977; Priest et al., 1977), 84 athletes were investigated (54 men; 30 women). The most striking findings concerned hypertrophy of the playing arm, showing an increase in humeral cortical thickness of 35% in males and

28% in females (Jones et al., 1977); significant asymmetrical differences were also found in the radius and ulna, but were less marked than those observed in the humerus (Priest et al., 1977).

As discussed above, further detailed analysis in which the crucial variables of duration of playing time, age at which X-rays were taken, and most especially age at which intense activity began, was performed by Ruff et al. (1994). Moreover, Ruff and colleagues also assessed external humeral (epiphyseal) dimensions, and also found significant hypertrophy on the playing side, but not as striking asymmetries as found in diaphyseal cross-sectional parameters. Moreover, the external epiphyseal dimensions were *not* age-related.

The findings from the array of expert analyses of this particular data set of elite tennis players have, understandably, had a wide influence for osteologists. Indeed, much of the more sophisticated analyses have been directly contributed by anthropologists (most notably, Christopher Ruff). The wider applicability of results derived from these studies, however, has not yet been clearly established. Carter (1984) noted, for example, from a general perspective that:

> Clinical investigations have shown that rigorous physical exercise can result in above average bone mass. These studies, however, have tended to be focused on individuals who engage in very rigorous activities and have begun these activities at an early stage of skeletal maturity (Carter, 1984: S22).

Similarly, Trinkaus and colleagues caution that it is "unclear to what extent nonhuman animal experimental protocols or professional athletes reflect *real-life situations*, either for the animals or for prehistoric humans, and the changes evident in clinical contexts, by their very nature, reflect abnormal situations" (Trinkaus et al., 1994: 2) [emphasis added].

In a balanced and concise fashion, these authors go on to conclude that, while the activities displayed by elite tennis players can "hardly be considered representative of a 'normal' subsistence-oriented population, they provide an extreme but nonpathological example of the nature of responses in humeral diaphyseal dimensions to marked unilateral activity patterns" (ibid.).

A few other contemporary samples of individuals participating in athletic activities have also been evaluated for consistent patterns of bone remodeling. Dalen and Olsson (1974) found in a sample of long-distance runners (aged 50–59, who had been running for at least 25 years) higher mineral content (as compared to controls), especially as reflected in trabecular bone. Interestingly, bone mineral was increased not just in the lower appendage, but in the upper appendage as well (humeral head, distal radius and ulna).

Lane et al. (1986) in a more controlled study, also of long-distance runners ($N=41$), found an even greater increase in BMC (+40%), as compared to controls. The assessment was performed using CT of L1, and thus provides a useful basis for potential further analysis. The bone content findings from this long-term longitudinal study (while mostly addressing OA; see Chapter 3), however, are not entirely unambiguous. Further analysis (Lane et al., 1990) has demonstrated an interesting phenomenon: there was a significant *loss* of bone mineral content in runners after they quit or reduced their activity.

Thus, yet another possible confounder is added to interpretation of this, and perhaps other, indicator(s) of bone status. The length of time prior to ascertainment that the activity ceased or significantly altered might be an important influence on eventual bone mineral status (in archaeological contexts this would translate as the time prior to death).

A few other focused clinical studies have provided provisional and potentially useful data. In a radiographic study of ballet dancers Schneider et al. (1974) found increased "cortical thickening" (especially in the metatarsals of female dancers). Claussen (1982) reported marked ulnar hypertrophy in professional rodeo performers (bareback riders); however, the condition here was attributed to "chronic irritation" (circumferential periostitis, secondary to trauma). Moreover, Bollen and Wright (1994) observed increased bone density on the proximal and middle hand phalanges of rock climbers, and, lastly, Daly et al. (1997) found "positive adaptive response" in the skeleton of young male gymnasts.

OSTEOLOGICAL STUDIES RELATING BONE GEOMETRY TO ACTIVITY

In the last decade considerable interest has developed among osteologists in evaluation of bone geometric features, and, most especially, in interpreting such osseous changes as relating to activity. A few of these studies have focused specifically on fossil hominids. An initial attempt, along these lines, was that by Kennedy (1985), who evaluated cortical bone thickness in *Homo erectus*. Most prominently, however, the collaborative research of Erik Trinkaus, Christopher Ruff, and their colleagues (Ruff et al., 1993; Trinkaus, 1993; Ruff et al., 1994; Trinkaus et al., 1994) has contributed innovative new applications to the investigation of activity in earlier hominids.

Trinkaus (1993), as in much of his other research, is interested in reconstructing behavioral patterns among Neandertals. In this particular study, femoral neck angle was examined and compared with several more recent populations (4 samples of foragers, 11 samples of agriculturists, and 6 samples from modern 'urban' contexts). Trinkaus explicitly recognized that developmental plasticity most likely influenced the eventual phenotype and, accordingly, argued the trait "should provide a relative indicator of differential *juvenile* and *early adolescent* activity levels, even if they inform us little about adult locomotor levels" (Trinkaus, 1993: 400). The pattern in Neandertals, with unusually low femoral neck angles, was interpreted by Trinkaus to indicate these hominids were highly active, i.e., mobile. At a general level, such a conclusion fits neatly with other data available regarding Neandertal adaptation. However, the general pattern for the varied populations of modern *H. sapiens* were not as clear. Trinkaus found considerable variation within and between groups in femoral neck angles, the only significant trend being an increase in angle among urban groups. Trinkaus concluded this pattern as well suggested "differences in activity level of immature individuals" (Trinkaus, 1993: 405). Again, such an interpretation is certainly a good hypothesis, but, beyond possible activity differences in subadults, contemporary urban groups differ systematically from the other samples in a variety of parameters (including diet, terrain, footwear, even, perhaps, customary infant carrying position). Thus, at present, this interpretation must be taken as speculative. It should be noted, however, that Trinkaus did systematically review the available clinical literature and substantiated his functional explanation to as strong a degree as was currently possible (see particularly, Houston and Zaleski, 1967).

Furthermore, the similarity found in the femoral neck angles of the early *H. sapiens* remains from Skhul and Qafzeh to that seen in modern urban groups is not easily explained. Bridges (1996b), however, argued the high neck angles in these ancient Near Easterners reflected low levels of activity during growth and development (a good explanation, assuming that neck angle variability results *primarily* from activity and has done so consistently across populations samples for the last 100,000 years).

In a further series of papers investigating cross-sectional geometry in fossil hominids, Ruff, Trinkaus, and Walker (Ruff et al., 1993; 1994; Trinkaus et al., 1994) presented systematic data contrasting different species of genus *Homo*. Rigorous allometric controls were utilized, and comparisons were made across species (i.e., covering considerable time). As in their other research, the investigators innovatively used these data to make broad generalizations, most notably that mechanical loading during hominid evolution has reduced through time. Ruff et al. (1993) suggested that shaft dimensions, as compared to articular size, were more likely responsive to mechanical loading, and thus argued that overall body size reduction was *not* the cause of the observed diaphyseal patterning; further, such changes in shaft dimensions resulted more likely from shifts in activity rather than from genetic changes within *Homo*. However, it could also be argued that diaphyseal dimensions and those of articular ends are under control of *different* genetic mechanisms and thus need not respond in concert, especially when considering long-term adaptive trends. It remains to be seen what influence natural selection might have contributed to these patterns, perhaps in tandem with, or partly independent of activity. Nevertheless, Ruff and colleagues have been quite firm in their assertions: "Whether observed variations in diaphyseal robusticity within *Homo* are entirely developmental or in part genetic in origin, it seems *almost certain* that the primary adaptive value of these changes was mechanical in nature" (Ruff et al., 1994: 47) [emphasis added].

Most of the energies of osteologists in utilizing bone geometric data to infer activity have concerned more recent human populations. In this sense, this perspective is highly analogous to the attempts by skeletal biologists to infer behavior from OA, enthesial reactions, and the other osseous manifestations discussed here. Compared to these earlier applications, in the bone geometry approach, the methods are more recently developed, have begun from a more concentrated research focus (of a restricted number of investigators), and, further, have attempted more rigorously to specify the theoretical/functional foundations. Nevertheless, many of the same limitations as with the other approaches also apply.

Techniques of analysis to assess cross-sectional geometry in archaeological bone include direct sections (Ruff and Hayes, 1983a,b; van Gerven et al., 1985; Ruff and Larsen, 1990), standard (biplanar) radiography (Ruff, 1987; Fresia et al., 1990), and CT scans (Brock and Ruff, 1988; Bridges, 1989a). In some cases external bone measurements have also been used (Ruff, 1987; Bridges, 1989a). Even more sophisticated techniques have been more recently attempted, beginning with gross bone sections, but then projecting and digitizing photographic images (Larsen and Ruff, 1994). In this last application, calculations could be performed directly, using a computer interfaced with the digitizer.

The techniques vary in accuracy, especially in standard radiography (even when using multiple views). However, Trinkaus et al. (1994)

have argued these inconsistencies are perhaps not such a problem, when the objective is to assess the degree of bilateral asymmetry; in this case the errors should produce a consistent, and thus manageable, bias.

As noted above, much of the stimulus for this research has come from the research by Ruff. In fact, working from the outset in close collaboration with clinicians (Ruff and Jones, 1981; Ruff and Hayes, 1983a,b), Ruff has demonstrated the obvious benefits of such productive professional interaction. Ruff's first report (Ruff and Jones, 1981) concerned the degree of bilateral asymmetry in the Ala-329 materials from Central California (the same sample as utilized by this author for OA and trauma; see Chapters 4 and 6). In this early work, the most notable finding was a greater degree of asymmetry in the upper as compared to the lower limb. These perspectives were further refined and expanded in much more detailed analysis of osteological remains from Pecos Pueblo (Ruff and Hayes, 1983a,b).

In these reports the first mention is made of the variability in femoral cross-sectional shape (A/P *vs.* M/L; i.e., in the degree of roundness of the femoral shaft). Initially, the interpretation was quite conservative; while the authors noted anecdotally that some activities, e.g., stair climbing and jogging, *should* increase A/P loading forces about the knee, and presumably making the femur less round, they concluded: "The behavioral implications of this apparent difference are complex and cannot be completely determined from a purely structural analysis" (Ruff and Hayes, 1983a: 375–376).

They further went on (Ruff and Hayes, 1983; 1983b) to delineate some other complexities, particularly the differential sex-based influence in pelvic orientation (i.e., a potentially species-wide factor). Nevertheless, from selective gleaning of ethnographic sources *hinting* at potentially more long-distance running in males, the authors postulated this behavior as a likely contributing ('causative') factor. The inadequacy of such ethnographic sources for establishing behavioral correlations is obvious and has been critiqued in prior chapters. Such data, by no means, provide an adequate *test* for any functional hypothesis.

What is perhaps most disturbing is that this very tentative interpretation of the etiology of altered femoral shaft shape, i.e., A/P expansion due *specifically* to long-distance travel, has been repeated numerous times elsewhere (e.g., Fresia et al., 1990; Larsen and Ruff, 1994; Bridges, 1996b). Given enough repetitions, the hypothesis now appears to be accepted (uncritically) as established. But, to my knowledge, there is not one single clinical study which has demonstrated a relationship between activity levels and consistent changes in the shape of the femoral diaphysis. Given this lack of even a general functional correlation, the claims that such changes are the result of a *specific* activity are even more speculative. Of course, the hypothesis may eventually be confirmed, and would prove the suggestion to be a brilliant (and most useful) insight. Clearly, however, until substantiation is forthcoming, judicious restraint is called for.

Ruff and Hayes (1983a) further cautiously suggested differences in female femoral diaphyseal shape "may indicate differential mechanical stressing of the lower limb bones during particular activities, *possibly* partially sex specific" (p. 399). Ruff (1987) later expanded his comparative samples and drew a somewhat firmer conclusion.

> ...these findings indicate that Pecos males and females engaged more often in sex-specific activities which placed different bending loads on the lower limb bones in the region around the knee than recent U.S. Whites. If the Pecos sex difference near the knee reflects more running and long-distance travel by males, than the reduction of such a difference in the modern U.S. sample may indicate less sexual dimorphism in the time spent in these activities (Ruff, 1987: 397).

Indeed, from consideration of eight other archaeological samples (and one from a dissecting room), Ruff (1987) postulated a quite widespread general relationship: that hunter-gatherers consistently displayed less round, but A/P more expanded, femoral diaphyses than agriculturists (or modern groups). Further, this variation could be explained by a simple behavioral shift, that is, hunter-gatherers engaged in more long-distance travel than in these other groups.

As above, this even wider behavioral hypothesis has attracted considerable support among other workers in the field (e.g., Fresia et al., 1990; Larsen and Ruff, 1994), but, obviously, it is based upon the same speculative foundation. Why the consistency in patterning? Is there a behavioral marker, amazingly consistent in its expression and beautifully elegant in its simplicity? Perhaps. But, perhaps, also it is a little too good to be true. It is difficult to imagine that activity pattern would be so consistently linked to such extremely rough categories of subsistence adaptation. Bridges has critically reviewed this topic (in her excellent discussion of OA) and made the point of the great variability of activities within the traditional categories of subsistence economy (Bridges, 1992). In fact, in my own research of so-called hunter-gatherers (or "foragers") the differences between the arctic-hunting Inuit, on the one hand, and the semi-sedentary Central California Indians on the other, could scarcely be more dramatic. That these groups would share activity-based similarities in bone shape would not be expected, and is not easily explained. Thus, to simply assume the most significant portion of the variance in bone diaphysis shape is accounted for by activity, is not obvious.

Ruff (1987) and Bridges (1989b) have noted that potential systemic effects (e.g., diet) also do not easily explain the observed pattern of shaft dimensions. Most particularly, they have argued that the changes are manifest most prominently in the distal femur, but *not* near the hip. In this view, more systemic factors, such as diet, *should* be expressed more uniformly throughout the skeleton; the more localized distribution of such shape patterning is thus thought to reflect mechanical influences.

However, it is not evident that all areas of the skeleton would necessarily respond similarly to dietary influences. As suggested by clinical researchers, bone remodeling can be highly site specific as well as frustratingly imprecise (Carter, 1984; Kuiper et al., 1997). Thus, it is quite possible that dietary influence, perhaps interacting with genetic influences and activity, could produce differential effects on the skeleton. The findings of extremely broad patterns of distal femoral diaphyseal shape in groups differing (evidently) in diet is, at minimum, highly suggestive.

It should be pointed out that Ruff has provided good *theoretical* support for his contentions, especially as documented in his 1987 article. Here, he described the probable biomechanical effects on certain muscle groups (and thus, ostensibly, on bone). What is lacking, however, is documentation in human samples with *known* behavior of the specific effects of differing amounts and types of mechanical loading (just as in the attempts to posit enthesial reactions to represent behavioral markers; see Chapter 5). Moreover, alternative hypotheses and multifactorial interpretations are frequently overlooked, in the pursuit of activity-based explanations.

Perhaps, because such contemporary documentation is still so elusive, analogies have been mined, instead, from the imprecise ethnographic record. Accordingly, Ruff was able confidently to conclude: "Therefore, it is apparent from ethnographic observations of living and recent peoples that sex differences in relative mobility are indeed greatest in hunter-gatherers and decline with the adoption and intensification of agriculture" (Ruff, 1987: 407).

Ruff also was well aware of the significant gap in finding adequate clinical data in commenting: "*comparable* data are not strictly available for industrial societies" (ibid.). Of course, *if* there were such contemporary data obtainable, they would not really be *comparable*. In well controlled studies they should be *much* better, even if they tend to show the process to be much more complex then hypothesized (see the varied clinical interpretations of OA in Chapter 3 for illustration of this point).

It is generally accepted that, excepting, of course, a few specialized sports-related studies (most notably of tennis players—see above), little relevant clinical data exist relating bone quality to normal ranges of behavior in humans. However, the available data base may not be quite so deficient as feared. There *are* some potential sources of highly useful data from investigations completed or still in progress (see below).

Another fairly early attempt in the interpretation of bone geometric features in archaeological materials was that by van Gerven and colleagues (1985). In this analysis of a Sudanese Nubian skeletal sample, the amount of cortical area as well as bone strength (as measured by 2nd moments of inertia) were calculated. The main objective, however, was *not* to relate these bone features to activity, but to assess biocultural stress at potentially sensitive stages of development; these results contributed to a broader perspective applied to this sample, which from other osseous indicators, such as growth rates, enamel hypoplasias, and porotic hyperostosis, has been interpreted as under considerable stress.

The functionally related hypothesis relating bone geometry to activity has, nevertheless, among some other osteologists attracted much interest. For example, Brock and Ruff (1988) utilized this approach to interpret behavioral patterns in three southwestern groups, and Bridges (1989a) did a similar analysis in a large sample from northwestern Alabama. In this latter study, especially, some of the logistical constraints in this type of analysis are readily apparent. From an initial sample of more than 1900 individuals, Bridges reduced the available materials (of known age and sex) to a smaller subsample ($N = 140$ for Archaic groups; $N = 126$ for Mississippian groups). Further, for the detailed analysis derived from CT scans, the final samples (combined) consisted of 49 humeri and 42 femora. Those osteologists who *begin* with limited materials should regard their potential venture into sampling oblivion with some caution.

Bridges' main objective was to contrast hunter-gatherers with agriculturists, and she stated her goal quite succinctly: "The objective was to reconstruct activity levels and, if possible, *specific* activities in different groups and to relate them to subsistence economy and sexual division of labor" (Bridges, 1989a: 385).

However, *contra* the results of Ruff, Bridges observed that agriculturists had larger (and "stronger") long bones than did the hunter-gatherers. Accommodating an activity-based interpretation, Bridges explained this pattern in northwestern Alabama as resulting from *increased* physical demands among hoe agriculturists. "These changes are related to alterations in both level and type of activity accompanying the transition to maize agriculture. *For this region, maize agriculture was more physically demanding than hunting and gathering*" (Bridges, 1989a: 392) [emphasis added].

What is immediately obvious here is that activity is *assumed* to explain all the variation present, regardless of the complexities of the patterns expressed. Moreover, what is being assumed is *exactly* what such analyses should be trying to demonstrate. In other words, it is not *known* what the activity patterns were in prehistoric Alabama (or, for that matter, anywhere else). Ostensibly, this is what one major aspect of skeletal analysis is trying to ascertain.

Imaginatively constructed "best fit" activity-based scenarios to explain skeletal variation can be used to accommodate and correlate *any* pattern. Exactly as critiqued for similar ventures in explaining the distribution of OA in archaeological groups, such reasoning becomes irresistibly circular.

It should be pointed out that Bridges has also been quite forceful in arguing that *specific* activities are most difficult to infer from skeletal manifestations. She has made this point in regard to OA (see Chapter 4) as well as for bone geometric features, reaffirming that "it is difficult to pinpoint specific activities" (Bridges, 1989a: 392).

Fresia et al. (1990) as well as Ruff and Larsen (1990) applied these same perspectives to interpretation of activity patterns in archaeological groups from the Georgia Coast. Here, as in the other studies, the variation in bone shape as well as patterns of asymmetry were uniformly interpreted as resulting from differing activity patterns. For example, Fresia and colleagues argued that, for asymmetry patterns, "these changes reflect alterations in mechanical loading, and thus activity patterns in this region" (p. 121). Moreover, agreeing with Bridges (1989a,b), these authors further suggested that reduced upper limb asymmetry in coastal Georgia females was likely due to greater grinding activity practiced by women. However, why the pre-agricultural group in this region was more asymmetric was not explained; moreover, why pre-agricultural males in Alabama were more bilaterally asymmetric than in Georgia also was not obviously related to straightforward behavioral explanations. Fresia and colleagues, nevertheless, provide yet another imaginative scenario: "... these individuals may not have been engaged in hunting to the same extent as the inland Tennessee River Valley group [i.e., N.W. Alabama], especially hunting involving use of the atlatl" (p. 129).

Fresia et al. (1990) further postulated that, since subsistence activities along the Georgia Coast included more "activities associated with the acquisition of marine resources . . . . Therefore it stands to reason that males of this population would have reduced levels of bone asymmetry relative to inland populations" (ibid.).

Such explanations are highly seductive in their seeming rationality and clarity. Yet, they are also based on speculation as well as highly simplistic assumptions. For example, the assumption was that *only* certain activities associated with terrestrial hunting would have produced asymmetries in bone geometry. Could not differing activities produce similar bone manifestations? What if the opposite patterning had been observed in Georgia as compared to Alabama, so that coastal Georgia males were more asymmetric than inland groups? It would then be easy to speculate regarding the purported influences of "throwing harpoons," "casting nets," etc. The inferred activity, in either case, is the product of a "just-so" story.

Ruff and Larsen (1990) also evaluated these same materials from coastal Georgia. It is worthy of note that this (and the above study by Fresia et al.) are part of a sophisticated, long-term, multidisciplinary effort by Larsen and colleagues to understand the bio-archaeology of this region. In addition to many other innovative applications of contemporary skeletal biology, some aspects of this research have been directed at reconstructing activity patterns, utilizing OA patterning (see Chapter 4) as well as bone geometry. It is to the credit of all the investigators involved in this research that the most recently developed methodologies were attempted. The accompanying risks in using pioneering, and oftentimes, untested, approaches, however, must also be recognized.

Ruff and Larsen clearly recognize many of these limitations, for example in commenting on the conflicting results obtained in studies of Georgia Coast individuals compared to those reported by Bridges (1989a; see above).

> Regardless, those results of the two studies that are contradictory serve as a warning against overly simplistic reconstructions of "global" changes occurring with environmental transitions which are based on an analysis of only a single geographic area (Ruff and Larsen, 1990: 119).

Ruff and Larsen further cautioned,

> These interpretations should be treated as *hypotheses which await further testing* using both additional samples and other techniques of skeletal analysis (p. 120) [emphasis added].

These conclusions are noteworthy for their circumspection and recognition of the need for further confirmation. Nevertheless, there were contained within this analysis some quite specific behavioral inferences. Most especially, Ruff and Larsen suggested from five post-European contact male skeletons (with increased femoral A/P dimensions as compared to other contemporary males):

> ...these results indicate that among some of the males in the contact period, long-distance travel had increased greatly from the average levels documented prior to contact, while in other males and all females long-distance travel either stayed the same or declined (p. 112).

From perusal of historical records Ruff and Larsen conjectured these males may have been forcefully recruited by the Spanish to make long distance trips. "Thus, changes in cross-sectional geometry are consistent with *certain* behavioral changes among these populations *known* to occur during the mission period" (ibid.) [emphasis added].

Once again, what is "known" here is not at all assured, and the selective culling of "certain" behaviors to accord with the skeletal pattern will almost always make for a *seemingly consistent* pattern. Moreover, it should also be noted that, given interpretations of the likely stages of the greatest skeletal plasticity, it would have required that these males would have been recruited as children or young adolescents; much of their latter lives would have then been spent in isolation as they journeyed (perhaps, ran) across the reaches of North America (irresistibly, images of *Forest Gump* come to mind). Sarcasm aside, the degree of specificity conjectured here is surprising and has been further employed by these researchers to explain the pattern of femoral diaphyseal geometry in some males from post-contact Florida (Larsen and Ruff, 1994).

In this latter study, the authors do acknowledge the likely contribution owing to changes in diet (on post-contact groups), suggesting a multifactorial explanation for changes in bone geometry, with results "likely represent(ing) a complex interplay of nutritional and mechanical factors" (Larsen and Ruff, 1994: 28). Along these same lines, Larsen and Ruff concluded:

> We do not argue that mechanical factors are solely responsible for the skeletal modifications observed in long bone cross sections reported here ... [Yet] the overall pattern is consistent with adaptations to *localized*, that is, mechanical factors, rather than *systemic* (nutritional) factors, *although the two certainly appear to be interrelated* (pp. 30–31).

The analysis of upper limb asymmetry done by Trinkaus et al. (1994) is among the most intriguing of any work yet done relating to interpretation of bone geometry. As discussed above, the orientation was an interpretation of behavior in fossil hominids, most especially among Neandertals. However, as relevant to the present discussion, data were also presented for five contemporary groups (archaeological, dissecting room, and tennis players). What stood out most dramatically was the extreme (humeral) asymmetry seen both in Neandertals (24–57% for different variables) and in the elite tennis players (almost an identical degree of asymmetry, 28–57%). All the other recent groups of *H. sapiens* showed only very moderate levels of such asymmetry, ranging between 5% and 14%.

It appears there may well be a baseline level of asymmetry uniformly manifested in most human populations. Trinkaus and colleagues, in fact, did argue this level of variation can be difficult to separate from underlying "developmental noise and fluctuating asymmetry." They do posit that lack of asymmetry *may* be due to bilaterally uniform activities. Only in extreme circumstances does the pattern appear to vary significantly, as in tennis players (who began playing rigorously early in life) and, fascinatingly, also among Neandertals. Of further interest, Aleut populations are also highly asymmetric in the upper limb (Berget and Churchill, 1994; Bridges, 1996b). They too may reflect remodeling effects of an extreme adaptation among arctic hunters, mirroring the unique OA patterning which has been well documented among the Inuit (see Chapter 4). Just what the Neandertals were up to has captured the imagination and creative talents of Trinkaus over the last several years.

For most contemporary groups, however, the biomechanical assessment of bilateral asymmetry might not contain the useful information that had been anticipated. Certainly, at present, this area of research has yet to demonstrate the *specificity* required to allow any confident degree of predictability.

In a more recent attempt further to understand the underlying contributing factors of bone shape remodeling, Ruff (1995) has considered the possible role of terrain. His results are most interesting, in that terrain appeared to influence femoral robusticity more than either sex or subsistence strategy. Ruff thus concluded: "These results suggest general levels of mechanical loading of the postcranial skeleton, at least in the lower limb, are more determined by the physical environment than by specific culturally linked behaviors" (p. 186). He, however, thinks "… orientations of loadings are more related to *sex-specific subsistence activities*" (ibid.) [emphasis added].

Appropriate restraint regarding the specificity of inferences is thus here emphasized by the leading advocate of this approach. While some other osteologists in recent publications have begun to reflect this same caution, there still remains a strong inclination expressed by several researchers to draw inferences that are nearly entirely limited to activity (e.g., Berget and Churchill, 1994; Larsen, 1995; Bridges, 1996b).

CONCLUSIONS

As noted at the beginning of the chapter, the study of bone (diaphyseal) geometric features is both innovative and potentially highly informative. Moreover, unlike much of the other research in paleopathology relating to activity/functional inferences, this method *begins* with sound biomechanical theoretical foundations. In addition, the relevant anatomical features are assessed systematically using established mathematical formulae, and results are thus presented quantitatively. Furthermore, new computer models are assisting in specifying key aspects of biomechanical loading and their likely influence on bone geometry (Mittelmeir et al., 1994; Carter et al., 1996). In all these respects, much of the confusion surrounding the reliability of assessment, as that seen in analysis of OA (and other purported behavioral indicators), is avoided.

The major limitations of this approach, in so far as it allows inferences relating to activity, are, firstly, the probable lack of both specificity and precision of bone response to activity. Secondly, whether even general levels of activity *might* be determined from geometric indicators, remains to be demonstrated. What is urgently required, of course, are studies that provide detailed data relating specific activities to specific changes in bone geometry (the same point belabored for OA studies, where so much has already been attempted, as well as for enthesial reaction investigations, etc.).

As an alternative to contemporary clinical documentation, some researchers have sought to use ethnographic data, again paralleling the history of anthropological applications of OA interpretations. For example, Ruff (1987) utilized ethnographic information to help support his contention that hunter-gatherers traveled further than agriculturists (and, in this sense, were thus more active than agriculturists). Bridges (1989a), on the other hand, used very similar ethnographic analogy to explain her contradictory osteological findings; here, she argued, for example, that agriculturists often worked *harder* than hunter-gatherers (as exemplified by the San). Experience with building such ethnographically based (and empirically limited) scenarios to explain OA patterning has provided osteologists with an unambiguous model of the ultimate non-productivity of this approach.

The only well-controlled contemporary clinical investigation providing at least generally reliable data on cross-sectional geometry in a group with known behavioral parameters is that of elite tennis players (Jones et al., 1977; Priest et al., 1977). This data set has stimulated much discussion and considerable insight, and it, understandably, is frequently referred to by osteologists. Indeed, anthropologically trained skeletal biologists have contributed much to a greater refinement in analyses of these data (Ruff et al., 1994; Trinkaus et al., 1994). Nevertheless, this is but one group of athletes performing one (extreme) type of activity. The lesson learned in OA investigations regarding the drawing of hasty conclusions from such restricted clinical data remind us all of the need for much more work of this nature.

Moreover, it is well-recognized by clinical researchers and osteologists that the tennis player sample represents an extreme case. What of the more typical ranges of activity practiced by most humans? Carter (1984) noted that such behavior may *not* be all that easily determined from skeletal remains: "Those activities may vary significantly from day to day or month to month without a strong influence on bone mass. There exists, in a sense, a physiologic 'band' of activity wherein bone tissue is fairly unresponsive to changes in loading history" (Carter, 1984: S20).

Besides mechanical loading, there are other potential influences on bone geometry that must be considered. These include, at minimum: sex (endocrine factors), diet, differential age influences, genetic factors, terrain, and even typical manner of locomotion. Regarding this last factor, a study by Lanyon et al. (1975) found (in one individual) lower limb strain measurements, while running, varied by up to 40%, depending on whether the subject was shod or unshod. The models needed to understand the ways in which geometric proportions of bone are modified must be more multifactorial than those typically presented. Certainly, a range of potential influences must initially be considered and adequate testing to control for these confounders performed, *before* the particular role of mechanical loading/activity can be seriously addressed.

Moreover, even given that mechanical factors are crucial, if not deterministic, they almost certainly do not function in a simple manner. Those clinical investigations thus far completed suggest that such factors as type of loading, duration, amplitude, age begun, age ceased, bone element involved, and even site within element all probably complicate interpretations of the observed pattern of variation.

Given these constraints, it appears that in many cases the confidence in behavioral inferences expressed by osteologists has gotten in front of the basic science. The motivations producing such enthusiasm are not difficult to understand. This ready willingness is frequently irresistible and has caught most of us at one time or another.

What can then be done to alleviate the current constraints? Firstly, there are some clinical data which *might* assist in providing some degree of confirmation relating to the role of activity. For example, there are several studies of OA of the knee in runners which have utilized radiography (sometimes from more than one view) (e.g., Puranen et al., 1975; Panush et al., 1986; Lane et al., 1990; Konradsen et al., 1990). Moreover, another knee investigation of individuals from a sports clinic context (compared to a hospital sample) utilized three radiographic views (Rose and Cockshott, 1982).

It has been noted that standard radiographic data can produce considerable error in deriving geometric parameters, especially 2nd moments of inertia. One solution (Trinkaus et al., 1994) has been to focus on bilateral asymmetry, thus systematically controlling for potential bias. Most of the focused epidemiological work mentioned above (and available elsewhere), however, most likely employ radiographic imaging from just one side.

Clearly, much work remains to be done. Osteologists have thus far been the innovators in establishing this new approach. Their continued contributions, in collaboration with clinical colleagues, are surely the best avenue to help substantiate its utility. Once sound confirmation of the reliability and specificity of geometric bone indicators is established, they could then become the best tool available to osteologists seeking to reconstruct activity patterns from ancient human skeletal remains.

Chapter 8

Conclusion: Defining the Limits of Interpretation

The Camel's hump is an ugly lump
  Which well you may see at the Zoo;
But uglier yet is the hump we get
  From having too little to do.

The cure for this ill is not to sit still
  Or frowst with a book by the fire;
But to take a large hoe and a shovel also,
  And dig till you gently perspire.

And then you will find that the sun and the wind
And the Djinn of the Garden too,
  Have lifted the hump—
  That horrible hump—
The hump that is black and blue!

I get it as well as you-oo-oo—
If I have n't enough to do-oo-oo—
  We all get hump—
  Camelous hump—
Kiddies and grown-ups too!

(Kipling, 1902: 29)

> Finally, we individually, and as a community, must stand firm about the limits of our abilities. We have not yet developed techniques which will provide us information on many of the interactions between the skeleton and the events in the person's life (Galloway, 1995: 14).

Osteologists have gone to great lengths in their attempts to derive behavioral inferences from skeletal material; most especially appealing are those presumed osseous indicators of activity. The desire, indeed, seeming compulsion, to make such activity-based interpretations is itself a major bias in contemporary osteological studies. Perhaps, this bias is what most stimulates ongoing and expanded attempts of this nature. Nevertheless, in striving to attain that seductive goal of intriguing behavioral inference, much is lost along the way.

Most fundamentally, perhaps, many osteologists choose to ignore much of the available clinical evidence, especially when it shows the etiopathogenesis of the relevant conditions to be more complex than would be convenient for arguing activity-based scenarios. In point of fact, those skeletal manifestations so central to these interpretive approaches result from complex, multi-factorial causation. Because of the mountains of detailed clinical documentation, osteoarthritis is perhaps the best example; but many of the other conditions discussed here are probably almost as complex. Certainly, *none* of these osseous "markers" result directly or simply from activity alone.

Yet, even in this seeming etiological confusion, there may well be some global patterning. For example, systemic (genotypic/endocrine) influences on bone remodeling probably have much to do with the later development of osteoarthritis, enthesial reactions, and, perhaps, diaphyseal geometry as well. The potential insight to be gained by addressing other possible influences on these skeletal manifestations is equally rewarding as focusing on behavior/activity. However, what I have termed an "activity-only myopia," has frequently acted to stifle broader-based interpretations of these skeletal conditions.

The over-willingness to leap to simple behavioral conclusions is probably most dramatically reflected in a variety of investigations dealing with small, poorly controlled samples. For example, in studies of osteoarthritis, numerous reports have made behavioral inferences without stratifying the respective samples by age (or, in some cases, by sex). Especially, as relating to poor age control, the imaginative investigations might still produce interesting stories, but contribute little in the way of scientific discipline.

Such inadequacies of the skeletal data are compensated by osteologists in a variety of ways. One approach is to provide more "detail" relating to behavioral components, particularly specific activities presumably practiced by earlier populations. The sources for such inference are typically ethnohistorical reports or historic forms of documentation. However, as made obvious by even a cursory inspection of such sources, they are woefully inadequate. The types of inferences made reflect highly selective culling of the behavioral "data," to create some correspondence with observed skeletal patterning. These attempts are highly reminiscent of "just-so stories," and like Kipling's explanation of dorsal humps in camelids and humans, could be used to *explain* most any feature.

It is difficult to see how such forms of inference provide *anything* but the most tentative of hypotheses. Osteologists might thus be inspired to look for certain types of bone lesions in certain populations. However, *verification* will have to come from elsewhere.

Another device to compensate for poor data, owing to inadequacies in sample size and/or age-control, or inherently weak specificity of individual osseous markers, is to use multiple indicators, each ostensibly providing some evidence of activity. Multiple indicators used in combination frequently include OA, enthesophytes, trauma, and more recently, bone geometry as well. Such an approach has been creatively applied by Merbs (1983), Larsen and colleagues (e.g., Ruff and Larsen, 1990; Larsen and Ruff, 1994; Larsen et al., 1995), Lai and Lovell (1995), and has also been recently advocated by Stuart-Macadam (1997).

Such more holistic interpretation is admirable, but does not necessarily contribute further rigor than the more restricted attempts using single markers. Firstly, adding more data, all of which are equally weak, does not produce greater precision. Some researchers may argue that, when several markers point in the same direction, i.e., indicating increased activity, this patterning provides some means of internal verification. Yet, there are other, perhaps even more obvious, explanations for such patterning. For example, the systemic effects of age influences both osteophyte and enthesophyte development. Moreover differential sex/endocrine factors further systemically influence such skeletal patterning. Lastly, genetically influenced rates of bone remodeling also probably produce systemic effects, and, in so doing, produce clear patterning (but *not* necessarily one related to activity).

Accordingly, it is argued that many of these osteological "markers" are probably not very informative of activity. Particularly, some indicators of OA (osteophytes) and most manifestations of enthesial reactions develop late in life and reflect primarily systemic influences. They separately, and in combination, are *poor* indicators of activity.

Yet *some* (extreme) types of activity might produce distinct skeletal patterns, particularly if they impact the skeleton early in life. For example, it has been suggested here that severe forms of elbow OA might be the product of extreme biomechanical stresses in the developing joint. The higher prevalence of this condition among the Inuit could, in part, reflect such an activity-related etiology. Clinical investigations of hip OA have imputed differential effects of early onset stress in the later development of degenerative changes in athletes and among farmers. Moreover, in the knee and, perhaps, elbow, skeletal evidence of osteochondritis dissecans could also indicate early onset of mechanical stress in developing joints.

Additionally, extreme mechanical loading commencing in childhood/adolescence has been shown to produce a number of stress fractures. Most notably, as demonstrated in both skeletal and clinical series, such lesions are manifested in the lumbar spine (i.e., spondylolysis). Contemporary sports data show similar types of bone changes in the elbow and ankle, suggesting there may be other potentially informative skeletal manifestations of this nature.

Another highly informative reflection of extreme biomechanical stress on the developing skeleton is seen in altered diaphyseal geometry. The clearest effect of activity on cross-sectional long bone geometric patterning has also been shown to result from unusually severe loading affecting very young individuals (indeed, perhaps < 10 years of age).

Lastly, it is even possible that Schmorl's nodes might indicate some component of mechanical loading placed on young spines. However, as this skeletal condition is so common, its degree of specificity is likely considerably less than the other "markers" just discussed. If Schmorl's nodes relate to *any* consistent functional consideration, it would be as part of an early onset manifestation. However, it is advised that these "lesions" be interpreted with great caution, as they are likely to obscure otherwise informative skeletal patterning.

Thus, there may well be a multiple-indicator approach, useful in suggesting certain types of functional overloading. However, activities relating to these skeletal expressions do not correspond to those engaged in by adults, at least not those performed *solely* by adults. The types of activities producing such osseous changes in certain joints, in certain geometric proportions, and at certain sites of likely bone fatigue, apparently all must begin early in life. Such conditions may, of course, continue throughout adulthood, but the distinctive skeletal involvement will not be easily interpreted, apparently unless the stresses begin before growth is completed. Indeed, if functional interpretation is the goal of osteologists, these skeletal changes *may* not be interpretable at all in those individuals who begin, even rigorous, activity later in life.

The situation relating to activity-related skeletal changes may thus not be entirely bleak. A combination of markers might occasionally be interpretable as they relate to what could be termed an "Early Onset Stress Syndrome." Several further limitations, must still be considered. It is probable such skeletal conditions are expressed but *rarely*—and, then, only in certain populations. Clinical evidence argues the activities must, not only begin early in life, but also be *extreme* in the amplitude, and perhaps repetivity/duration, of biomechanical loading, e.g., analogous to adolescent elite tennis players.

In archaeological contexts, the Inuit are probably the best example of a cultural group participating in the types and timing of activities necessary to produce potentially identifiable skeletal markers. How often such patterning can be identified in other populations is, at present, not known. We should, however, not expect too much.

Additionally, there is no clear link between these osseous changes and particular activities. Since many different activities, or combination of activities, can produce similar bone involvement, it follows that skeletal data offer little hope of inferring *specific* behavior from *specific* lesions.

As a last caution, there is the always-necessary step of substantiation using *known* analogs. Thus, more attention needs to be addressed to clinical investigations, especially those involving young athletes. Correlative, multi-site, longitudinal studies would go a long way towards providing better substantiation to the *hypotheses* presented here.

Activities in past populations which primarily involved adults may be mostly invisible to skeletal biologists. One notable exception, however, concerns trauma. Many traumatic lesions do result from behavior/activity, and some of these are interpretable from skeletal evidence. Moreover, most of the best markers, excluding stress fractures, occur in adults.

Some traumatic lesions are quite unambiguous in terms of causation. Such lesions relate most especially to weapon wounds and patterning of craniofacial injury. Postcranial evidence of healed fractures frequently is somewhat less obvious, but these fractures *usually* result from accidental falls. This evidence, too, can be informative, especially its patterning in cross-populational perspective. For example, the prevalence and distribution of healed postcranial fractures can sometimes be correlated with risk components relating to terrain and potentially other biocultural parameters as well.

Most of the interest concerning interpretation of traumatic lesions has focused on interpersonal aggression. However, interpretations relying on forearm ("parry") fractures are usually ambiguous. Craniofacial involvement is a much better indicator of patterned aggression, and, in its absence, so-called parry fractures are best attributed to a non-aggressive etiology.

Osteological evidence obviously has much to offer in a broader understanding of many conditions which impact skeletal tissue. The most productive avenue has been displayed in those efforts where skeletal biologists have worked collaboratively with a variety of clinicians. In this way, the skeletal data are more directly linked to real morbid changes, and the osteological evidence becomes a collateral and supportive means of interpretation. The Paleopathology Association has consistently promoted cooperation between anthropologists and clinicians, and there has been in recent years a number of investigations which exemplify the benefits of such coordinated research. These include studies of osteoarthritis and enthesial reactions by Rogers and colleagues in the United Kingdom as well as research in the U.S. on these same conditions by Rothschild and associates. Similarly, long-term assistance by clinical colleagues has contributed much to Merbs' interpretations of spondylolysis. Moreover, the intensive research by Ruff and colleagues on patterning of diaphyseal geometry has from its outset involved clinical collaboration. Lastly, the innovative interpretation of traumatic lesions by Trinkaus and colleagues has also been enriched by detailed clinical support.

The most basic message relating to behavioral reconstruction is that *some* form of verification must be available. Otherwise, reasoning is perpetually circular, and hypotheses remain open-ended, and merely "suggestive." Worse yet, the stories and scenarios become ends in themselves. This is not to say there cannot be some role for imagination and some use for anecdote. But these essentially non-scientific avenues should be utilized to stimulate research which can then be substantiated by more rigorous methodologies. The inherent limitations of such evidence, including most of that obtained from ethnohistorical sources, must always be clearly recognized.

The temptations and limitations of undisciplined "storytelling" have also been recognized as a component of paleoanthropology (Landau, 1984; Fedigan, 1986). Anecdotal evidence has also long played a role in primatological research. For example, Goodall (1986) explicitly detailed many anecdotal events in the lives of the Gombe chimpanzees, but also supported conclusions with a wealth of systematic, quantifiable observations. Of further interest to skeletal biologists, certain documented events in the lives of the Gombe chimpanzees are reflected in the skeletons of these animals (Jurmain, 1989; 1997).

In recent years, cultural anthropologists have come to regard the individual as the most significant focus of research. This "post-modernist" perspective thus treats anecdotal life history information as central. Indeed, post-modernists, not only exalt anecdotal evidence, they, at the same time, disdain empirically based attempts at broader generalization. Such an orientation has done little to bolster academic cooperation within anthropology, but, more disturbingly, it surely weakens already confused perceptions of scientific methodology among the general public. Healthy skepticism is, of course, a fundamental part of intellectual discourse, and *must* always be a component of good science. But some of our colleagues in cultural anthropology, perhaps bewildered in the confusing fog inherent in "behavioral science," have apparently lost their intellectual compass.

Paleopathologists should recognize that post-modernist critiques sometimes strike a little too closely for comfort. Story-telling and scenario-building may have some (limited) utility. However, without clear empirical controls and application of rigorous means of verification, skeletal biologists can become lost in the same intellectual fog. What is needed is the steadfast recognition of the differences between sound scientific method and less well controlled approaches. Physical anthropologists are usually still trained to allow this recognition, even if students in other parts of anthropology are not.

Unless we are to adopt a type of "post-modern osteology," we have no choice but to be constantly vigilant and critical of our *own* biases and methods. In a sense, we must straddle a 'pyramid' of evidence, method, and inference (Figure 8-1). In this way, we not only talk and

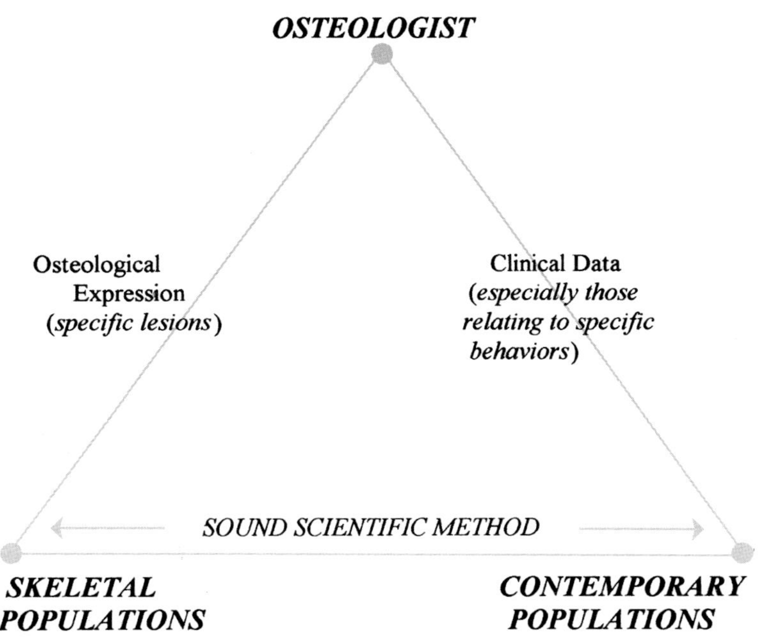

Figure 8-1 "Pyramid" of bioarchaeological behavioral inference.

teach about scientific method and good critical thinking, we also display them in our own research.

As a final alternative, osteologists could remain oblivious to the inherent pitfalls, and, consequently, tumble from the precarious intellectual edifice we have constructed. The trauma may leave tell-tale signs on the skeleton, and thus osteologists, as a group, could then join our colleagues in the legal profession in manifesting yet another marker of occupational stress:

> "You are old," said the youth, "and your jaws are too weak
> For anything tougher than suet;
> Yet you finished the goose, with the bones and the beak—
> Pray, how did you manage to do it?"
> "In my youth," said his father, "I took to the law,
> And argued each case with my wife;
> And the muscular strength, which it gave to my jaw
> Has lasted the rest of my life."
>
> (Carroll, 1865).

Appendix A

Age-Specific Prevalence of Osteoarthritis in Skeletal Samples

	Right		Left	
	mod	sev	mod	sev
Shoulder joint—Males				
Terry White[1]				
21–30	11.1	0.0	12.5	0.0
31–40	18.8	0.0	18.5	0.0
41–50	23.5	0.0	35.3	0.0
51–60	52.6	0.0	55.6	0.0
61–70	68.8	0.0	72.2	0.0
71+	93.8	6.3	88.2	5.9
Terry Black[1]				
21–30	5.0	0.0	0.0	0.0
31–40	16.7	0.0	18.8	0.0
41–50	40.0	0.0	40.0	0.0
51–60	88.9	0.0	87.5	0.0
61–70	72.2	11.1	94.1	5.9
71+	88.2	11.8	88.2	11.8
Pecos[1]				
21–30	0.0	0.0	0.0	0.0
31–40	0.0	0.0	0.0	0.0
41–50	21.4	0.0	18.8	0.0
51–60	58.3	8.3	45.5	0.0
61–70	57.1	0.0	44.4	0.0
71+	66.7	0.0	57.1	14.3
Inuit[1]				
21–30	30.0	0.0	9.5	0.0
31–40	61.1	0.0	53.3	0.0
41–50	71.4	0.0	63.6	0.0
51–60	xx	xx	100.0	0.0
61–70	xx	xx	xx	xx
71+	xx	xx	xx	xx

(continued)

	Right		Left	
	mod	sev	mod	sev
Calif (Ala-329)[2] (males and females combined)				
21–30	0.0	0.0	0.0	0.0
31–40	2.0	0.0	0.0	0.0
41+	12.9	0.0	6.7	0.0
Kulubnarti[3]	(mod+sev)		(mod+sev)	
20–29	0.0		0.0	
30–39	0.0		5.3	
40+	21.0		33.3	
Averbuch[4] (L and R combined males and females)		(mod)	(sev)	
19–29		77.0	0.0	
30–39		80.0	10.0	
40+		88.2	11.8	
Indian Knoll[4] (L and R combined males and females)				
19–29		52.6	0.0	
30–39		83.3	5.8	
40+		82.9	9.8	
Shoulder joint—Females				
Terry White[1]				
21–30	0.0	0.0	14.3	0.0
31–40	18.2	0.0	16.7	0.0
41–50	36.4	0.0	39.1	0.0
51–60	50.0	0.0	44.4	0.0
61–70	68.4	5.3	66.7	0.0
71+	55.0	35.0	75.0	25.0
Terry Black[1]				
21–30	15.8	0.0	5.9	5.9
31–40	38.9	5.6	50.0	0.0
41–50	50.0	5.0	60.0	0.0
51–60	66.7	11.1	64.7	5.9
61–70	64.7	17.6	68.8	12.5
71+	88.2	11.8	88.2	11.8
Pecos[1]				
21–30	0.0	0.0	0.0	0.0
31–40	0.0	0.0	14.3	0.0

(*continued*)

	Right		Left	
	mod	sev	mod	sev
41–50	7.7	0.0	0.0	0.0
51–60	50.0	0.0	70.0	0.0
61–70	80.0	0.0	66.7	0.0
71+	xx	xx	75.0	0.0
Inuit[1]				
21–30	6.7	0.0	0.0	0.0
31–40	37.5	0.0	40.0	0.0
41–50	16.7	0.0	33.3	33.3
51–60	xx	xx	xx	xx
61–70	xx	xx	xx	xx
71+	xx	xx	xx	xx
Kulubnarti[3]	(mod+sev)		(mod+sev)	
20–29	0.0		0.0	
30–39	9.5		0.0	
40+	41.0		42.1	
Elbow joint—Males				
Terry White[1]				
21–30	0.0	0.0	0.0	0.0
31–40	11.1	0.0	0.0	0.0
41–50	5.9	0.0	5.6	0.0
51–60	15.0	5.0	9.5	4.8
61–70	20.0	5.0	25.0	15.0
71+	15.0	20.0	21.1	10.5
Terry Black[1]				
21–30	0.0	0.0	0.0	0.0
31–40	10.5	0.0	5.3	0.0
41–50	5.3	0.0	5.3	0.0
51–60	31.6	5.3	11.1	0.0
61–70	40.0	5.0	30.0	5.0
71+	31.6	21.1	26.3	15.8
Pecos[1]				
21–30	0.0	8.3	0.0	0.0
31–40	5.9	0.0	0.0	0.0
41–50	12.5	0.0	5.6	0.0
51–60	13.3	0.0	6.3	0.0
61–70	16.7	8.3	30.8	0.0
71+	28.6	14.3	22.2	11.1
Inuit[1]				
21–30	4.8	14.3	22.7	9.1
31–40	42.9	14.3	38.5	0.0
41–50	53.3	20.0	35.7	21.4

(continued)

	Right		Left	
	mod	sev	mod	sev
51–60	xx	xx	0.0	100.0
61–70	xx	xx	xx	xx
71+	xx	xx	xx	xx
Calif (Ala-329)[2] (males and females combined)				
21–30	0.0	0.0	0.0	0.0
31–40	5.8	0.0	6.4	2.1
41+	8.8	2.9	9.1	0.0
Kulubnarti[3]	(mod+sev)		(mod+sev)	
20–29	0.0		0.0	
30–39	0.0		0.0	
40+	13.6		17.4	
Averbuch[4] (L and R combined males and females)	(mod)		(sev)	
19–29	80.4		2.2	
30–39	45.4		45.4	
40+	71.4		28.6	
Indian Knoll[4] (L and R combined males and females)				
19–29	62.9		0.0	
30–39	83.3		2.7	
40+	93.3		6.7	
Elbow joint—Females				
Terry White[1]				
21–30	0.0	0.0	0.0	0.0
31–40	0.0	0.0	0.0	0.0
41–50	0.0	0.0	0.0	0.0
51–60	5.0	0.0	5.0	0.0
61–70	23.8	4.8	20.0	0.0
71+	35.0	0.0	20.0	0.0
Terry Black[1]				
21–30	0.0	0.0	0.0	0.0
31–40	15.8	0.0	5.3	0.0
41–50	23.8	0.0	4.8	0.0
51–60	23.8	4.8	14.3	9.5
61–70	17.6	0.0	17.6	5.9
71+	50.0	0.0	26.3	5.3

(*continued*)

	Right		Left	
	mod	sev	mod	sev
Pecos[1]				
21–30	0.0	8.3	0.0	0.0
31–40	0.0	0.0	0.0	0.0
41–50	6.7	0.0	11.8	0.0
51–60	28.6	0.0	13.3	6.7
61–70	28.6	0.0	80.0	0.0
71+	xx	xx	50.0	0.0
Inuit[1]				
21–30	0.0	0.0	7.1	0.0
31–40	42.9	14.3	36.4	18.2
41–50	33.3	0.0	0.0	40.0
51–60	xx	xx	xx	xx
61–70	xx	xx	xx	xx
71+	xx	xx	xx	xx
Kulubnarti[3]	(mod + sev)		(mod + sev)	
20–29	0.0		0.0	
30–39	0.0		5.0	
40+	5.6		8.1	
Hip joint—Males				
Terry White[1]				
21–30	0.0	0.0	11.1	0.0
31–40	16.7	0.0	27.8	0.0
41–50	41.2	0.0	33.3	0.0
51–60	57.1	0.0	45.0	5.0
61–70	90.0	5.0	70.0	5.0
71+	70.6	11.8	88.2	11.8
Terry Black[1]				
21–30	10.0	0.0	0.0	0.0
31–40	21.1	0.0	33.3	0.0
41–50	15.8	0.0	33.3	0.0
51–60	70.6	0.0	55.6	5.6
61–70	84.2	5.3	75.0	10.0
71+	88.9	5.0	88.9	5.6
Pecos[1]				
21–30	7.7	0.0	7.7	0.0
31–40	5.9	0.0	5.6	0.0
41–50	5.6	0.0	5.6	0.0
51–60	18.8	0.0	26.3	0.0
61–70	61.5	0.0	38.5	0.0
71+	40.0	20.0	43.8	12.5

(*continued*)

	Right		Left	
	mod	sev	mod	sev
Inuit[1]				
21–30	13.0	0.0	13.0	0.0
31–40	39.1	4.3	40.0	4.0
41–50	50.0	5.6	57.9	5.3
51–60	66.7	0.0	80.0	0.0
61–70	xx	xx	xx	xx
71+	xx	xx	xx	xx
Calif (Ala-329)[2] (males and females combined)				
21–30	0.0	0.0	0.0	0.0
31–40	2.1	0.0	0.0	0.0
41+	0.0	0.0	2.9	0.0
Kulubnarti[3]	(mod+sev)		(mod+sev)	
20–29	0.0		0.0	
30–39	0.0		0.0	
40+	43.5		36.0	
Averbuch[4] (L and R combined males and females)		(mod)	(sev)	
19–29		67.4	6.5	
30–39		60.9	30.4	
40+		62.5	37.5	
Indian Knoll[4] (L and R combined males and females)				
19–29		9.1	0.0	
30–39		27.0	0.0	
40+		63.9	11.3	
Hip joint—Females				
Terry White[1]				
21–30	16.7	0.0	11.1	0.0
31–40	8.3	0.0	27.8	0.0
41–50	39.1	0.0	33.3	0.0
51–60	36.8	10.5	45.0	5.0
61–70	47.4	21.1	20.0	5.0
71+	55.0	25.0	88.2	11.8
Terry Black[1]				
21–30	22.0	0.0	0.0	0.0
31–40	15.8	0.0	33.3	0.0

(*continued*)

	Right		Left	
	mod	sev	mod	sev
41–50	47.6	4.8	33.3	0.0
51–60	66.7	4.8	55.6	5.6
61–70	61.1	16.7	75.0	10.0
71+	72.2	22.2	88.9	5.6
Pecos[1]				
21–30	4.8	0.0	0.0	0.0
31–40	5.9	0.0	5.6	0.0
41–50	27.8	0.0	15.8	0.0
51–60	15.0	0.0	22.2	0.0
61–70	54.5	0.0	50.0	0.0
71+	60.0	0.0	83.3	0.0
Inuit[1]				
21–30	4.8	0.0	5.3	0.0
31–40	28.6	0.0	19.0	0.0
41–50	27.0	0.0	53.8	0.0
51–60	60.0	20.0	75.0	25.0
61–70	xx	xx	xx	xx
71+	xx	xx	xx	xx
Kulubnarti[3]	(mod+sev)		(mod+sev)	
20–29	0.0		0.0	
30–39	20.0		4.8	
40+	41.7		41.0	
Knee joint—Males				
Terry White[1]				
21–30	0.0	0.0	0.0	0.0
31–40	5.3	0.0	10.5	0.0
41–50	23.5	5.9	27.8	0.0
51–60	26.3	0.0	6.3	0.0
61–70	42.1	0.0	40.0	0.0
71+	50.0	11.1	42.1	10.5
Terry Black[1]				
21–30	0.0	0.0	0.0	0.0
31–40	10.5	0.0	10.5	0.0
41–50	17.6	0.0	27.8	0.0
51–60	55.6	0.0	6.3	0.0
61–70	68.4	10.5	40.0	0.0
71+	73.7	15.8	42.1	10.5
Pecos[1]				
21–30	0.0	0.0	0.0	0.0
31–40	11.1	0.0	10.0	0.0

(continued)

	Right		Left	
	mod	sev	mod	sev
41–50	18.2	0.0	11.1	0.0
51–60	36.4	0.0	45.5	0.0
61–70	66.7	0.0	57.1	0.0
71+	37.5	12.5	66.7	11.1
Inuit[1]				
21–30	8.3	0.0	0.0	10.0
31–40	54.5	9.1	22.2	0.0
41–50	40.0	30.0	37.5	12.5
51–60	xx	xx	xx	xx
61–70	xx	xx	xx	xx
71+	xx	xx	xx	xx
Calif (Ala-329)[2] (males and females combined)				
21–30	0.0	0.0	0.0	0.0
31–40	10.4	2.1	4.1	0.0
41+	14.8	3.7	20.0	4.0
Kulubnarti[3]	(mod + sev)		(mod + sev)	
20–29	0.0		0.0	
30–39	7.1		0.0	
40+	50.0		45.0	
Averbuch[4] (L and R combined males and females)	(mod)	(sev)		
19–29	65.8	5.3		
30–39	84.6	0.0		
40+	xx	xx		
Indian Knoll[4] (L and R combined males and females)				
19–29	27.9	0.0		
30–39	44.8	0.0		
40+	33.3	25.0		
Todd White[5]	(mod + sev)			
31–40	4.1			
41–50	14.4			
51–60	31.7			
61–70	38.3			
71–80	52.6			

(*continued*)

	Right			Left	
	mod	sev		mod	sev
Todd Black[5]					
31–40			10.1		
41–50			25.3		
51–60			43.8		
61–70			55.9		
71–80			57.9		
Knee joint—Females					
Terry White[1]					
21–30	14.3	0.0		14.3	0.0
31–40	16.7	0.0		8.3	0.0
41–50	26.1	0.0		18.2	0.0
51–60	47.4	10.5		45.0	10.0
61–70	38.1	19.0		50.0	15.0
71+	52.6	36.3		55.0	25.0
Terry Black[1]					
21–30	10.5	0.0		10.5	0.0
31–40	26.3	0.0		21.1	5.3
41–50	23.8	14.3		38.1	9.5
51–60	45.0	20.0		60.0	10.0
61–70	52.9	29.3		55.6	27.8
71+	35.3	52.9		50.0	31.3
Pecos[1]					
21–30	0.0	0.0		0.0	0.0
31–40	0.0	0.0		0.0	0.0
41–50	0.0	0.0		0.0	0.0
51–60	16.7	0.0		33.3	0.0
61–70	xx	xx		80.0	0.0
71+	xx	xx		xx	xx
Inuit[1]					
21–30	25.0	0.0		0.0	0.0
31–40	27.3	9.1		55.6	0.0
41–50	40.0	0.0		40.0	0.0
51–60	xx	xx		xx	xx
61–70	xx	xx		xx	xx
71+	xx	xx		xx	xx
Kulubnarti[3]	(mod + sev)			(mod + sev)	
20–29	10.0			9.1	
30–39	22.2			23.5	
40+	55.2			50.0	

(continued)

	Right		Left	
	mod	sev	mod	sev
Todd White[5]		(mod + sev)		
31–40		3.3		
41–50		18.8		
51–60		27.6		
61–70		42.9		
71–80		22.2		
Todd Black[5]		(mod + sev)		
31–40		15.5		
41–50		31.0		
51–60		53.8		
61–70		40.0		
71–80		62.5		

xx = insufficient data.
[1]Data from Jurmain, 1975; 1977a.
[2]Data from Jurmain, 1990 (males and females combined).
[3]Data from Kilgore, 1984 (moderate and severe involvement combined).
[4]Data from Pierce, 1987 (males and females combined; left and right combined).
[5]Data from Woods, 1995 (moderate and severe involvement combined).

Bibliography

Aalund, O., L. Danielsen, and R.O. Sanhueza (1990) Injuries due to deliberate violence in Chile. *Forensic Sci Inter*, *46*: 189–202.

Achinson, R.M., Y.K. Chan, and A.R. Clement (1970) New Haven survey of joint diseases. Distribution and symptoms of osteoarthritis of the hands with reference to handedness. *Ann Rheum Dis*, *29*: 275–286.

Acheson, R.M. and A.B. Collart (1975) New Haven survey of joint disease XVII. Relationship between systemic characteristics and osteoarthrosis in a general population. *Ann Rheum Dis*, *34*: 379–387.

Acsádi, Gy. and J. Nemeskéri (1970) *History of Human Life Span and Mortality*. Budapest: Akadémiai Kiadó.

Adams, I.D. (1976) Osteoarthritis and sport. *Clin Rheum Dis*, *2*: 523–541.

Adams, I.D. (1979) Osteoarthritis and sport. *J Royal Soc Med*, *72*: 185–187.

Adams, J.E. (1965) Injury to the throwing arm. *Calif Medicine*, *102*: 127–132.

Adams, W.S. (1951) The etiology of swimmers' exostoses of the external auditory canal. Parts I & II. *J Laryngol*, *64*: 133–153; *65*: 232–250.

Aggrawal, N.D., R. Kaur, S. Kumar et al. (1979) A study of the changes in the spine of weight-lifters and other athletes. *Brit J Sports Med*, *13*: 58–61.

Aiola, J., S.H. Cohen, J.A. Ostuni et al. (1978) Prevention of involutional bone loss in exercise. *Ann Intern Med*, *89*: 356.

Albright, J.A., P. Jokl, R. Shaw et al. (1978) Clinical study of baseball pitchers: correlation of injury to throwing arm and method of delivery. *Am J Sports Med*, *6*: 15–20.

Alekel, L., J.L. Clasey, P.C. Fehling et al. (1995) Contribution of exercise, body composition, and age to bone mineral density in premenopausal women. *Med Sci Sports Exerc*, *27*: 1477–1485.

Ali-Gombe, A., P.R. Croft, and A.J. Silman (1996) Osteoarthritis of the hip and acetabular dysplasia in Nigerian men. *J Rhematol*, *23*: 512–515.

Allander, E., O.J. Björnsson, O. Olafsson et al. (1974) Normal range of joint movements in shoulder, hip, wrist, and thumb with special reference to side: a comparison between two populations. *Int J Epidemiology*, *3*: 253–261.

Altman, R. (1991) Classification of Disease: Osteoarthritis. *Seminar Arthritis Rheum*, *20*(*Suppl 2*): 40–47.

Altman, R., E. Asch, D. Bloch et al. (1986) Development of criteria for the classification and reporting of osteoarthritis. Classification of osteoarthritis of the knee. *Arthritis Rheum*, *29*: 1039–1049.

Ambre, T. and B.E. Nilsson (1978) Degenerative changes in the first metatarsophalangeal joint of ballet dancers. *Acta Orthop Scan*, *49*: 317–319.

Andersen, B.C. (1986) Parturition as a consequence of flexible pelvic architecture. Ph.D. thesis, Simon Fraser University, Vancouver, B.C.

Andersen, S. and F. Winckler (1979) The epidemiology of primary osteoarthrosis of the knee in Greenland. *Arch Orthop Traumat Surg*, *93*: 91–94.

Anderson, J.A.D. (1974) Occupation as a modifying factor in the diagnosis and treatment of rheumatic diseases. *Curr Med Res Opin*, *2*: No. 9.

Anderson, J.A.D. (1984) Arthrosis and its relation to work. *Scand J Work Environ Health*, *10*: 429–433.

Anderson, J.A.D. and J.J. Duthie (1963) Rheumatic complaints in dockyard workers. *Ann Rheum Dis*, *22*: 401–409.

Anderson, J.A.D., J.J.R. Duthie, and B.P. Moody (1962) Social and economic effects of rheumatic diseases in mining populations. *Ann Rheum Dis*, *21*: 342–351.

Anderson, J.E. (1963) *The People of Fairty. An Osteological Analysis of a Iroquois Ossuary*. National Museum of Canada, Bull. No. 193, Contributions to Anthropology 1961–1962.

Anderson, J.J. and D.T. Felson (1988) Factors associated with osteoarthritis of the knee in the First National Health and Nutritional Survey (HANES I): Evidence for an assocation of overweight, race, and physical demands of work. *Am J Epidemiol*, *128*: 179–189.

Anderson, L.D. (1984) Fractures of the shafts of the radius and ulna. In: Rockwood, C.A. and D.P. Green (eds.), *Fractures in Adults*. Philadelphia: J.B. Lippincott, pp. 511–558.

Anderson, T. (1996) Cranial weapon injuries from Anglo-Saxon Dover. *Int J Osteoarchaeol*, *6*: 10–14.

Angel, J.L. (1964) The reaction area of the femoral neck. *Clin Orthop*, *32*: 130–142.

Angel, J.L. (1966) Early skeletons from Tranquility, California. *Smithsonian Contrib Anthropol*, *2*: 1–19.

Angel, J.L. (1969) The bases of paledemography. *Am J Phys Anthropol*, *30*: 427–437.

Angel, J.L. (1971) *The People of Lerna*. Washington, D.C.: Smithsonian Institution Press.

Angel, J.L. (1976) Colonial to modern skeletal changes in the U.S.A. *Am J Phys Anthropol*, *45*: 723–735.

Angel, J.L., J.O. Kelley, M. Parrington et al. (1987) Life stresses of the Free Black Community as represented by the First African Babtist Church, Philadelphia, 1823–1841. *Am J Phys Anthropol*, *74*: 213–229.

Asadi, S.J. and Z. Asadi (1996) Site of the mandible prone to trauma: A two year retrospective study. *Int Dent J*, *46*: 171–173.

Bagneres, H. (1967) Lesions osteo-articulaires chroniques des sportifs. *Rheum Haneologia Allergiologia*, *19*: 41–50.

Bard, C.C., J.J. Sylvester, and R.G. Dussault (1984) Hand osteoarthropathy in pianists. *J Can Assoc Radiol*, *35*: 154–158.

Barnett, C.H., D.V. Davies, and M.A. MacConaill (1961) *Synovial Joints: Their Structure and Mechanics*. Springfield, IL: Charles C. Thomas.

Barsi, C. and R. Rossaro (1963) Osteoartopatie da vibrazioni. *Rassegna di Medicina Industrale e di Ingriene del Lavoro*, *32*: 592–599.

Bateman, J.E. (1969) *American Academy of Orthopaedic Surgeons Symposium on Sports Medicine*. St Louis: C.V. Mosby.

Begg, A.C. (1954) Nuclear herniations of the intervertebral disk: Their radiological manifestation and significance. *J Bone Joint Surg*, *36B*: 180–193.

Bennett, G.A. and W. Bauer (1937) Joint changes resulting from patellar displacement and their relation to degenerative joint disease. *J Bone Joint Surg*, *19A*: 667–682.

Bennett, G.A., H. Waine, and W. Bauer (1942) *Changes in the Knee Joint at Various Ages*. New York: Commonwealth Fund.

Bennett, G.E. (1941) Shoulder and elbow lesions in the professional baseball pitcher. *JAMA*, *117*: 510–514.

Bennett, P.H. and T.A. Burch (1968) Osteoarthrosis in the Blackfoot and Pima Indians. In: Bennett, P.H. and P.H.N. Wood (eds.), q.v., pp. 407–412.

Bennike, Pia (1985) *Palaeopathology of Danish Skeletons*. Copenhagen: Akademisk Forlag.

Bennike, Pia (1988) Causes of death in the early Neolithic period in Denmark. *Revista di Anthropologia*, Vol. 66, *Supplement*, pp. 205–214.

Berger, T.D. and E. Trinkaus (1995) Patterns of trauma among the Neandertals. *J Archaeol Sci*, *22*: 841–852.

Berget, K.A. and S.E. Churchill (1994) Subsistence activity and humeral hypertrophy among western Aleutian islanders (abstract). *Am J Phys Anthropol* (*Suppl*), *18*: 55.

Bergmann, G.H. Kniggendorf, F. Graichen et al. (1995) Influence of shoes and heel strike on the loading of the hip joint. *J Biomech*, *28*: 817–827.

Binford, L.R. (1981) *Bones: Ancient Men and Modern Myths*. New York: Academic Press.

Binford, L.R. (1984) *Faunal Remains from Klasies River Mouth*. New York: Academic Press.

Birket-Smith, K.A.J. (1929) *The Caribou Eskimos. Report of the Fifth Thule Expedition 1921–24*. Vol. V, Parts 1 & 2. Copenhagen.

Black, T.K. (1979) The biological and social analysis of a Mississippian cemetery from Southeast Missouri: the Turner Site, 23BU21A. Ann Arbor: Museum of Anthropology, University of Michigan Anthropological Papers, No. 68.

Blumberg, B.S., K. Bloch, R.L. Black et al. (1961) A study in the prevalence of arthritis in Alaskan Eskimos. *Arth Rheum*, *4*: 325–341.

Boas, F. (1888) *The Central Eskimo*. Washington, D.C.: *Smithsonian Inst Bur Ethnol*, 6th Annual Report, 1884–1885, Vol. 6.

Bollen, S.R. and V. Wright (1994) Radiographic changes in the hands of rock climbers. *Brit J Sports Med*, *28*: 185–186.

Boström, L. (1997) Injury panorama and medical consequences for 1158 persons assaulted in the central part of Stockholm. *Arch Orthop Trauma Surg*, *116*: 315–320.

Bourke, J.B. (1967) A review of the paleopathology of the arthritic diseases. In: Brothwell, D.R. and A.T. Sandison (eds.), *Diseases in Antiquity*. Springfield, IL: Charles C. Thomas, pp. 302–370.

Bourke, J.B. (1969) Trauma and degenerative disease in ancient Egypt and Nubia. *J Human Evol*, *1*: 225–232.

Bouxsein, M.L., K.H. Myburgh, M.C. van der Meulen et al. (1994) Age-related differences in cross-sectional geometry of the forearm bones in healthy women. *Calcif Tissue Int*, *54*: 113–118.

Bovenzi, M., L. Petronio, and F. Di Marino (1980) Epidemiological survey of shipyard workers exposed to hand-arm vibration. *Int Arch Occup Environ Health*, *46*: 251–266.

Boyd, H.B. and J.C. Boals (1969) The Monteggia lesion. A review of 159 cases. *Clin Orthop*, *66*: 94–100.

Bradtmiller, B. (1983) The effect of horseback riding on Arikara arthritis patterns. Paper, American Anthropological Assoc., 82nd Annual Meetings, Chicago.

Bramblett, C. (1967) Pathology in the Darajani baboon. *Am J Phys Anthropol*, *21*: 331–340.

Brandt, K.D. and R.S. Fife (1986) Ageing in relation to the pathogenesis of osteoarthritis. *Clin Rheum Dis*, *12*: 117.

Breiting, V.B., O. Aalund, S.B. Albreksten et al. (1989) Injuries due to deliberate violence in areas of Denmark. I. The extent of violence. *Forensic Sci Int*, *40*: 183–199.

Bremner, J.M., J.S. Lawrence, and W.E. Miall (1968) Degenerative joint disease in a Jamaican rural population. *Ann Rheum Dis*, *27*: 326–332.

Bridges, P.S. (1989a) Changes in activity with the shift to agriculture in the Southeast United States. *Curr Anthropol*, *30*: 385–394.

Bridges, P.S. (1989b) Bone cortical area in the evaluation of nutrition and activity levels. *Am J Human Biology*, *1*: 785–792.

Bridges, P.S. (1989c) Spondylolysis and its relationship to degnerative joint disease in the prehistoric southeastern United States. *Am J Phys Anthropol*, *79*: 321–329.

Bridges, P.S. (1990) Osteological correlates of weapon use. In: J.E. Buikstra (ed.), *A Life in Science: Papers in Honor of J. Lawrence Angel*. Center for American Archaeology Scientific Papers, pp. 87–98.

Bridges, P.S. (1991) Degenerative joint disease in hunter-gatherers and agriculturists from the southeastern United States. *Am J Phys Anthropol*, *85*: 379–391.

Bridges, P.S. (1992) Prehistoric arthritis in the Americas. *Ann Rev Anthropol*, *21*: 67–91.

Bridges, P.S. (1993) The effect of variation in the methodology and outcome of osteoarthritic studies. *Intl J Osteoarchaeol*, *3*: 289–295.

Bridges, P.S. (1994) Vertebral arthritis and physical activities in the prehistoric southeastern United States. *Am J Phys Anthropol*, *93*: 83–93.

Bridges, P.S. (1996a) Warfare and mortality at Koger's Island, Alabama. *Int J Osteoarchaeol*, *6*: 66–75.

Bridges, P.S. (1996b) Skeletal biology and behavior in ancient humans. *Evol Anthropol*, *4*: 112–120.

Brock, S.L. and C.B. Ruff (1988) Diachronic patterns of change in the structural properties of the femur in the prehistoric American Southwest. *Am J Phys Anthropol*, *75*: 113–127.

Brodelius, A. (1961) Osteoarthritis of the talar joints in footballers and ballet dancers. *Acta Orthop Scan*, *30*: 309–314.

Brook, I.M. and N. Wood (1983) Aetiology and incidence of facial fractures in adults. *Intl J Oral Surg*, *12*: 293–298.

Brothwell, D. (1961) The paleopathology of early British man: An essay on the problem of diagnosis and analysis. *J Roy Anth Inst Gr Brit*, *91*: 318–345.

Brown, R.A. and J.B. Weiss (1988) Neovascularization and its role in the osteoarthritic process. *Ann Rheum Dis*, *47*: 881–885.

Bruns, J. and H. Kilma (1993) Osteochondrosis dessicans genus und Sport. *Sportverletz Sportschaden*, *7*: 68–72.

Buikstra, J.E. (1965) Healed fractures in *Macca mulatta*: Age, sex, and asymmetry. *Folia Primatol*, *23*: 140–148.

Buikstra, J.E. and D.C. Cook (1980) Palaeopathology: An American account. *Ann Rev Anthropol*, *9*: 433–470.

Buikstra, J.E. and D.H. Ubelaker (eds.) (1994) *Standards for Data Collection from Human Skeletal Remains*. Fayetteville, AK: Arkansas Archeological Survey, Research Series No. 44.

Burke, M.J., E.C. Fear, and V. Wright (1977) Bone and joint changes in pneumatic drillers. *Ann Rheum Dis*, *36*: 276–279.

Burr, D.B. (1980) The relationship among physical, geometrical, and mechanical properties of bone, with a note on the properties of non-human primate bone. *Yrbk Phys Anthrop*, *23*: 109–146.

Burr, D.B., R.B. Martin, R.D. Jurmain et al. (1983) Osteoarthrosis: Sex-specific relationship to osteoporosis. *Am J Phys Anthropol*, *61*: 299–303.

Burr, D.B. and E.L. Radin (1990) Trauma as a factor in the initiation of osteoarthritis. In: Brandt, K.D. (ed.), *Cartilage Changes in Osteoarthritis*. Indianapolis: Indiana University, School of Medicine, Ciba-Geigy, pp. 73–80.

Burrell, L.L., M.C. Maas, and D.P. van Gerven (1986) Patterns of long-bone fracture in two Nubian cemeteries. *Human Evol*, *1*: 495–506.

Butchart, A. and D.S.O. Brown (1991) Non-fatal injuries due to interpersonal violence in Johannesburg-Soweto: Incidence, determinants and consequences. *Forensic Sci Inter*, *52*: 35–51.

Caine, D. (1992) Stress changes of the distal radial growth plate. *Am J Sports Med*, *20*: 290–298.

Cardy, A. (1994) Whithorn: The late Medieval cemetery. Unpublished skeletal report.

Carlson, C.S., R.F. Loeser, and C.B. Purser (1996) Osteoarthritis in cynomolgus macaques. III. Effects of age, gender, and subchondral bone thickness on the severity of disease. *J Bone Miner Res*, *11*: 1209–1217.

Carroll, L. (1865) *Alice's Adventures in Wonderland*. First published as, *Alice's Adventures Underground*. New York: Macmillan.

Carter, D.R. (1984) Mechanical loading histories and cortical bone remodeling. *Calcif Tissue Int*, *36*: S19–S24.

Carter, D.R., M.C. Van Der Meulen, and G.S. Beaurple (1996) Mechanical factors in bone growth and development. *Bone*, *18*: 5S–10S.

Carter, S.R., M.J. Aldridge, R. Fitzgerald et al. (1988) Stress changes in the wrist in adolescent gymnasts. *Br J Radiol*, *61*: 109–112.

Cassidy, C.M. (1979) Arthritis in dry bones: Diagnostic problems. *Henry Ford Hosp Med J*, *27*: 68–69.

Centers for Disease Control (1995) Prevalence and impact of arthritis among women—United States, 1989–1991. *Morbidity and Mortality Weekly Report*, *44*: 329–334.

Centers for Disease Control (1996a) Prevalence and impact of arthritis by race and ethnicity—United States, 1989–1991. *Morbidity and Mortality Weekly Report*, *45*: 373–378.

Centers for Disease Control (1996b) Factors associated with prevalent self-reported arthritis and other rheumatic conditions—United States, 1989–1991. *Morbidity and Mortality Weekly Report*, *45*: 487–491.

Chagnon, N. (1983) *Yanomamo*: *The Fierce People*. 3rd. ed. New York: Holt, Rinehart, and Winston.

Chantraine, A. (1985) Knee joint in soccer players: Osteoarthritis and axis deviation. *Med Sci Sports Exerc*, *17*: 434–439.

Chapman, F.H. (1962) Incidence of arthritis in a prehistoric Indian population. *Proc Indiana Acad Sci*, *72*: 59–62.

Chapman, F.H. (1964) Comparison of osteoarthritis in three aboriginal populations. *Proc Indiana Acad Sci*, *74*: 84–85.

Chapman, F.H. (1972) Vertebral osteophytosis in prehistoric populations of central and southern Mexico. *Amer J Phys Anthropol*, *36*: 31–38.

Chinn, C.J., J.D. Priest, and B.E. Kent (1974) Upper extremity range of motion, grip strength, and girth in highly skilled tennis players. *Phys Therapy*, *54*: 474–482.

Cicuttini, F.M. and T.D. Spector (1996) Genetics of osteoarthritis. *Ann Rheum Dis*, *55*: 665–667.

Cicuttini, F.M., T. Spector, and J. Baker (1997) Risk factors for osteoarthritis in the tibiofemoral and patellofemoral joints of the knee. *J Rheumatol*, *24*: 1164–1167.

Claessens, A.A., J.S. Schouten, A. van den Ouwerland et al. (1990) Do clinical findings associate with radiographic osteoarthritis of the knee? *Ann Rheum Dis*, *49*: 771–774.

Claussen, B.F. (1982) Chronic hypertrophy of the ulna in the professional rodeo cowboy. *Clin Orthop*, *164*: 45–47.

Clement, D.B. et al. (1984) Achilles tendinitis and peritendinitis: etiology and treatment. *Am J Sports Med*, *12*: 179–184.

Cobb, S. (1971) *The Frequency of the Rheumatic Diseases*. Cambridge: Harvard University Press.

Cobb, S., W.R. Merchant, and T. Rudin (1957) The relation of symptoms to osteoarthritis. *J Chronic Dis*, *5*: 197–204.

Cockburn, A., H. Duncan, and J.M. Riddle (1979) Arthritis, ancient and modern: guidelines for field workers. *Henry Ford Hosp Med J*, *27*: 74–79.

Cohen, M.N. (1989) *Health and the Rise of Civilization*. New Haven: Yale University Press.

Cohen, M.N. and G.J. Armelagos (1984a) Paleopathology and the origins of agriculture: Editors' summation. In: Cohen, M.N. and G.J. Armelagos (eds.), q.v., pp. 585–601.

Cohen, M.N. and G.J. Armelagos (eds.) (1984b) *Paleopathology at the Origins of Agriculture*. Orlando, FL: Academic Press.

Cohen, M.N., K. O'Connor, M. Danforth et al. (1994) Health and disease at Tipu. In: C.S. Larsen and G.R. Milner (eds.), q.v., pp. 121–133.

Collins, D.H. (1950) *The Pathology of Articular and Spinal Diseases*. Liverpool: Edward Arnold and Co.

Cooper, C.T. McAlindon, D. Coggon et al. (1994) Occupational activity and osteoarthritis of the knee. *Ann Rheum Dis*, *53*: 90–93.

Cooke, T.D.V. (1983) The polyarticular features of osteoarthritis requiring hip and knee surgery. *J Rheumatol*, *10*: 288–290.

Courville, C.B. (1962) Forensic neuropathology II: Mechanisms of craniocerebral injury and their medicolegal significance. *J Forensic Sci*, *7*: 1–28.

Cox, M. and A. Scott (1992) Evaluation of the obstetric significance of some pelvic characters in an 18th Century British sample of known parity status. *Am J Phys Anthropol*, *89*: 431–440.

Croft, P.D. (1996) The occurrence of osteoarthritis outside Europe. *Ann Rheum Dis*, *55*: 661–664.

Croft, P.D. Coggan, M. Cruddas et al. (1992) Osteoarthritis of the hip: An occupational disease in farmers. *Brit Med Journal*, *304*: 1269–1272.

Crosby, A.C. (1985) The hands of karate experts. Clinical and radiological findings. *Br J Sports Med*, *19*: 41–42.

Cuccurullo, G.D.G. and F. Croce (1980) Macroscopic and microscopic aspects of osteophytes in advanced osteoarthritis of the hip. *Ital J Orthop Traumatology*, *6*: 117–122.

Currey, J. (1984) *The Mechanical Adpatations of Bone*. Princeton: Princeton University Press.

Dahl, W., K. Matthews, J. Midthun et al. (1981) "Musher's Knee" and "Hooker's Elbow" in the arctic. Correspondence. *New Eng J Med*, *304*: 737.

Dalen, N. and K.E. Olsson (1974) Bone mineral content and physical activity. *Acta Orthop Scan*, *45*: 170.

Daly, R.M., P.A. Rich, and R. Klein (1997) Influence of high impact loading on ultrasound bone measurements in children: A cross-sectional report. *Calcif Tissue Int*, *60*: 401–404.

Danielsen, L., O. Aalund, P.H. Mazza et al. (1989) Injuries due to deliberate violence in areas of Argentina. II. Lesions. *Forensic Sci Inter*, *42*: 165–175.

Danielsson, L.G. (1964) Incidence of osteoarthritis of the hip. *Acta Orthop Scand*, *Suppl 66*: 1–114.

Danielsson, L.G. and J. Hernborg (1970) Clinical and roentgenologic study of knee joints with osteophytes. *Clin Orthop Rel Res*, *69*: 302–312.

Davis, M.A., W.M. Ettinger, and J.M. Neuhaus (1988) The role of metabolic factors and blood pressure in the association of obesity with osteoarthritis of the knee. *J Rheumatol*, *15*: 1827–1832.

Deacon, A., K. Bennell, Z.S. Kiss et al. (1997) Osteoarthritis in the knee of retired, elite Australian Rules footballers. *Med J Aust*, *166*: 187–190.

Deleyiannis, F.W., B.D. Cockcroft, and E.F. Pinczower (1996) Exostoses of the external auditory canal in Oregon surfers. *Am J Otolarynol*, *17*: 303–307.

Demster, D.W. (1992) Bone remodeling. In: Coe, F.L. and M.J. Favus (eds.), *Disorders of Bone and Mineral Metabolism*. New York: Raven Press, pp. 355–380.

Dequeker, J.A. (1975) Bone and aging. *Ann Rheum Dis*, *34*: 100–115.

Dequeker, J., A. Burggens, and G. Creytens (1977) Are osteoarthrosis and osteoporosis the end result of normal ageing or two different disease entities? *Acta Rheum Belgiae*, *1*: 46–57.

Dequeker, J., L. Mokassa, J. Aersseus, and S. Boonen (1997) Bone density and local growth factors in generalized osteoarthritis. *Microsc Res Tech*, *37*: 358–371.

De Rousseau, C.J. (1988) *Osteoarthritis in Rhesus Monkeys and Gibbons: A Locomotor Model of Joint Degeneration*. Contributions to Primatology, Vol. 25, Basel: Karger.

Di Bartolomeo, J. (1979) Exostoses of the external auditory canal. *Ann Otol Rhinol Laryngol* (*Suppl 66*), *88*: 1–17.

Dieppe, P. (1987) Osteoarthritis and related disorders. In: Weatherall, D.J. and J.G.G. Ledingham (eds.), *Oxford Textbook of Medicine*. New York: Oxford University Press, pp. 16.76–16.84.

Dieppe, P. (1990) Osteoarthritis: A review. *J Royal Coll Phy London*, *24*: 262–267.

Dieppe, P. and I. Watt (1985) Crystal deposition in osteoarthritis: an opportunistic event? *Clin Rheum Dis*, *11*: 367–392.

Dobyns, J.H. and R.L. Linscheid (1984) Fractures and dislocations of the wrist. In: Rockwood, C.A. and D.P. Green (eds.), *Fractures in Adults*, 2nd ed. Philadelphia: J.B. Lippencott, pp. 411–509.

Doherty, M. and P. Dieppe (1986) Crystal deposition disease in the elderly. *Clin Rheum Dis*, *12*: 97–116.

Donaldson, C.L., S.B. Hulley, J.M. Vogel et al. (1970) Effects of prolonged bed rest on bone mineral. *Metabolism*, *19*: 1071–1084.

Douglas, D. (1993) Auditory exostoses at CA-Ala-329. Unpublished report, San Jose State University, Anthropology Laboratory.

Duncan, H. (1979) Osteoarthritis. *Henry Ford Hosp Med J*, *27*: 6–9.

Duncan, H. (1983) Cellular mechanisms of bone damage and repair in the arthritic joint. *J Rheumatol* Suppl., *11*: 29–37.

Du Toit, F.P. and R.P. Gräbe (1979) Isolated fractures of the shaft of the ulna. *South African Med J*, *56*: 21–25.

Dutour, O. (1986) Enthesopathy (lesions of muscle insertions) as indicators of activities of Neolithic Saharan populations. *Am J Phys Anthropol*, *71*: 221–224.

Dymond, I.W.D. (1984) The treatment of isolated fractures of the distal ulna. *J Bone Joint Surg*, *66B*: 408–410.

Ebong, W.W. (1985) Osteoarthritis of the knee in Nigerians. *Ann Rheum Dis*, *44*: 682–684.

Edynak, G.J. (1976) Life-styles from skeletal material. In: Giles, E. and J. Friedlaender (eds.), *The Measures of Man*. Cambridge: Peabody Museum, pp. 408–432.

Ehrlich, G.E. (1981) Osteoarthritis before, during, and after rheumatoid arthritis. *Semin Arthritis Rheum*, Vol. XI (1), Suppl 1: 123–124.

Eichner, E. (1989) Does running cause osteoarthritis? An epidemiological perspective. *The Physician and Sports Med*, *17*: 147–154.

Eisenberg, L. and D. Huttchinson (1996) Introduction. *Int J Osteoarchaeol*, *6*: 1.

Elliot-Smith, G. and F. Wood Jones (1910) *The Archaeological Survey of Nubia Report for 1907–1908*, Vol. II. *Report on the Human Remains*. Cairo: National Printing Dept.

Ende, L.S. and J. Wickstrom (1982) Ballet injuries. *Physician and Sportsmedicine*, *10*: 101–118.

Epstein, J.A. and L.S. Lavine (1964) Herniated lumbar intervertebral discs in teen-age children. *J Neurosurg*, *21*: 1070–1075.

Evans, E.M. (1949) Pronation injuries of the forearm with special reference to the anterior Monteggia fracture. *J Bone Joint Surg*, *31B*: 578–588.

Fedigan, L.M. (1986) The changing role of women in models of human evolution. *Ann Rev Anthropol*, *15*: 25–66.

Feik, S.A., C.D. Thomas, and J.G. Clement (1996) Age trends in remodeling of the femoral midshaft differ between the sexes. *J Orthop Res*, *14*: 590–597.

Felson, D.T. (1988) Epidemiology of hip and knee osteoarthritis. *Epidemiol Rev*, *10*: 1–28.

Felson, D.T., J.J. Anderson, A. Naimark et al. (1988) Obsesity and knee osteoarthritis. The Framingham Study. *Ann Intern Med*, *109*: 18–24.

Felson, D.T., J.J. Anderson, J.J. Naimark et al. (1989a) The prevalence of chondrocalcinosis in the elderly and its association with knee osteoarthritis. *J Rheumatol*, *16*: 1241–1245

Felson, D.T., J.J. Anderson, A. Naimark et al. (1989b) Does smoking protect against osteoarthritis? *Arthritis Rheum*, *32*: 166–172.

Felson, D.T., A. Naimark, J. Anderson et al. (1987) The prevalence of knee osteoarthritis in the elderly: The Framingham Study. *Arth Rheum*, *30*: 914–918.

Felson, D.T., M.T. Hannan, A. Naimark et al. (1991) Occupational physical demands, knee bending, and knee osteoarthritis: results from the Framingham study. *J Rheumatol*, *18*: 1587–1592.

Ferguson, C. (1980) Analysis of skeletal remains. In: Cordell (ed.), *Tijeras Canyon*: *Analysis of the Past*. Albuquerque: University of New Mexico.

Field, G.P. (1878) A case of ivory exostoses in both ears. *Lancet*, *2*: 81.

Fischer, A. (1932) Rheumatismus als Berufskrankheit. *Acta Rheumatol*, *4*: 24–28.

Fitzgerald, B. and G.R. McClatchie (1980) Degenerative joint disease in weight-lifters. Fact or fiction? *Brit J Sports Med*, *14*: 97–101.

Forman, M., R. Malamet, and D. Kaplan (1983) A survey of osteoarthritis of the knee in the elderly. *J Rheumatol*, *10*: 283–287.

Fowler, E.P. and P.M. Osmun (1942) New bone growth due to cold water in ears. *Arch Otolaryngol*, *36*: 455–466.

Fowles, J.V., N. Silman, and M.T. Kassab (1983) The Monteggia lesion in children. *J Bone Joint Surg*, *65A*: 1276–1283.

Frayer, D.W. (1980) Sexual dimorphism and cultural evolution in the late Pleistocene and Holocene of Europe. *J Hum Evol*, *9*: 399–415.

Frayer, D.W. (1997) Ofnet: Evidence for a Mesolithic massacre. In: Frayer, D.W. and D.L. Martin (eds.), q.v.

Frayer, D.W. and D.C. Martin (eds.) (1997) *Troubled Times: Osteological and Archaeological Evidence of Violence*. Amsterdam: Gordon and Breach.

Fresia, A.E., C.B. Ruff, and C.S. Larsen (1990) Temporal decline in bilateral asymmetry of the upper limb on the Georgia coast. In: Larsen, C.S. (ed.), *The Archaeology of Mission Santa Catalina de Guale: 2 Biocultural Interactions of a Population in Transition. Anthropol Papers, Am Mus Nat Hist*, *68*: 121–132.

Froimson, M.J., K.A. Athanasion, M.A. Kelley et al. (1990) Mismatch of cartilagenous properties at the patellofemoral joint. *Trans Orthop Res Soc*, *15*: 152.

Frost, H.M. (1991) A new direction for osteoporosis research: A review and proposal. *Bone*, *12*: 429–437.

Galloway, A. (1995) Determination of parity from the maternal skeleton; an appraisal. *Revista di Anthropologia*, *73*: 83–98.

Garden, R.S. (1961) Tennis elbow. *J Bone Joint Surg*, *43B*: 100–106.

Gardner, D.L. (1983) The nature and causes of osteoarthritis. *Brit Med J*, *286*: 418–424.

Gainor, B.J., G. Piotrowski, J. Puhl et al. (1980) The throw: biomechanics and acute injury. *Am J Sports Med*, *8*: 114–118.

Garn, S.M. (1970) *The Earlier Gain and Later Loss of Cortical Bone in Nutritional Perspective*. Springfield, IL: Charles C. Thomas.

Garn, S.M., E. Pao, and M. Ruhl (1964) Compact bone in Chinese and Japanese. *Science*, *143*: 1439–1440.

Gejvall, N.-G. (1983) Wear-and-tear: chronic traumatic lesions. In: Hart, G. (ed.), *Diseases in Ancient Man*. Toronto: Clarke Irwin, pp. 84–96.

Genti, G. (1989) Occupation and osteoarthritis. *Baillieres Clin Rheumatol*, *3*: 193–204.

Geist, E.S. (1931) The intervertebral disk. *JAMA*, *96*: 1676–1679.

Gevers, G., J. Dequeker, P. Geusens et al. (1989) Physical and histomorphological characteristics of the iliac crest bone differ according to the grade of osteoarthritis of the hand. *Bone*, *10*: 173–177.

Giffin, N.M. (1930) *The Roles of Men and Women in Eskimo Culture*. Chicago: University of Chicago Press.

Gilbert, B.M. and T.W. McKern (1973) A method for aging the female *os pubis*. *Am J Phys Anthropol*, *38*: 31–38.

Gilmor, M.K. (1990) A new approach to the use of pubic scars to estimate fertility. M.A. thesis, California State University, Fullerton, Fullerton, CA.

Glick, R. and N. Parhami (1979) Frostbite arthritis. *J Rheumatol*, *6*: 456–460.

Goldstein, M.S. (1957) Skeletal pathology of early Indians in Texas. *Am J Phys Anthropol*, *15*: 299–307.

Goodall, J. (1986) *The Chimpanzees of Gombe. Patterns of Behavior*. Cambridge: Harvard University Press.

Goodman, A.H., J. Lallo, G.J. Armelagos et al. (1984a) Health changes at Dickson Mounds (AD 950-1300). In: Cohen, M.N. and G.J. Armelagos (eds.), q.v., pp. 271–305.

Goodman, A.H., D. Martin, G.J. Armelagos et al. (1984b) Indications of stress from bone and teeth. In: Cohen, M.N. and G.J. Armelagos (eds.), q.v., pp. 13–49.

Goodship, A.E., L.E. Lanyon, and H. McFie (1979) Functional adaptation of bone to increased stress. *J Bone Joint Surg*, *61A*: 539–546.

Goulding, A., E. Gold, R. Cannan et al. (1996) Changing femoral geometry in growing girls: A cross-sectional DEXA study. *Bone*, *19*: 645–649.

Grauer, A.L. and C.A. Roberts (1996) Paleoepidemiology, healing, and possible treatment of trauma in a Medieval cemetery population of St. Helens-on-the-Walls, York, England. *Am J Phys Anthropol*, *100*: 531–544.

Gremion, G. and A. Chantraine (1990) Sport et arthrose. *Schweiz Z Sportmed*, *38*: 143–149.

Gresham, G.E. and W.K. Rathey (1975) Osteoarthritis in knees of aged persons. Relationship between roentegenographic and clinical manifestations. *JAMA*, *233*: 168–170.

Griffin, C.J., R. Powers, and R. Kruszynski (1979) The incidence of osteoarthritis of the temporomandibular joint in various cultures. *Austr Dent J*, *24*: 94.

Grimston, S.K., N.D. Willows, and D.A. Hanley (1993) Mechanical loading regime and its relationship to bone mineral density in children. *Med Sci Sports Exerc*, *25*: 1203–1209.

Guinness-Hay, M. (1980) The Koniag Eskimo presacral vertebral column: variation, anomalies, and pathologies. *Ossa*, *7*: 99–118.

Hadler, N.M. (1977) Industrial rheumatology: Clinical investigation into the influence of the pattern of useage on the pattern of regional musculoskeletal disease. *Arthritis Rheum*, *20*: 1019–1025.

Hadler, N.M. (1985) Osteoarthritis as a public health problem. *Clin Rheum Dis*, *11*: 175–185.

Hadler, N.M., D.B. Gillings, R. Imbus et al. (1978) Hand structure and function in an industrial setting. Influence of three patterns of stereotyped repetitive usage. *Arthritis Rheum*, *21*: 210–220.

Hamerman, D. (1989) The biology of osteoarthritis. *New Eng J Med*, *320*: 1322–1330.

Han, Z.H., S. Palnitkar, D.S. Rao et al. (1996) Effect of ethnicity and age on menopause on the structure and geometry of the iliac bone. *J Bone Miner Res*, *11*: 1967–1975.

Hannan, M.T., J.J. Anderson, T. Pincus et al. (1992) Educational attainment and osteoarthritis: differential associations with radiographic changes and symptom reporting. *J Clin Epidemiol*, *45*: 139–147.

Hannan, M.T., D.T. Felson, J.J. Anderson et al. (1993) Habitual physical activity is not associated with knee osteoarthritis in the Framingham study. *J Rheumatol*, *20*: 704–709.

Hansen, N.M.J. (1982) Epiphyseal changes in the proximal humerus of an adolescent baseball pitcher. *Am J Sports Med*, *10*: 380–384.

Hamanishi, C., T. Kawabata, T. Yosii et al. (1994) Schmorl's nodes on magnetic resonance imaging. Their incidence and clinical relevance. *Spine*, *19*: 450–453.

Hamerman, D. and M. Klagsbrun (1985) Osteoarthritis. Emerging evidence for cell interactions in the breakdown and remodeling of cartilage. *American J Med*, *78*: 495–499.

Hamilton, M.E. (1982) Sexual dimorphism in skeletal samples. In: Hall, R.L. (ed.), *Sexual Dimorphism in* Homo sapiens: *A Question of size*. New York: Praeger, pp. 107–163.

Harman, M., T.I. Molleson, and J.L. Price (1981) Burials, bodies, and beheadings in Romano-British and Anglo-Saxon cemeteries. *Bull Br Mus Nat Hist*, *35*: 145–188.

Harrison, D.F.N. (1951) Exostosis of the external auditory meatus. *J Laryngol Otol*, *65*: 704–714.

Harrison, D.F.N. (1962) The relationship of osteomata of the external auditory meatus to swimming. *Ann R Coll Surg*, *31*: 187–201.

Hawkes, S.C. (1989) Weapons and warfare in Anglo-Saxon England. Oxford: Oxford University Committee for Archaeology Monograph No. 21.

Hawkey, D. and C. Merbs (1995) Activity-induced musculoskeletal stress markers (MSM) and subsistence strategy changes among Hudson Bay Eskimos. *Intl J Osteoarchaeol*, *5*: 324–338.

Healy, J.F., B.B. Healy, W.H. Wong et al. (1996) Cervical and lumbar MRI in asymptomatic older male lifelong athletes: frequency of degenerative findings. *J Comput Assist Tomogr*, *1*: 107–112.

Hellman, D.B., C.A. Helms, and H.K. Genant (1983) Chronic repetition trauma: A cause of atypical degenerative joint disease. *Skeletal Radiol*, *10*: 236–242.

Hermann, B. and T. Bergfelder (1978) Über den diagnostischen Wert des songenannten Geburtstrauma am schambein die der Identifikation. *Zeitschrift für Rechtsmedizin*, *81*: 73–78.

Hernborg, J. and B.E. Nilsson (1973) The relationship between osteophytes in the knee joint, osteoarthritis and aging. *Acta Orthop Scand*, *44*: 69–74.

Hernborg, J. and B.E. Nilsson (1977) The natural course of untreated osteoarthritis of the knee. *Clin Orthop*, *123*: 130–137.

Herring, S.A. and K.L. Nilson (1987) Introduction to overuse injuries. *Clin Sports Med*, *6*: 225–239.

Hershkowitz, I., L. Bedford, L.M. Jellema et al. (1996) Injuries to the skeleton due to prolonged activity in hand-to-hand combat. *Intl J Osteoarchaeol*, *6*: 167–178.

Heyman, J. and A. Lundquist (1932) The symphysis pubis in pregnancy and parturition. *Acta Obstet Gynecol*, *12*: 191–226.

Hilton, R.C., J. Ball, and T.T. Benn (1976) Vertebral endplate lesions (Schmorl's nodes) in the dorsolumbar spine. *Ann Rheum Dis*, *35*: 127–132.

Hoaglund, F.T., C.S. Oishi, and G.G. Gialamas (1995) Extreme variations in racial rates of total hip arthroplasty for primary coxarthrosis: A population-based study in San Francisco. *Ann Rheum Dis*, *54*: 107–110.

Hoaglund, F.T., A.C. Yau, and W.L. Wong (1973) Osteoarthritis of the hip and other joints in southern Chinese in Hong Kong. Incidence and related factors. *J Bone Joint Surg*, *55A*: 545–557.

Hochberg, M.C. (1984) Chondrocalcinosis articularis of the knee: Prevalence and association with osteoarthritis of the knee (abstract). *Arthritis Rheum*, *27*: S49.

Hochberg, M.C., M. Lethbridge-Cejko, M. Plato et al. (1991) Factors associated with osteoarthritis of the hand in males: Data from the Baltimore Longitudinal Study of Aging. *Am J Epidemiol*, *134*: 1121–1127.

Hoffman, D.F. (1993) Arthritis and exercise. *Prim Care*, *20*: 895–910.

Holt, C. (1978) A re-examination of parturition scars on the human female pelvis. *Am J Phys Anthropol*, *49*: 91–94.

Hooton, E.A. (1920) Indian Village site and cemetery, Maddisonville, Ohio. *Papers of the Peabody Museum of American Archaeology and Ethnology*, *8*: 83–104.

Hooton, E.A. (1930) *The Indians of Pecos Pueblo.* New Haven: Yale University Press.

Houghton, P. (1974) The relationship of the pre-auricular groove of the ilium to pregnancy. *Am J Phys Anthropol*, *41*: 381–390.

Hough, A.J., Jr. (1993) Pathology of osteoarthritis. In: McCarty, D.J. and W.J. Coopman (eds.), *Arthritis and Allied Conditions 12th ed.* Philadelphia: Lea and Febiger, pp. 1699–1721.

Houghston, J.C. (1957) Fracture of the distal radial shaft. *J Bone Joint Surg*, *39A*: 249–264.

Houston, C.S. and W.A. Zaleski (1967) The shape of vertebral bodies and femoral necks in relation to activity. *Radiology*, *89*: 59–66.

Howell, D.S. and J.P. Pelletier (1993) Etiopathogenesis of osteoarthritis. In: McCarthy, D.J. and W.J. Coopman (eds.), *Arthritis and Allied Conditions, 12th ed.* Philadelphia: Lea and Febiger, pp. 1723–1734.

Howell, D.S., J.F. Woessner, Jr., S. Jimenez et al. (1979) A view on the pathogenesis of osteoarthritis. *Bull Rheum Dis*, *29*: 996–1001.

Hrdlička, A. (1914) Special notes on some of the pathological conditions shown by the skeletal materials of the ancient Peruvians. *Smithson Miscell Coll*, *61*: 57–69.

Hrdlička, A. (1935) Ear exostoses. *Smithson Miscell Coll*, *93*: 1–100.

Hulkko, A., S. Orova, and P. Nikula (1986) Stress fractures of the olecranon in juvenile throwers. *Int J Sports Med*, *7*: 210.

Hunter, D., A.I.G. McLaughlin, and K.M.A. Parry (1945) Clinical effects of the use of pneumatic tools. *Brit J Industr Med*, *2*: 10–16.

Huthchinson, D.L. (1996) Brief encounters: Tatham Mound and the evidence for Spanish and Native American confrontation. *Intl J Osteoarchaeol*, *6*: 51–65.

Hutchinson, D.L., C.B. Denise, H.J. Daniel et al. (1997) A reevaluation of the cold water etiology of external auditory exostoses. *Am J Phys Anthropol*, *103*: 417–422.

Imeokparia, R., J. Barret, M. Arrieta et al. (1994) Physical activity as a risk factor for osteoarthritis of the knee. *Ann Epidemiol*, *4*: 221–230.

Inglemark, B.E., V. Møller-Christensen, and O. Brinch (1959) Spinal joint changes and dental infections. *Acta Anatomics Suppl*, *36*: *Vol. 38.*

Jacobs, K.H. (1985) Evolution of the postcranial skeleton of late Glacial and early Postglacial Europeans hominids. *Z Morph Anthopol*, *75*: 307–326.

Jacobsson, B., N. Dalen, and B. Tjornstrand (1987) Coxarthrosis and labor. *Int Orthop*, *11*: 311–313.

Jasin, H.E. (1989) Immune mechanisms in osteoarthritis. *Seminar Arthritis Rheum*, *18*: 81–89.

Johnson, L.C. (1966) Discussion. In: Jarcho, S. (ed.), *Human Paleopathology*. New Haven: Yale University Press, pp. 68–81.

Johnson, L.C. (1959) Kinetics of osteoarthritis. *Laboratory Invest*, *8*: 1223–1241.

Jonck, L.M. (1961) The distribution of osteophytes in the lumbar spine in the Bantu. *African Lab Clin Med*, *8*: 71–80.

Jones, H.H., J.D. Priest, W.C. Hayes et al. (1977) Humeral hypertrophy in response to exercise. *J Bone Joint Surg*, *59A*: 204–208.

Jorgensen, V., S. Sonne-Holm, F. Lauridsen et al. (1987) Long-term follow-up of meniscectomy in athletes. A prospective longitudinal study. *J Bone Joint Surg*, *69B*: 80–83.

Judd, M. (1994) Fracture patterns from two Medieval populations in Britain. Unpublished M.Sc. thesis, University of Bradford, Department of Archaeological Sciences.

Julkenen, I., J. Kichila, and H. Julkenen (1981) Etiological, social, and therapeutical aspects of osteoarthrosis and soft-tissue rheumatism in a Finnish health centre. *Scan J Rheum*, *10*: 215–221.

Jurmain, R.D. (1975) The distribution of degnerative joint disease in skeletal populations. Ph.D. thesis, Harvard University, Cambridge, MA.

Jurmain, R.D. (1977a) Stress and the etiology of osteoarthritis. *Am J Phys Anthropol*, *46*: 353–366.

Jurmain, R.D. (1977b) Paleoepidemiology of degenerative knee disease. *Medical Anthropol*, *1*: 1–14.

Jurmain, R.D. (1978) Paleoepidemiology of degenerative joint disease. *Med Coll Virg Quart*, *14*: 45–56.

Jurmain, R.D. (1980) The pattern of involvement of appendicular degenerative joint disease. *Am J Phys Anthropol*, *53*: 143–150.

Jurmain, R.D. (1989) Trauma, degenerative disease, and other pathologies among the Gombe chimpanzees. *Am J Phys Anthropol*, *80*: 229–237.

Jurmain, R.D. (1990) Paleoepidemiology of a central California prehistoric population from CA-Ala-329. II. Degenerative Disease. *Am J Phys Anthropol*, *83*: 83–94.

Jurmain, R.D. (1991a) Paleoepidemiology of trauma in a central California population. In: Ortner, D.J. and A.C. Aufderhiede (eds.), *Paleopathology: Current Synthesis and Future Options.* Washington, D.C.: Smithsonian Institution Press, pp. 241–248.

Jurmain, R.D. (1991b) Degenerative disease as an indicator of occupational stress; opportunities and limitations. *Intl J Osteoarchaeol*, *1*: 247–252.

Jurmain, R.D. (1997) Skeletal evidence of trauma in African apes, with special reference to the Gombe chimpanzees. *Primates*, *38*: 1–14.

Jurmain, R.D. and V.I. Bellifemine (1997) Patterns of cranial trauma in a prehistoric population from Central California. *Intl J Osteoarchaeol*, *7*: 43–50.

Jurmain, R.D. and L. Kilgore (1995) Skeletal evidence of osteoarthritis: a paleopathological perspective. *Ann Rheum Dis*, *54*: 443–450.

Jurmain, R.D. and L. Kilgore (n.d.) Sex-linked patterns of trauma in humans and African apes. In: Grauer, A. and P. Stuart-Macadam (eds.), *Sex and Gender in Paleopathological Perspective.* Cambridge: Cambridge University Press. In preparation.

Kallman, D.A., F.M. Wigley, W.W. Scott et al. (1990) The longitudinal course of hand osteoarthritis in a male population. *Arthritis Rheum*, *33*: 1323–1332.

Karegeannes, J.C. (1995) Incidence of bony outgrowths of the external ear canal in U.S. Navy divers. *Undersea Hyperb Med*, *22*: 301–306.

Keeley, L.H. (1996) *War before Civilization. The Myth of the Peaceful Savage*. New York: Oxford University Press.

Keen, R.W., D.J. Hart, J.S. Lanchbury, and T.d. Spector (1997) Association of biosteoarthritis of the knee with a Taq I polymorphism of the Vitamin D receptor gene. *Arthritis Rheum*, *40*: 1444–1449.

Kellgren, J.H. (1961) Osteoarthrosis in patients and populations. *Brit Med J*, *2*: 1–6.

Kellgren, J.H. and J.S. Lawrence (1957) Radiological assessment of osteoarthritis. *Ann Rheum Dis*, *16*: 494–502.

Kellgren, J.H., and J.S. Lawrence (1952) Rheumatism in miners. Part II: X-ray study. *Brit J Industr Med*, *9*: 197–207.

Kellgren, J.H., J.S. Lawrence, and F. Bier (1964) Genetic factors in osteoarthritis. *Ann Rheum Dis*, *23*: 109–122.

Kelley, M.A. (1979) Parturition and pelvic changes. *Am J Phys Anthropol*, *51*: 541–546.

Kelley, M.A. (1982) Intervertebral osteochondrosis in ancient and modern populations. *Am J Phys Anthropol*, *59*: 271–279.

Kelley, J.O. and J.L. Angel (1987) Life stresses of slavery. *Am J Phys Anthropol*, *74*: 199–211.

Kennedy, G.E. (1985) Bone thickness in *Homo erectus*. *J Hum Evol*, *14*: 699–708.

Kennedy, G.E. (1986) The relationship between auditory exostoses and cold water: A latitudinal analysis. *Am J Phys Anthropol*, *71*: 401–415.

Kennedy, K.A.R. (1983) Morphological variations in ulnar supinator crests and fossae as identifying markers of occupational stress. *J Forensic Sci*, *28*: 871–876.

Kennedy, K.A.R. (1989) Skeletal markers of occupational stress. In: Iscan, M.Y. and K.A.R. Kennedy (eds.), *Reconstruction of Life from the Skeleton*. New York: Wiley-Liss, pp. 129–160.

Kern, D., M.B. Zlatking, and M.K. Dalinka (1988) Occupational and post-traumatic arthritis. *Radiol Clin North Am*, *26*: 1349–1358.

Kilgore, L. (1984) Degenerative joint disease in a Medieval Nubian population. Ph.D. thesis, University of Colorado, Boulder, CO.

Kilgore, L. (1989) Pathologies in ten free-ranging chimpanzees from Gombe National Park, Tanzania. *Amer J Phys Anthropol*, *90*: 219–227.

Kilgore, L. (1990) Biomechanical relationships in the development of degenerative joint disease of the spine. Paper presented at Eighth European Meeting of the Paleopathology Association, Cambridge, England.

Kilgore, L. (1993) Pathological conditions of the dentition of free-ranging chimpanzees. *Am J Phys Anthropol Suppl*, *16*: 125 (abstract).

Kilgore, L. (n.d.) Biomechanical factors in the development of spinal arthropathies. In preparation.

Kilgore, L., R. Jurmain, and D.P. van Gerven (1997) Paleoepidemiological patterns of trauma in a Medieval Nubian skeletal population. *Intl J Osteoarchaeol*, *7*: 103–114.

King, J.W., H.J. Brelsford, and H.S. Tullos (1969) Analysis of the pitching arm of the professional pitcher. *Clin Orthop Rela Res*, *67*: 116–123.

Kipling, R. (1902) *Just So Stories*. New York: Doubleday.

Klünder, K.B., B. Rud, and J. Hansen (1980) Osteoarthritis of the hip and knee joint in retired football players. *Acta Orthop Scan*, *51*: 925–927.

Knüsel, C.J., S. Goggel, and D. Lucy (1997) Comparative degenerative joint disease of the vertebral column in the Medieval monastic cemetery of the Gilbertine Priory of St. Andrew, Fishergate, York, England. *Am J Phys Anthropol*, *103*: 481–495.

Knüsel, C.J., C.A. Roberts, and A. Boylston (1996) Brief Communication: When Adam delved ... An activity-related lesion in three human skeletal populations. *Am J Phys Anthropol*, *100*: 427–434.

Kohatsu, N.D. and D.J. Schurman (1990) Risk factors for the development of osteoarthritis of the knee. *Clin Orthop*, *261*: 242–246.

Konradsen, L., E.M. Hansen, and L. Sundergaard (1990) Long distance running and osteoarthrosis. *Am J Sports Med*, *18*: 379–381.

Kotani, P.T., N. Ichikawa, W. Wakabayashi et al. (1970) Studies of spondylolysis found among weight-lifters. *Br J Sports Med*, *6*: 4–8.

Kouba, R. (1967) Pressluftschäden bei Steinarbeiten. *Zbl Arbeitsmed*, *17*: 67–73.

Krogman, W.M. (1940) The skeletal and dental pathology of an early Iranian site. *Bull Hist Med*, *8*: 28–48.

Krølner, B. and B. Toft (1983) Vertebral bone loss: And unheeded side effect of therapeutic bed rest. *Clin Sci*, *64*: 537–540.

Krølner, B., B. Toft, S.P. Nielsen et al. (1983) Physical exercise as prophylaxis against involutional vertebral bone loss: A controlled trial. *Clin Sci*, *64*: 541.

Kuiper, J.W., C. Van Kuijk, and J.L. Grashuis (1997) Distribution of trabecular and cortical bone related to geometry. A quantitative computed tomography study of the femoral neck. *Invest Radiol*, *32*: 83–89.

Kujala, U.M., J. Kaprio, and S. Sarna (1994) Osteoarthritis of weight bearing joints of lower limbs in former elite athletes. *Brit Med J*, *308*: 231–234.

Kujala, U.M., J. Kettunen, H. Paananen et al. (1995) Knee osteoarthritis in former runners, soccer players, weight lifters, and shooters. *Arthritis Rheum*, *38*: 539–546.

Kumlin, T., M. Wilkeri, and P. Sumari (1973) Radiologic changes in carpal and metacarpal bones and phalanges caused by chain saw vibration. *Br J Industr Med*, *30*: 71.

Kurtz, A.D. (1917) Apophysitis of the os calcis. *Am J Orthop Surg*, *15*: 659–664.

Lai, P. and N.C. Lovell (1992) Skeletal markers of occupational stress in the fur trade: A case study from a Hudson's Bay Company fur trade post. *Intl J Osteoarchaeol*, *2*: 221–234.

Lambert, P.M. (1994) War and peace on the western front: A study of violent conflict and its correlates in prehistoric hunter-gatherer societies of coastal southern California. Ph.D. thesis, University of California, Santa Barbara: Santa Barbara, CA.

Lambert, P.M. (1995) Projectile injuries as evidence for violent conflict in prehistoric populations (abstract). *Am J Phys Anthropol*, *Suppl 20*: 128.

Lambert, P.M. (1997) Patterns of violence in prehistoric hunter gatherer societies of coastal southern Caifornia. In: Frayer, D.W. and D.L. Martin (eds.), q.v. (in press).

Lambert, P.M. and P.L. Walker (1991) Physical anthropological evidence for the evolution of social complexity in coastal southern California. *Antiquity*, *65*: 963–973.

Landau, M. (1984) Human evolution as narrative, *Amer Scientist*, *72*: 262–268.

Lane, N.E. (1995) Exercise: a cause of osteoarthritis. *J Rheumatol, Suppl*, *43*: 3–6.

Lane, N.E. (1996) Physical activity at leisure and risk of osteoarthritis. *Ann Rheum Dis*, *55*: 682–684.

Lane, N.E., D.A. Bloch, H.B. Hubert et al. (1990) Running, osteoarthritis, and bone density: Initial 2-year longitudinal study. *Am J Med*, *88*: 452–459.

Lane, N.E., D.A. Bloch, H.H. Jones et al. (1986) Long-distance running, bone density, and osteoarthritis. *JAMA*, *255*: 1147–1151.

Lane, N.E., D.A. Bloch, H.H. Jones et al. (1989) Osteoarthritis in the hand: a comparison of handedness and hand use. *J Rheumatol*, *16*: 637–642.

Lane, N.E., D.A. Bloch, P.D. Wood et al. (1987) Aging, long-distance running and the development of musculoskeletal disability. A controlled study. *Am J Med*, *82*: 772–780.

Lane, N.E., B. Michel, and A. Bjorkengren (1993) The risk of osteoarthritis with running and aging: A 5-year longitudinal study. *J Rheumatol*, *20*: 461–468.

Lanyon, L.E. (1984) Functional strain as a determinant for bone remodeling. *Calcif Tissue Int*, *36*: S56–S61.

Lanyon, L.E. and D.G. Baggott (1976) Mechanical function as an influence on the structure and form of bone. *J Bone Joint Surg*, *58B*: 436–443.

Lanyon, L.E., W.G.H. Hamson, A.G. Goodship et al. (1975) Bone deformation recorded *in vivo* from strain gauges attached to the human tibial shaft. *Acta Orthop Scand*, *45*: 256–258.

Lanyon, L.E., P.T. Magee, and D.G. Baggott (1979) The relationship of functional stress and strain to the process of bone remodeling. An experimental study of the sheep radius. *J Biomechan*, *12*: 593–600.

Laros, G.S., C.M. Tipton, and R.R. Cooper (1971) Influence of physical activity on ligament insertions in the knees of dogs. *J Bone Joint Surg*, *53A*: 275–285.

Larsen, C.S. (1982) The Anthropology of St. Catherine's Island: 3. Prehistoric Human Biological Adaptation. *Anthrop Pap, Am Mus Nat History*, *57*: 155–270.

Larsen, C.S. (1984) Health and disease in prehistoric Georgia: the transition to agriculture. In: Cohen, M.N. and G.J. Armelagos (eds.), q.v., pp. 367–392.

Larsen, C.S. (1987) Bioarchaeological interpretations of subsistence economy and behavior from human skeletal remains. *Adv Archaeol Meth Theor*, *10*: 339–445.

Larsen, C.S. (ed.), (1990) *The Archaeology of Mission Santa Catalina de Guale: 2 Biocultural Interactions of a Population in Transition. Anthropol Papers Am Mus Nat Hist*, *68*: 121–132.

Larsen, C.S. (1995) Biological changes in human populations with agriculture. *Ann Rev Anthropol*, *24*: 185–213.

Larsen, C.S., J. Craig, and L.E. Sering (1995) Cross Homestead: Life and death on the Midwestern frontier. In: Grauer, A. (ed.), *Bodies of Evidence*. New York: Wiley-Liss pp. 139–159.

Larsen, C.S. and G.R. Milner (eds.) (1994) *In the Wake of Contact. Biological Responses to Conquest*. New York: Wiley-Liss.

Larsen, C.S. and C.B. Ruff (1994) The stresses of conquest in Spanish Florida: structural adaptation and change before and after conquest. In: Larsen, C.S. and G.R. Milner (eds.) q.v., pp. 21–34.

Larson, R.L., K.M. Singer, R. Bergstrom et al. (1976) Little League survey: The Eugene study. *Am J Sports Med*, *4*: 201–209.

Lau, E.M., F. Lin, D. Silman et al. (1995) Hip osteoarthritis and dysplasia in Chinese men. *Ann Rheum Dis*, *54*: 965–969.

Lawrence, J.S. (1955) Rheumatism in coal miners. Part III: Occupational factors. *Brit J Industr Med*, *12*: 249–261.

Lawrence, J.S. (1961) Rheumatism in cotton operatives. *Brit J Industr Med*, *18*: 270–276.

Lawrence, J.S. (1969) Generalized osteoarthritis in a population sample. *Am J Epidemiol*, *90*: 381–389.

Lawrence, J.S. and J. Aitken-Swan (1952) Rheumatism in miners. Part I: Rheumatic complaints. *Brit J Industr Med*, *9*: 1–18.

Lawrence, J.S., J.M. Bremner, and F. Bier (1966) Osteoarthrosis. Prevalence in the population and relationship between symptoms and X-ray changes. *Ann Rheum Dis*, *25*: 1–24.

Lawrence, J.S. and M. Sebo (1980) The geography of osteoarthrosis. In: Nuki, G. (ed.), *The Aetiology of Osteoarthrosis*. Kent: Pitman Medical Pub. Co., pp. 155–183.

Lawrence, R.C., M.C. Hochberg, J.L. Kelsey et al. (1989) Estimates of the prevalence of selected arthritic and musculoskeletal diseases in the United States. *J Rheumatol*, *16*: 427–441.

Layani, F., J. Roeser, and M. Nadeud (1960) Les lesion osteoarticulaires des catcheurs. *Rev Rheum Malad Osteoartic*, *7/8*: 244–248.

Lee, E.J., K.A. Long, and W.L. Risser (1995) Variations in bone status of contralateral and regional sites in young athletic women. *Med Sci Sports Exercise*, *27*: 1354–1361.

Lee, P., P.J. Rooney, R.D. Stonock et al. (1971) The etiology and pathogenesis of osteoarthritis: a review. *Sem Arth Rheum*, *3*: 189–218.

Lehman, W.L., Jr. (1984) Overuse syndromes in runners. *Am Family Phys*, *29*: 157–161.

Levy, L.F. (1967) Lumbar intervertebral disc disease in Africans. *J Neurosurg*, *26*: 31.

Levy, L.F. (1968) Porter's neck. *Brit Med J*, *2*: 16–19.

Lewin, R. (1986) Pressure measured in live hip joint. Research News, *Science*, *232*: 1192–1193.

Lindberg, H. and L.G. Danielsson (1984) The relation between labor and coxarthrosis. *Clin Orthop Rela Res*, *191*: 159–161.

Lindberg, H. and F. Montgomery (1987) Heavy labor and the occurrence of gonarthrosis. *Clin Orthop*, *21*: 235–236.

Linde, F. (1932) Uber Schädigung der Ellenbogengelenke durch Arbeit mit Pressluftwerkzeugen, eine besonders bei Bergleuten häufige Erkrankung. *Münch Med Wehnehr*, *79*: 2009–2011.

Lindholm, C., K. Hagenfeldt, and H. Ringertz (1995) Bone mineral content of young female former gymnasts. *Acta Paediatr*, *84*: 1109–1112.

Liston, M.A. and B.J. Baker (1996) Reconstructing the massacre at Fort William Henry, New York. *Intl J Osteoarchaeol*, *6*: 28–41.

Lovejoy, C.O. and K.G. Heiple (1981) Analysis of fractures in skeletal populations with an example from the Libben Site, Ottawa County, Ohio. *Am J Phys Anthropol*, *55*: 529–541.

Lovell, N.C. (1990) *Patterns of Injury and Illness in Great Apes. A Skeletal Analysis.* Washington, D.C.: Smithsonian Institution Press.

Lovell, N.C. (1991) An evolutionary framework for assessing injury in nonhuman primates. *Yrbook Phys Anthropol*, *34*: 117–155.

Lovell, N.C. (1994) Spinal arthritis and physical stress at Bronze Age Harappa. *Am J Phys Anthropol*, *93*: 149–164.

Lovell, N.C. (1997) Trauma analysis in paleopathology. *Yearbook Phys Anthropol*, *40*: 39–170.

Louyot, P. and R. Savin (1966) La coxarthrose chez l'agriculteur *Revue des Rheumatisme*, *33*: 625–632.

Lynch, F.W. (1920) The pelvic articulations during pregnancy, labor and puerperium. *Surg Gynecol Obstet*, *30*: 575–580.

Mack, P.B., P.A. LaChance, G.P. Vase et al. (1967) Bone demineralization of foot and hand of Gemini-Titan IV, V, and VII astronauts during orbital flight. *Am J Roentgen*, *100*: 503–511.

Maetzel, A., M. Mäkelä, G. Hawker, and C. Bombardier (1997) Osteoarthritis of the hip and mechanical occupational exposure—A systematic overview of the evidence. *J Rheumatol*, *24*: 1599–1607.

Manchester, K. and O.E.C. Elmhirst (1980) Forensic aspects of an Anglo-Saxon injury. *Ossa*, *7*: 179–188.

Mankin, H.J. (1974) The reaction of articular cartilage to injury in osteoarthritis. Parts I and II. *New Engl J Med*, *29*: 1285–1292; 1335–1340.

Mankin, H.J., K.D. Brandt, and L.E. Shulman (1986) Workshop on etiopathogenesis of osteoarthritis. Proceedings and recommendations. *J Rheumatol*, *13*: 1130–1160.

Mankin, H.J. and E. Radin (1985) Stucture and function of joints. In: McCarty, D.J. (ed.), *Arthritis and Allied Conditions*, 10th ed. Philadelphia: Lea and Febiger, pp. 179–195.

Mannienen, P. H. Riihimäck, M. Heliövaara et al. (1996) Overweight, gender, and knee osteoarthritis. *Int J Obes Relat Metab Disord*, *20*: 595–597.

Manzi, G., A. Sperduti, and P. Passarello (1991) Behavior induced auditory exostoses in Imperial Roman society: Evidence from urban and rural communities near Rome. *Am J Phys Anthropol*, *85*: 253–260.

Marcus, R. (1992) Secondary forms of osteoporosis. In: Coe, F.L. and M.J. Favus (eds.), *Disorders of Bone and Mineral Metabolism*. New York: Raven Press, pp. 355–380.

Marti, B. (1991) Health effects of recreational running in women. *Sports Med*, *11*: 20–51.

Marti, B., M. Knobloch, A. Tschopp et al. (1989) Is excessive running predictive of degenerative hip disease? Controlled study of former elite athletes. *BMJ*, *289*: 91–93.

Martin, D.L. (1997) Violence against women in the La Plata River Valley. In: Frayer, D.W. and D.L. Martin (eds.), q.v. (in press).

Martin, D.L., G.J. Armelagos, and J.R. King (1979) Degenerative joint disease of the long bones in Dickson Mounds. *Henry Ford Hosp Med J*, *27*: 60–63.

Martin, R.B. and P.J. Atkinson (1977) Age and sex-related changes in the structure and strength of the human femoral shaft. *J Biomechan*, *10*: 223–231.

Massardo, L., I. Watt, J. Cushnaghan et al. (1989) Osteoarthritis of the knee joint: An eight year prospective study. *Ann Rheum Dis*, *48*: 893–897.

Matsuda, J.J., R.F. Zernicke, R.F. Vailas et al. (1986) Structural and mechanical adaptation of immature bone to strenuous exercise. *J Appl Physiol*, *60*: 2028.

Matthew, P.K., F. Kapua, P.J. Soaki et al. (1996) Trauma admissions in the southern highlands of Papua, New Guinea. *Aust NZ Surg*, *66*: 659–663.

Mays, S.A. (1996) Healed limb amputations in human osteoarchaeology and their causes: A case study from Ipswich, U.K. *Intl J Osteoarchaeol*, 6: 101–113.

McAlindon, T.E., P. Jacques, Y. Zhang et al. (1996) Do antioxidant micronutrients protect against the development and progression of knee osteoarthritis? *Arthritis Rheum*, *39*: 648–658

McAlindon, T.E., D.T. Felson, Y. Zhang et al. (1996) Relation of dietary intake and serum levels of Vitamin D to progression of osteoarthritis of the knee among participants of the Framingham study. *Ann Intern Med*, *125*: 353–359.

McAlindon, T.E., S. Snow, C. Cooper et al. (1992) Radiographic patterns of osteoarthritis of the knee joint in the community: The importance of the patellofemoral joint. *Ann Rheum Dis*, *51*: 844–849.

McAlindon, T.E., S. Snow, K. Vires et al. (1994) Mechanical and constitutional risk factors for symptomatic knee osteoarthritis: differences between medial tibiofemoral and patellofemoral disease. *J Rheumatol*, *21*: 307–313.

McDermott, M. and P. Freyne (1983) Osteoarthrosis in runners with knee pain. *Brit J Sports Med*, *17*: 84–87.

McDonald, E. and C. Marino (1990) Manual labor metacarpophalangeal arthropathy in a baker. *NY State J Med*, *90*: 268–269.

McKeag, D.B. (1992) The relationship of osteoarthritis and exercise. *Clin Sports Med*, *11*: 471–487.

McKinley, J.I. (1993) A decapitation from the Romano-British cemetery at Baldock, Hertfordshire. *Intl J Osteoarchaeol*, *3*: 41–44.

Meade, J.B. (1989) The adaptation of bone to mechanical stress: experimentation and current concepts. In: Cowin, S.C. (ed.), *Bone Mechanics*. Boca Raton, FL: CRC Press, pp. 211–251.

Mebane, W.N. (1981) Dog-walker's elbow. Correspondence. *New Engl J Med*, *304*: 613–614.

Merbs, C.F. (1983) Patterns of activity-induced pathology in a *Canadian* Inuit population. *Archaeological Survey of Canada, Paper, No. 119.* Ottawa: National Museums of Canada.

Merbs, C.F. (1989a) Trauma. In: Iscan, M.Y. and K.A.R. Kennedy (eds.), *Reconstruction of Life from the Skeleton*. New York: Alan R. Liss, pp. 161–189.

Merbs, C.F. (1989b) Spondylolysis; its nature and anthropological significance. *Int J Anthropol*, *4*: 163–169.

Merbs, C.F. (1995) Incomplete spondylolysis and healing. A study of ancient Canadian Eskimo skeletons. *Spine*, *20*: 2328–2334.

Merbs, C.F. (1996a) Spondylolysis of the sacrum in Alaskan and Canadian Inuit skeletons. *Am J Phys Anthropol*, *100*: 357–367.

Merbs, C.F. (1996b) Spondylolysis and spondylolisthesis: a cost of being an erect biped or a clever adaptation? *Yrbook Phys Anthropol*, *39*: 201–228.

Michael, R.H. and L.E. Holder (1985) The soleus syndrome: A cause of medial tibial stress (shin splints). *Am J Sports Med*, *13*: 87–94.

Micheli, L.J. and M.L. Ireland (1987) Prevention and management of calcaneal apophysitis in children: An overuse syndrome. *J Pediatr Orthop*, 7: 34–38.

Micheli, L.J. and J.D. Klein (1991) Sports injuries in children and adolescents. *Br J Sports Med*, *25*: 6–9.

Micheli, L.J. and R. Wood (1995) Back pain in young athletes. Significant differences from adults in causes and patterns. *Arch Pediatr Adolesc Med*, *149*: 15–18.

Mickic, Z.D.J. (1975) Galeazzi fracture dislocation. *J Bone Joint Surg*, *57A*: 1071–1080.

Mieklejohn, C., C. Schentag, A. Venema, and P. Key (1983) Socioeconomic change and patterns of pathology and variation in the Mesolithic and Neolithic of Western Europe: Some suggestions. In: Cohen, M.N. and G.J. Armelagos (eds.), q.v., pp. 75–100.

Milgram, J.W. (1983) Morphologic alterations of the subchondral bone in advanced degenerative arthritis. *Clin Orthop Rel Res*, *173*: 293–312.

Miller, E. (1992) The effect of horseback riding on the human skeleton. Paper presented at the Paleopathology Asssociation Annual Meeting, Las Vegas, NV.

Miller, J.E. (1960) Javelin thrower's elbow. *J Bone Joint Surg*, *42B*: 788.

Miller, R.J. (1985) Lateral epicondylitis in a prehistoric Central Arizona Indian population from Nuvakwewtagu. In: Merbs, C.F. and R.J. Miller (eds.), *Health and Disease in the Prehistoric Southwest*. Tucson: Arizona State University Press, pp. 391–400.

Miles, J.S. (1966) Diseases encountered at Mesa Verde, Colorado. In: Jarcho, S. (ed.), *Human Paleopathology*. New Haven: Yale University Press, pp. 91–97.

Miles, J.S. (1975) Orthopedic problems of the Wetherill Mesa population in archaeology: Wetherill Mesa Studies. U.S. Department of Interior: National Park Service.

Milner, G.R., E. Anderson, and V.G. Smith (1991) Warfare in late prehistoric West-central Illinois. *Amer Antiquity*, *56*: 581–603.

Mintz, G. and A. Fraga (1973) Severe osteoarthritis of the elbow in foundry workers. *Arch Environ Health*, *27*: 78–80.

Mittelmeir, T., C. Mattheck, and F. Dietrich (1994) Effects of mechanical loading on the profile of human femoral diaphyseal geometry. *Med Eng Phys*, *16*: 75–81.

Molleson, T. (1989) Seed preparation in the Mesolithic: The osteological evidence. *Antiquity*, *63*: 356–362.

Montoye, H.J., E.L.Smith, D.F. Fardon et al. (1980) Bone mineral in senior tennis players. *Scan J Sports Sci*, *2*: 26–32.

Morel, P. and J.L. Demetz (1961) *Pathologic Overuse du Haut Moyes Age*. Paris: Masson.

Moretz, J.A., S.D. Harlan, J. Goodrich et al. (1984) Long-term follow-up of the injuries in high school football players. *Am J Sports Med*, *12*: 298–300.

Morey, E.R. and D.J. Baylink (1978) Inhibition of bone formation during space flight. *Science*, *201*: 1138–1141.

Morse, D. (1969) Ancient diseases in the Midwest. Illinois State Museum, Reports of Investigations, No. 15.

Moskowitz, R.W. (1987) Primary osteoarthritis: epidemiology, clinical aspects, and general management. *Am J Med*, *83*: 5–10.

Moskowitz, R.W. (1993) Clinical and laboratory findings in osteoarthritis. In: McCarty, D.J. and W.J. Coopman (eds.), *Arthritis and Allied Conditions*, 12th ed. Philadelphia: Lea and Febinger, pp. 1735–1760.

Murad, T.A. and D. Mertz (1982) A forensic analysis of a prehistoric vertebral column from northeastern California. *Am J Phys Anthropol*, *57*: 212.

Murray, R.O. and C. Duncan (1975) Athletic activity in adolescence as an etiological factor in degenerative hip disease. *J Bone Joint Surg*, *53B*: 406–419.

Murray-Leslie, C.F., D.J. Lintott, and V. Wright (1977) The knees and ankles in sports and veteran parachutists. *Ann Rheum Dis*, *36*: 327–331.

Myburgh, K.H., S.S. Charette, S. Arnaud et al. (1991) The influence of activity and strength on ulnar bending stiffness in adult men (abstract). *Med Sci Sports Med* (*Suppl*) *23*: S121.

Nagy, B. (1996) The utility of osteoarthritis in behavioral reconstructions. Paper presented at 23rd Annual Meeting, Paleopathology Association, Durham, N.C.

Nakamura, R., Y. Ono, E. Harii et al. (1993) The aetiological significance of work-load in the development of osteoarthritis of the distal interphalangeal joint. *J Hand Surg*, *18*: 540–542.

Nathan, H. (1962) Osteophytes of the vertebral column: an anatomical study of their development according to age, race, and sex with considerations as to their etiology and significance. *J Bone Joint Surg*, *44A*: 243–268.

Nelson, R.K. (1969) *Hunters of the Northern Ice*. Chicago: University of Chicago Press.

Newell, R.L. (1995) Spondylolysis. An historical review. *Spine*, *20*: 1950–1956.

Newman, R.W. (1957) A comparative analysis of prehistoric skeletal remains from the Lower Sacramento Valley. Berkeley: University of California Archaeological Survey Reports, 39. University of California.

Newton, P.M., V.C. Mow, T.R. Gardner et al. (1997) The efect of life-long exercise on canine articular cartilage. *Am J Sports Med*, *25*: 282–287.

Nilsson, B.E., L.G. Danielsson, and S.A.J. Hernborg (1982) Clinical features and natural course of coxarthrosis and gonarthrosis. *Scan J Rheum* (*Suppl*) *43*: 13–21.

Nilsson, B.E. and N.E. Westlin (1971) Bone density in athletes. *Clin Orthop Rel Res*, *77*: 179–182.

Noble, P.C., G.G. Box, E. Kamaric et al. (1995) The effect of aging on the shape of the proximal femur. *Clin Orthop*, *316*: 31–44.

Nordström, P., G. Nordström, and R. Lorentzen (1997) Correlation of bone density to strength and physical activity in young with a low or moderate level of physical activity. *Calcif Tissue Int*, *60*: 332–337.

Noyes, F.R., P.J. Torvik, W.B. Hyde et al. (1974) Biomechanics of ligament failure. *J Bone Joint Surg*, *56A*: 1406–1418.

Nuber, G.W. and M.T. Diment (1992) Olecranon stress fractures in throwers. *Clin Orthop*, *273*: 58–61.

Oddis, C.V. (1996) New perspectives on osteoarthritis. *Am J Med*, *26*: 10–15.

O'Donoghue, D.H. (1970) *Treatment of Injuries to Athletes. 2nd ed.* London: W.B. Saunders.

Olsen, O., E. Vingård, M. Koster et al. (1994) Etiologic fractions for physical work load, sports, and overweight in the occurrence of coxarthrosis. *Scand J Work Environ Health*, *20*: 184–188.

O'Neill, D.B. and L.J. Micheli (1988) Overuse injuries in the young athlete. *Clin Sports Med*, *7*: 591–610.

Ord, R.A. and R.M. Benian (1995) Baseball bat injuries to the maxillofacial region caused by assault. *J Oral Maxofac Surg*, *53*: 514–517.

Ortner, D.J. (1968) Description and classification of degenerative bone changes in the distal joint surfaces of the humerus. *Am J Phys Anthropol*, *28*: 139–156.

Ortner, D.J. and W.G.J. Putschar (1981) *Identification of Pathological Conditions in Human Skeletal Remains*. Washington, D.C.: Smithsonian Institution Press.

Owsley, D.W., H.E. Berryman, and W.M. Bass (1977) Demographic and osteological evidence for warfare at the Larson site, South Dakota. *Plains Anthropol Memoir*, *13*: 119–131.

Owsley, D.W., G.W. Gill, and S.D. Owsley (1994) Biological effects of European contact on Easter Island. In: Larsen, C.S. and G.R. Milner (eds.), q.v., pp. 161–177.

Owsley, D.W., J. Orser, C.E. Mann et al. (1987) Demography and pathology of an urban slave population from New Orleans. *Am J Phys Anthropol*, *74*: 185–197.

Panush, R.S. and D.G. Brown (1987) Exercise and arthritis. *Sports Med*, *4*: 54–64.

Panush, R.S. and J.D. Inzinna (1994) Recreational activities and degenerative joint disease. *Sports Med*, *17*: 1–5.

Panush, R.S. and N.E. Lane (1994) Exercise and the musculoskeletal system. *Baillierés Clin Rheumatol*, *8*: 79–102.

Panush, R.S., C. Schmidt, J.R. Caldwell et al. (1986) Is running associated with degenerative joint disease? *JAMA*, *255*: 1152–1154.

Pappas, A.M. (1982) Elbow problems associated with baseball during childhood and adolescence. *Clin Orthop*, *164*: 30–41.

Partridge, R.E.H., J.A.D. Anderson, M.A. McCarthy et al. (1968) Rheumatic complaints among workers in iron foundries. *Ann Rheum Dis*, *27*: 441–453.

Partridge, R.E.H. and J.J.R. Duthie (1968) Rheumatism in dockers and civil servants: A comparison of heavy manual and sedentary workers. *Ann Rheum Dis*, *27*: 559–568.

Pearce, M.S., Y.E. Buttery, and R.N. Brueton (1996) Knee pathology among seafarers: A review of 299 patients. *Occup Med*, *46*: 137–140.

Petersson, C.J. (1983) Degeneration of the gleno-humeral joint. *Acta Orthop Scand*, *54*: 277.

Peyron, J.G. (1986) Osteoarthritis: The epidemiologic viewpoint. *Clin Orthop*, *213*: 13–19.

Pfeiffer, S. (1977) The skeletal biology of Archaic populations of the Great Lakes region. *Archaeol Surv Can Pap*, *64*: 1–384.

Pfeiffer, S. (1985) Paleopathology of Archaic peoples of the Great Lakes. *Canad Rev Phys Anthropol*, *4*: 1–7.

Pickering, R.B. (1979) Hunter-gatherer/agriculturist arthritic patterns: a preliminary investigation. *Henry Ford Hosp Med J*, *27*: 50–53.

Pickering, R.B. (1984) An examination of patterns of arthritis in Middle Woodland, Late Woodland, and Mississippian skeletal series from the Lower Illinois Valley. Ph.D. thesis, Northwestern University, Evanston, IL.

Pierce, L.C. (1987) A comparison of the pattern of involvement of degenerative joint disease between an agricultural and non-agricultural skeletal series. Ph.D. thesis, University of Tennessee, Knoxville, Tennessee.

Pietrusewsky, M. and M.T. Douglas (1994) An osteological assessment of health and disease in precontact and historic (1778) Hawaii. In: Larsen, C.S. and G.R. Milner (eds.), q.v., pp. 179–196.

Pommer, G. (1920) Die functionelle Theorie der Arthritis deformans vor den Forum das Tierversuchen und der Pathologischen Anatomie. *Arch f Orthop U Unfallchir*, *17*: 573–593.

Preidler, K.W. and D. Resnick (1996) Imaging of osteoarthritis. *Radiol Clin North Am*, *34*: 259–271.

Priest, J.D., H.H. Jones, and D.A. Nagel (1974) Elbow injuries in highly skilled tennis players. *J Sports Med*, *2*: 137–149.

Priest, J.D., H.H. Jones, C.J.C. Tichenor et al. (1977) Arm and elbow changes in expert tennis players. *Minnesota Med*, *60*: 399–404.

Puranen, J., L. Ala-Ketola, P. Peltokallio et al. (1975) Running and primary osteoarthritis of the hip. *Brit Med J*, *2*: 424–425.

Putschar, W. (1927) Zür Kenntniss der Knorpelinseln in den Wirbelkörpern. *Beitr Path Anat*, *79*: 150.

Putschar, W. (1931) *Entwicklung, Wachstrum und pathologie der Beckenverbindungen des Menschen, mit besonders Beruchsichtigung von Schwangerschaft, Gerburt und ihren folgen.* Jenai: Gutav Fischer.

Putschar, W. (1976) The structure of the human symphysis pubis with special consideration of parturtion and its sequelae. *Am J Phys Anthropol*, *45*: 589–594.

Radin, E.L. (1975) Mechanical aspects of osteoarthrosis. *Bull Rheum Dis*, *26*: 862–865.

Radin, E.L. (1976) Etiology of osteoarthritis. *Clin Rheum Dis*, *2*: 509–522.

Radin, E.L. and D.B. Burr (1984) Hypothesis: Joints can heal. *Semin Arth Rheum*, *13*: 293–302.

Radin, E.L., D. Eyre, J.L. Kelman et al. (1979) Effect of prolonged walking on concrete on the joints of sheep (abstract). *Arthritis Rheum*, *22*: 649.

Radin, E.L. and I.L. Paul (1971) Responses of joints to impact loading. 1. *In vitro* wear. *Arthritis Rheum*, *14*: 356–362.

Radin, E.L., I.C. Paul, and R.M. Rose (1972) Mechanical factors in osteoarthritis. *The Lancet*, *1*: 519–522.

Radin, E.L. and R.M. Rose (1986) Role of subchondral bone in the initiation and prognosis of cartilage damage. *Clin Orthop Rel Res*, *213*: 34–40.

Rall, K.L., G.L. McElroy, and T.E. Keats (1964) A study of long term effects of football injury to the knee. *Missouri Med*, *61*: 435–438.

Radosevich, S.C. (1993) The six deadly sins of trace element analysis: A case of wishful thinking in science. In: Sandford, M.K. (ed.),

Investigations of Ancient Human Tissue. Clinical Analyses in Anthropology. Newark: Gordon and Breach.

Rasmussen, O.V. (1990) Medical aspects of torture. Torture forms and their relation to symptoms and lesions in 200 victims, followed by a description of the medical profession in relation to torture. Thesis. Logeforenivseus Forlag, University of Copenhagen.

Rathbun, J.B. and I. McNab (1970) The microvascular pattern of the rotator cuff. *J Bone Joint Surg*, *52B*: 540–553.

Rathbun, T.A. (1984) Skeletal pathology from the Paleolithic through the Metal Ages in Iran and Iraq. In: Cohen, M.N. and G.J. Armelagos (eds.), q.v., pp. 137–167.

Rathbun, T.A. (1987) Health and disease at a South Carolina plantation: 1840–1870. *Am J Phys Anthropol*, *74*: 239–253.

Reed, M.F.T. (1981) Stress fractures of the distal radius in adolescent gymnasts. *Br J Sports Med*, *15*: 272–276.

Reinhard, K.J., L. Tieszen, K.L. Sandness et al. (1994) Trade, contact, and female health in northeast Nebraska. In: Larsen, C.S. and G.R. Milner (eds.), *In the Wake of Contact: Biological Responses to Conquest*. New York: Wiley-Liss, pp. 63–74.

Resnick, D. and G. Niwayama (1978) Intravertebral herniation: Cartilagenous (Schmorl's) nodes. *Radiology*, *126*: 57–65.

Resnick, D. and G. Niwayama (1981) *Diagnosis of Bone and Joint Disorders*. Vol. 2. Philadelphia: W.B. Saunders.

Resnick, D. and G. Niwayama (1983) Entheses and enthesopathy: Anatomical, pathological and radiological correlation. *Radiology*, *146*: 1–9.

Resnick, D. and G. Niwayama (eds.) (1988) *Diagnosis of Bone and Joint Disorders*. 2nd ed., Vol. 3. Philadelphia: W.B. Saunders.

Resnick, D., R.F. Shapiro, K.B. Wiesner et al. (1978) Diffuse idopathic skeletal hyperostosis (DISH). *Semin Arthritis Rheum*, *7*: 153–187.

Roach, K.E., V. Persky, T. Miles et al. (1994) Biomechanical aspects of occupation and osteoarthritis of the hip: A case control study. *J Rheumatol*, *21*: 2334–2340.

Robb, J. (1994) Skeletal signs of activity in the Italian Metal Ages: Methodological and interpretive notes. *Human Evol*, *9*: 215–229.

Robb, J. (1997) Violence and gender in early Italy. In: Frayer, D.W. and D.L. Martin (eds.), q.v. (in press).

Roberts, C.A. (1987) Case reports in paleopathology, Number 9. Possible case of scurvy. *Paleopath Newsletter*, *57*: 14–15.

Roberts, C.A. (1991) Trauma and treatment in the Bristish Isles. In: Ortner, D.J. and A.C. Aufderheide (eds.), *Human Paleopathology: Current Synthesis and Future Options*. Washington, D.C.: Smithsonian Inst. Press, pp. 225–240.

Roberts, C.A. and K. Manchester (1995) *The Archaeology of Disease*. 2nd ed. Ithaca, N.Y.: Cornell University Press.

Roche, M.B. (1957) Incidence of osteophytosis and ostearthritis in 419 skeletonized vertebral columns. *Am J Phys Anthropol*, *15*: 433–434.

Roche, M.M.L., J. Maitrepierre, E. Le Jeune et al. (1961) Les atteintes du membre supérieur chez les ouvriers travaillant au marteau pneumatique. *Arch Mal Prop J*, *22*: 57–61.

Rogers, J. and P. Dieppe (1993) Ridges and grooves on the bony surfaces of osteoarthritic joints. *Osteoarthritis and Cartilage*, *1*: 167–170.

Rogers, J., L. Shepstone, and P. Dieppe (1997) Boneformers: Osteophyte and enthesophyte formation are postively associated. *Ann Rheum Dis*, *56*: 85–90.

Rogers, J. and T. Waldron (1989) The paleopathology of enthesopathy. In: Capasso, L. (ed.), *Advances in Paleopathology*. Proceedings of the VII European Meeting of the Paleopathology Association, Lyon, Sept. 1988. *J Paleopathology*, Monographic Pub. No. 1. Chieti: Marino Solfanelli Editore, pp. 169–172.

Rogers, J. and T. Waldron (1995) *A Field Guide to Joint Diseases in Archaeology*. New York: John Wiley.

Rogers, J., T. Waldron, P. Dieppe et al. (1987) Arthropathies in paleopathology: The basis of classification according to most probable cause. *J Archaeol Sci*, *14*: 179–193.

Rogers, J., I. Watt, and P. Dieppe (1981) Arthritis in Saxon and Medieval skeletons. *Brit Med J*, *283*: 1666–1670.

Rogers, J., I. Watt, and P. Dieppe (1990) Comparison of visual and radiographic detection of bony changes at the knee joint. *BMJ*, *300*: 367–368.

Rogers, L.F. (1992) *Radiology of Skeletal Trauma*. New York: Churchill Livingstone.

Roney, J.G. (1966) Palaeoepidemiology: An example from California. In: Jarcho, S. (ed.), *Human Paleopathology*. New Haven: Yale University Press, pp. 99–107.

Rose, C.P. and W.P. Cockshott (1982) Anterior femoral erosion and patello-femoral osteoarthritis. *J Canad Assoc Radiologists*, *33*: 32–34.

Rose, J.C., B.A. Burnett, M.S. Nassaney, and M.N. Blauer (1984) Palaeopathology and the origin of maize agriculture in the Lower Missippi Valley. In: Cohen, M.N. and G.J. Armelagos (eds.), q.v., pp. 393–424.

Rosenberg, B.N., J.C. Richmond, and W.N. Levine (1995) Long-term followup of Bankart reconstruction. Incidence of late degenerative glenohumeral arthrosis. *Am J Sports Med*, *23*: 538–544.

Rostock, P. (1936) Gelenkenschaden durch Arbeiten mit Prestluftwerkenzeugen und andere schwere Röperliche Arbeit. *Medizinische Klinik*, *11*: 341–343.

Rothschild, B.M. (1990) Radiologic assessment of osteoarthritis in dinosaurs. *Ann Carnegie Mus*, *59*: 295–301.

Rotshchild, B.M. (1993) Skeletal paleopathology of rheumatic diseases; The sub-*Homo* connection. In: McCarty, D.J. and W.J. Koopman (eds.), *Arthritis and Allied Conditions*, 6th ed. Philadelphia: Lea & Febiger, pp. 3–7.

Rothschild, B.M. (1995) Paleopathology, its character and contribution to understanding and distinguishing among rheumatic diseases: Perspectives on rheumatoid arthritis and spondyloarthropathy. *Clin Exper Rheumatol*, *13*: 657–662.

Rothschild, B.M. (1997) Porosity: A curiosity without diagnostic significance. *Am J Phys Anthropol*, *104*: 529–533.

Rothschild, B.M. and C.R. Rothschild (1994) No laughing matter; Spondyloarthropathy and osteoarthritis in Hyaenidae. *J Zoo Wildlife Med*, *25*: 259–263.

Rothschild, B.M., K.R. Turner, and M.A. DeLuca (1988) Symmetrical erosive peripheral polyarthritis in the Late Archaic Period of Alabama. *Science*, *241*: 1498–1501.

Rothschild, B.M. and R.J. Woods (1992) Osteoarthritis, calcium pyrophosphate deposition disease, and osseous infection in Old World primates. *Am J Phys Anthropol*, *87*: 341–347.

Roy, T.A., C.B. Ruff, and C.C. Plato (1994) Hand dominance and bilateral asymmetry of the second metacarpal. *Am J Phys Anthropol* *94*: 203–211.

Rubin, C.T. and L.E. Lanyon (1984a) Dynamic strain similarity in vertebrates: An alternative to allometric limb bone scaling. *J Ther Biol*, *107*: 321–327.

Rubin, C.T. and L.E. Lanyon (1984b) Regulation of bone formation by applied dynamic loads. *J Bone Joint Surg*, *66*A: 397–402.

Ruff, C.B. (1987) Sexual dimorphism in human lower limb bone structure: relationship to subsistence strategy and sexual division of labor. *J Human Evol*, *16*: 391–416.

Ruff, C.B. (1995) Limb bone structure: Influence of sex, subsistence, and terrain (abstract). *Am J Phys Anthropol* (*Suppl 20*): 186.

Ruff, C.B. and W.C. Hayes (1983a) Cross-sectional geometry of Pecos Pueblo femora and tibiae—a Biomechanical investigation: I. Method and general patterns of variation. *Am J Phys Anthropol*, *60*: 359–381.

Ruff, C.B. and W.C. Hayes (1983b) Cross-sectional geometry of Pecos Pueblo femora and tibiae—a biomechanical investigation: II. Sex, age and side differences. *Am J Phys Anthropol*, *60*: 383–400.

Ruff, C.B. and W.C. Hayes (1984) Age changes in geometry and mineral content of the lower limb bones. *Ann Biomed Eng*, *12*: 573–584.

Ruff, C.B. and H.H. Jones (1981) Bilateral asymmetry in cortical bone of the humerus and tibia—age and sex factors. *Human Biology*, *53*: 69–86.

Ruff, C.B. and C.S. Larsen (1990) Postcranial biomechanical adaptations to subsistence strategy changes on the Georgia coast. In: Larsen, C.S. (ed.), q.v., pp. 94–120.

Ruff, C.B. and J.A. Runestad (1992) Primate limb bone structural adaptation. *Ann Rev Anthropol*, *21*: 407–433.

Ruff, C.B., E. Trinkaus, A.Wallker et al. (1993) Postcranial robusticity in *Homo*. I. Temporal trends and mechanical interpretation. *Am J Phys Anthropol*, *91*: 35–54.

Ruff, C.B., A. Walker, and E. Trinkaus (1994) Postcranial robusticity in *Homo*. II. Humeral bilateral asymmetry and bone plasticity. *Am J Phys Anthropol*, *93*: 1–34.

Ruffer, M.A. and A. Rietti (1912) On osseous lesions in ancient Egyptians. *J Path Bact*, *16*: 439–465.

Russell, M.D. (1987a) Bone breakage in the Krapina collection. *Am J Phys Anthropol*, *72*: 373–379.

Russell, M.D. (1987b) Mortuary practices at the Krapina Neandertal site. *Am J Phys Anthropol*, *72*: 381–397.

Sairanen, E., L. Brushaber, and M. Keskinen (1981) Felling work, low back pain, and osteoarthritis. *Scand J Work Environ Health*, *7*: 18–30.

Sakalinskas, V. and R. Jankanskas (1993) Clinical otosclerosis and auditory exostoses in ancient Europeans (investigation of Lithuanian paleoosteological samples). *J Laryngol Otol*, *107*: 489–491.

Sandford, M.K. (ed.) (1993) *Investigations of Ancient Human Tissue*. Newark: Gordon and Breach.

Sandness, K.L. and K.J. Reinhard (1992) Vertebral pathology in prehistoric and historic skeletons from northeast Nebraska. *Plains Anthropol*, *37*: 299–309.

Schell, L.M., F.E. Johnson, D.R. Smith et al. (1985) Directional asymmetry of body dimensions among White adolescents. *Am J Phys Anthropol*, *67*: 317–322.

Scher, A.T. (1978) Injuries to the cervical spine sustained while carrying loads on the head. *Paraplegia*, *16*: 94–101.

Schlomka, G., G. Schroter, and A. Ocherval (1955) Über der bedeutung der beruflischer Belastung für die entschung der degenerativen Gelankleiden. *Z Gesamte Inn Med*, *10*: 993–999.

Schneider, H.J., A.Y. King, J.L. Brunson et al. (1974) Stress injuries and developmental change of lower extremities in ballet dancers. *Radiology*, *113*: 627–632.

Schmorl, G. (1927) Über die an den Wirbelbandscheiben vorkommenden ausdehnungs—und zerreisungs—vorgänge und die dadurch an ihnen und der wirbelspongosia hervorgerufenen veränderungen. *Verh dtsch path Ges*, *22*: 250.

Schmorl, G. and H. Junghanns (1971) *The Spine in Health and Disease*. 2nd ed. New York: Grune and Stratton.

Schouten, J.S., F.A. van den Ovweland, and H.A. Valkenburg (1992) A 12 year follow up study in the general population on prognostic factors of cartilage loss in osteoarthritis of the knee. *Ann Rheum Dis*, *51*: 932–937.

Scott, R. (n.d.) Data survey of competitive surfers from Stubbies Classic Surf Contest, San Onofre, CA, March 1977. (reported in Douglas, 1993, q.v.)

Schultz, A.H. (1939) Notes on diseases and healed fractures in wild apes. *Bull Hist Med*, *7*: 571–582.

Schultz, A.H. (1941) Growth and development of the orangutan. *Contr Embryol*, *29*: 57–110.

Schultz, A.H. (1950) Morphological observations on gorillas. In: Gregory, W.K. (ed.), *The Anatomy of the Gorilla*. New York: Columbia University Press, pp. 227–254.

Schultz, R. (1990) *The Language of Fractures*. 2nd ed. Baltimore: Williams and Wilkins.

Seedhom, B.B., T. Takeda, M. Tsubuku et al. (1979) Mechanical factors and patellofemoral osteoarthritis. *Ann Rheum Dis*, *38*: 307–316.

Seftel, D.M. (1977) Ear canal hyperostosis—surfer's ear. An improved surgical technique. *Arch Otolarygnol*, *103*: 58–60

Shaibani, A., R. Workman, and B.M. Rothschild (1993) The significance of enthesopathy as a skeletal phenomenon. *Clin Exper Rheumatol*, *11*: 399–403.

Sheperd, J.P., M. Shapland, N.X. Pearce et al. (1990) Pattern, severity, and aetiology of injuries in victims of assault. *J Roy Soc Med*, *83*: 75–78.

Shermis, S. (1982) Domestic violence in two skeletal populations. *Ossa*, *9–11*: 143–151.

Shumacher, H.R., C. Argudelo, and R. Labowitz (1972) Jackhammer arthropathy. *J Occup Med*, *14*: 563–564.

Silberberg, M., S.F. Jarrett, and R. Silberberg (1956) Obesity and degnerative joint disease in 'yellow' mice. *Arch Path*, *61*: 280–288.

Silman, A.J. and M.C. Hochberg (1993) *Epidemiology of the Rheumatic Diseases*. New York: Oxford University Press.

Slemenda, C.W., C.H. Turner, M. Peacock et al. (1996) The genetics of proximal femur geometry, distribution of bone mass and bone mineral denisty. *Osteoporosis Int*, *6*: 178–182.

Smith, D.H. and F.P. Sage (1957) Medullary fixation of forearm fractures. *J Bone Jt Surg*, *39(A)*: 91.

Smith, E.L., P.E. Smith, C.J. Ensign et al. (1984) Bone involution decrease in exercising middle-aged women. *Calcif Tissue Int*, *36*: S129–S138.

Smith, M.O. (1996) 'Parry' fractures and female-directed interpersonal violence: Implications from the Late Archaic Period of west Tennessee. *Intl J Osteoarchaeol*, *6*: 84–91.

Smith, M.O. (1997) Osteological indicators of warfare in the Archaic period of the Western Tennessee Valley. In: Frayer, D.W. and D.L. Martin (eds.), q.v. (in press).

Snow, C.E. (1948) Indian Knoll Skeletons of Site Oh 2 Ohio County, Kentucky. Knoxsville: University of Kentucky Reports in Anthrop. 4, pp. 321–554.

Sohn, R.S and L.J. Micheli (1985) The effect of running on the pathogenesis of osteoarthritis of the hips and knees. *Clin Orthop*, *198*: 106–109

Sokoloff, L. (1985) Endemic forms of osteoarthritis. *Clin Rheum Dis*, *11*: 187–202.

Sokoloff, L. (1987) Osteoarthritis as a remodeling process. *J Rheumatol*, *14(Suppl)*: 7–10.

Solomon, L., P. Beighton, and J.S. Lawrence (1976) Osteoarthrosis in a rural South African Negro population. *Ann Rheum Dis*, *35*: 274–278.

Solomon, L.P., C.M. Schnitzler, and J.P. Browett (1982) Osteoarthritis of the Hip: The patient behind the disease. *Ann Rheum Dis*, *41*: 118–125.

Solonen, K.A. (1966) The joints of the lower extremities of football players. *Ann Chir Gyn Fenn*, *55*: 176–180.

Sowers, M.F., M. Hochberg, and J.P. Crabbe (1996) Association of bone mineral denisty and sex hormone levels with osteoarthritis of the hand and knee in premenopausal women. *Am J Epidemiol*, *143*: 38–47.

Spector, T.D., F. Cicuttini, F. Baker et al. (1996) Genetic influences on osteoarthritis in women: a twin study. *BMJ*, *312*: 940–943.

Spector, T.D., J.E. Dacre, P.A. Harris et al. (1992) Radiological progression of osteoarthritis: An 11 year follow up study of the knee. *Ann Rheum Dis*, *51*: 1107–1110.

Spector, T.D., P.A. Harris, D.J. Hart et al. (1996) Risk of osteoarthritis associated with long-term weight bearing sports: A radiologic survey of the hips and knees in female ex-athletes and population controls. *Arthritis Rheum*, *39*: 988–995.

Spector, T.D. and D.J. Hart (1992) How serious is knee osteoarthritis? *Ann Rheum Dis*, *51*: 1105–1106.

Spencer, R.F. (1969) *The North Alaskan Eskimo*. Washington, D.C.: Smithsonian Inst. Press

Stäbler, A., M. Bellan, M. Weiss et al. (1997) MR imaging of enhancing interosseous disk herniation (Schmorl's nodes). *Am J Roentgenol*, *168*: 933–938.

Standen, V.G. and B.T. Arriaza (1996) External auditory exostoses in prehistoric Chilean populations: A test of chronology of geographic distribution. Paper presented at Papeopathology Association Annual Meetings, Durham, N.C.

Standen, V.G., B.T. Arriaza, and C.M. Santoro (1997) External auditory exostoses in prehistoric Chilean populations: A test of the cold water hypothesis. *Am J Phys Anthropol*, *103*: 119–129.

Stanitski, C.L. (1994) Overuse syndromes. In: Satnitski, C.L., J.C. Delee, and D. Piez (eds.), *Pediatric and Adolescent Sports Medicine*. Philadelphia: W.B. Saunders, pp. 94–109.

Stecher, R.M. (1940) Heberden's nodes: The incidence of hypertrophic arthritis of the fingers. *New Engl J Med*, *222*: 300–308.

Stecher, R.M. (1959) Heredity of the joint diseases. *Rheumatismo*, *11*: 1–17.

Steen, S.L, S.R. Street, D.E. Hawkey et al. (1996) An application of a comprehensive strategy for recording activity-related stress markers on the skeletal remains of Alaskan Eskimos. Poster presentation, Annual Meetings, Amer. Assoc of Physical Anthropologits, Durham, N.C.

Steensby, H.P. (1910) Contributions to the ethnography and anthropogeography of the Polar Eskimo. *Medd om Grønland*, *34*: 255–405.

Steinbock, R.T. (1976) *Paleopathological Diagnosis and Interpretation*. Springfield, IL: Charles C. Thomas.

Stenlund, B. (1993) Shoulder tendinitis and osteoarthrosis of the acromioclavicular joint and their relation to sports. *Br J Sports Med*, *27*: 125–130.

Stenlund, B., I. Goldie, M. Hagberg et al. (1992) Radiographic osteoarthrosis in the acromioclavicular joint resulting from manual work or exposure to vibration. *Br J Indust Med*, *49*: 588–591.

Stewart, T.D (1947) Racial patterns in vertebral osteoarthritis. *Am J Phys Anthropol*, *5*: 230–231.

Stewart, T.D. (1956) Examination of the possibility that certain skeletal characteristics predispose to defects in neural arches. *Clin Orthop*, *8*: 44–60.

Stewart, T.D. (1957) Distortion of the pubic symphseal surface in females and its effect on age determination. *Am J Phys Anthropol*, *15*: 9–18.

Stewart, T.D. (1958) The rate of development of vertebral osteoarthritis in American Whites and its significance in skeletal age identification. *The Leech*, *28*: 144–151.

Stewart, T.D. (1966) Some problems in human paleopathology In: Jarcho, S. (ed.), *Human Paleopathology*. New Haven: Yale University Press, pp. 43–55.

Stewart, T.D. (1970) Identification of the scars of parturtion in the skeletal remains of females. In: Stewart, T.D. (ed.), *Personal Identification in Mass Disasters*. Washington, D.C.: Smithsonian Inst. Press, pp. 127–135.

Stewart, T.D. (1974) Nonunion of fractures in antiquity, with description of five cases from the New World involving the forearm. *Bull N.Y. Acad Med*, *50*: 875–891.

Stewart, T.D. and L.G. Quade (1969) Lesions of the frontal bone in American Indians. *Am J Phys Anthropol*, *30*: 89–110.

Stirland, A. (1991) The diagnosis of occupationally related paleopathology. Can it be done? In: Ortner, D.J. and A.C. Aufderhiede (eds.), *Paleopathology*: *Current Synthesis and Future Options*. Washington, D.C.: Smithsonian Inst. Press, pp. 40–47.

Stirland, A. (1996) Patterns of trauma in a unique Medieval parish cemetery. *Intl J Osteoarchaeol*, *6*: 92–100.

Stirland, A. (1997) Musculoskeletal evidence for use of the Medieval longbow from King Henry VIII's flagship, the *Mary Rose* (abstract). *Am J Phys Anthropol* (*Suppl 24*): 220.

Straus, W.L. and A.J.E. Cave (1957) Pathology and the posture of Neanderthal man. *Quart Rev Biol*, *32*: 348–363.

Stroud, G. and R. Kemp (1993) Cemeteries of St Andrew, Fishergate. The archaeology of York. The Medieval cemeteries 12/2. York: Council for British Archaeology for York Archaeological Trust.

Ström, C., G. Johanson, and A. Nordenram (1992) Facial injuries due to criminal violence: a retrospective study of hospital attenders. *Med Sci Law*, *32*: 345–353.

Stuart-Macadam, P. (1985) Porotic hyperostosis: representative of a childhood condition. *Am J Phys Anthropol*, *66*: 391–398.

Stuart-Macadam, P. (1997) Interpreting activity patterns: an integrative approach (abstract). *Am J Phys Anthropol* (*Suppl 24*): 223.

Stulberg, S.D., K. Schulman, S. Stuart et al. (1980) Breaststroker's knee: pathology, etiology, and treatment. *Am J Sports Med*, *8*: 164–171.

Suchey, J.M and L.C. Pierce (1992) Assessment of the Gilmor technique for estimating births from the *os pubis* (abstract). *Am J Phys Anthropol* (*Suppl 14*): 159–160.

Suchey, J.M., D.V. Wesley, R.F. Green et al. (1979) Analysis of dorsal pitting in the os pubis in an extensive sample of modern American females. *Am J Phys Anthropol*, *51*: 517–540.

Swann, A.C. and B.B. Seedhöm (1993) The stiffness of normal articular cartilage and the predominant acting stress levels: Implications for the aetiology of osteoarthritis. *Br J Rheumatol*, *32*: 16–25.

Swedborg, I. (1974) *Degenerative Changes in the Human Spine*. Stockholm: Osteological Research Laboratory, University of Stockholm.

Tague, R.G. (1988) Bone resoprtion of the pubis and preauricular area in humans and nonhuman mammals. *Am J Phys Anthropol*, *76*: 251–267.

Tainter, J.A. (1980) Behavior and status in a Middle Woodland mortuary population from the Illinois Valley. *Amer Antiquity*, *45*: 308–313.

Takahashi, K., T. Miyazaki, H. Ohnari et al. (1995) Schmorl's nodes and low back pain. Analysis of magnetic resonance imaging findings in symtomatic and asymtomatic individuals. *Eur Spine J*, *4*: 56–59.

Tenney, J. (1986) Trauma among early California populations (abstract). *Am J Phys Anthropol*, *69*: 271.

Thelin, A. (1990) Hip joint arthrosis: an occupational disorder among farmers. *Am J Ind Med*, *18*: 339–343.

Thould, A.K. and B.T. Thould (1983) Arthritis in Roman Britain. *BMJ*, *287*: 1909–1911.

Tipton, C.M., R.D. Matthes, J.A. Maynard et al. (1975) The influence of physical activity on ligaments and tendons. *Med Sci Sports*, *7*: 165–175.

Trinkaus, E. (1976) The evolution of the hominid femoral diaphysis during the Upper Pleistocene in Europe and the Near East. *Z Morphol Anthropol*, *67*: 291–319.

Trinkaus, E. (1983) *The Shanidar Neandertals*. New York: Academic Press.

Trinkaus, E. (1985) Pathology and posture of the La Chapelle-aux-Saints Neandertal. *Am J Phys Anthropol*, *67*: 19–41.

Trinkaus, E. (1993) Femoral neck-shaft angles of Qafzeh-Skhul early modern humans, and activity levels among immature Near Eastern Middle Paleolithic hominids. *J Human Evol*, *25*: 393–416.

Trinkaus, E., S.E. Churchill, and C.B. Ruff (1994) Postcranial robusticity in *Homo*. III. Ontogeny. *Am J Phys Anthropol*, *93*: 35–54.

Tullos, H.S. and J.W. King (1973) Throwing mechanism in sports. *Orthop Clin North Am*, *4*: 709–721.

Turner, C. (1993) Cannibalism in Chaco Canyon: the charnel pit excavated in 1926 at Small House by Frank H.H. Roberts Jr. *Am J Phys Anthropol*, *91*: 421–439.

Tyson, R.A. (1977) Historical accounts as aids to physical anthropology. Examples of head injury in Baja California. *Pacif Coast Archaeol Soc Q*, *13*: 52–58.

Uitterlinden, A.G., H. Burger, Q. Huang et al. (1997) Vitamin D receptor genotype is associated with radiographic osteoarthritis at the knee. *J Clin Invest*, *100*: 259–263.

University of Manchester, Department of Rheumatology and Medical Illustration (1973) The epidemiology of chronic rheumatism: Vol. II, *Atlas of Standard Radiographs of Arthritis*. Philadelphia: F.A. Davis, pp. 1–15.

U.S. Department of Health, Education, and Welfare (1976) Arthritis: out of the Maze. II. Osteoarthritis. Report to Congress of the United States, April, 1976. N.I.H. Pub. # 76–1151, pp. 249–251.

U.S. Department of Health, Education, and Welfare (1979) Basic data on arthritis of the knee, hip, and sacroiliac joints in adults ages 25–74 years, United States, 1971–1975. DHEW Pub. No (PHS) 79–1661. Hyattsville, MD: National Center for Health Statistics.

Valkenburg, H.A. (1983) Osteoarthritis in some developing countries. *J Rheumatol* (*Suppl*), *10*: 20–24.

van Gerven, D.P., J.R. Hummert, and D.B. Burr (1985) Cortical bone maintenance and geometry of the tibia in prehistoric children from Nubia's Batn el Hajar. *Am J Phys Anthropol*, *66*: 275–280.

Van Gilse, P.H.G. (1938) Des observations ulterieures sur la genèse des exostoses du conduit externe par l'irritation d'eau froide. *Acta Otolaryngol*, *26*: 343–352.

Van Saase, J.L., L.K. Van Romunde, A. Cats et al. (1989) Epidemiology of osteoarthritis: Zoetermeer Survey. Comparison of radiological ostearthritis in a Dutch population with that in 10 other countries. *Ann Rheum Dis*, *48*: 271–280.

Videman, T. (1982) The effect of running on the osteoarthritic joint: A experminetal matched-pair study with rabbits. *Rheum Rehab*, *21*: 1–8.

Videman, T., M. Nurminen, and J.D. Troup (1990) 1990 Volvo award in clinical sciences. Lumbar spinal pathology in cadavaric material in relation to history of back pain, occupation, and physical loading. *Spine*, *15*: 728–740.

Villa, P. (1990) Torralba and Aridos: Elephant exploitation in Middle Pleistocene Spain. *J Human Evol*, *19*: 299–309.

Vincelette, P., C.A. Laurin, and H.P. Lévesque (1972) The footballer's ankle and foot. *CMA Journal*, *107*: 872–877.

Vingård, E., L. Alfredsson, I. Goldie et al. (1991) Occupation and osteoarthrosis of the hip and knee: A register-based cohort study. *Int J Epidemiol*, *20*: 1025–1031.

Vingård, E., L. Alfredsson, I. Goldie et al. (1993) Sports and osteoarthritis of the hip. An epidemiological study. *Am J Sports Med*, *21*: 195–200.

Vingård, E., L. Alfredsson, and H. Malchau (1997) Osteoarthritis of the hip in women and its relation to physical load at work and in the home. *Ann Rheum Dis*, *56*: 593–598.

Vingård, E., C. Hogstedt, L. Alfredsson et al. (1991) Coxarthrosis and physical work load. *Scand J Work Environ*, *17*: 104–109.

Vossenaar, A.H. (1936) Affections articulaires dues au maniement des outils pneumatiques. *Echo Méd Nord*, *6*: 949–960.

Waldron, H.A. (1997) Association between osteoarthritis of the hand and knee in a population of skeletons from London. *Ann Rheum Dis*, *56*: 116–118.

Waldron, H.A. and M. Cox (1989) Occupational arthropathy. Evidence from the past. *Br J Ind Med*, *46*: 420–422.

Waldron, T. (1992) Osteoarthritis in a Black Death cemetery in London. *Intl J Osteoarchaeol*, *2*: 235–240.

Waldron, T. (1993) The health of adults. In: *The Spitalfields Project*, Vol. 2. York: Council for British Archaeology, Research Report 86.

Waldron, T. (1994a) *Counting the Dead. The Epidemiology of Skeletal Populations.* New York: John Wiley.

Waldron, T. (1994b) The human remains. In: Evison, V. (ed.), *An Anglo-Saxon Cemetery from Great Chesterford, Essex*. York: Council for British Archaeology, Reserach Report 91, pp. 52–66.

Waldron, T. (1996) Legalized trauma. *Intl J Osteoarchaeol*, *6*: 114–118.

Waldron, T. and J. Rogers (1991) Inter-observer variation in coding osteoarthritis in human skeletal remains. *Intl J Osteoarchaeol*, *1*: 49–56.

Wakely, J. (1996) Limits to interpretation of skeletal trauma—two case studies from Medieval Abindton, England. *Intl J Osteoarchaeol*, *6*: 76–83.

Walker, P.L. (1989) Cranial injuries as evidence of violence in prehistoric southern California. *Am J Phys Anthropol*, *80*: 313–323.

Walker, P.L. (1997) Wife beating, boxing, and broken noses: Skeletal evidence for the cultural patterning of violence. In: Frayer, D.W. and D.L. Martin (eds.), q.v. (in press).

Walker, P.L., D.C. Cook, and P.M. Lambert (1997) Skeletal evidence for child abuse: A physical anthropological perspective. *J Forensic Sci*, *42*: 196–207.

Walker, P.L. and S.E. Holliman (1989) Changes in osteoarthritis associated with the development of maritime economy among southern Californian Indians. *Intl J Anthropology*, *4*: 171–183.

Walker, P.L. and P.M. Lambert (1989) Skeletal evidence for stress during a period of cultural change in prehistoric California. *J Paleopathology, Monographs*, *1*: 207–212.

Warren, R.F. and J.L. Marshall (1978) Injuries of the anterior and medial collateral ligaments of the knee: A long-term follow-up of 86 cases II. *Clin Orthop Rel Dis*, *136*: 198–211.

Washburn, S.L. (1962) The strategy of physical anthropology. In: Tax, S. (ed.), *Anthropology Today*. Chicago: University of Chicago Press, pp. 1–14.

Washington, E.L. (1978) Musculskeletal injuries in theatrical dancers: site frequency and severity. *Amer J Sports Med*, *2*: 75–98.

Webb, S. (1995) *Palaeopathology of Aboriginal Australians*. Cambridge: Cambridge University Press.

Webb, W.S. and C.E. Snow (1945) *The Adena People*. Lexington: Publications of the Department of Anthropology and Archaeology, Vol VI, University of Kentucky.

Weightman, B. (1976) Tensile fatigue of human articular cartilage. *J Biomechan*, *9*: 193–199

Wells, C. (1962) Joint pathology in ancient Anglo-Saxons. *J Bone Joint Surg*, *44B*: 948.

Wells, C. (1963) Hip disease in ancient man. Report of three cases. *J Bone Joint Surg*, *45B*: 790–791.

Wells, C. (1964) *Bones, Bodies, and Disease. Evidence of Disease and Abnormality in Early Man*. New York: Praeger.

Wells, C. (1965) Disease of the knee in Anglo-Saxons. *Med Biol Illustra*, *15*: 100–107.

Wells, C. (1967) Weaver, tailor, or shoemaker? An osteological detective story. *Med Biol Illustra*, *17*: 39–47.

Wells, C. (1972) Ancient arthritis. *M&B Pharm Bull* (*December*), pp. 1–4.

Wells, C. (1982) The human burials. In: McWhirr, A., L. Viner, and C. Wells (eds.), *Romano-British Cemeteries at Cirenster*. Cirenster Excavation Committee, pp. 135–201.

Wendorf, F. (1968) Site 117: A Nubian Final Paleolithic graveyard near Jebel Shaba, Sudan. In: Wendorf, F. (ed.), *The Prehistory of Nubia*. Vol. II. Southern Methodist University Contribution to Anthropology, No. 2, pp. 954–985.

Wenham, J. (1989) Anatomical interpretation of Anglo-Saxon weapon injuries. In: Hawkes, S.C. (ed.), *Weapons and Warfare in Anglo-Saxon England*. Oxford: Oxford University Committee for Archaeology, Monograph 21, pp. 123–139.

Weston, W.J. (1956) Clay shoveller's disease in adolescents (Schmitt's Disease). A report on two cases. *Radiol*, *30*: 378–380.

White, J.A., V. Wright, and A.M. Hudson (1993) Relationship between habitual physical activity and osteoarthrosis in ageing women. *Public Health*, *107*: 459–470.

White, T.D. (1986) Cut marks on the Bodo cranium: A case of prehistoric defleshing. *Am J Phys Anthropol*, *69*: 503–509.

White, T.D. (1992) *Prehistoric Cannibalism at Mancos 5MTUMR-2346*. Princeton: Princeton University Press.

White, T.D. and N. Toth (1991) The question of ritual cannibalism at Grotta Guatteri. *Curr Anthropol*, *32*: 118–138.

White, W. (1988) *The Cemetery of St. Nicholas Shambles*. London: Museum of London and the London and Middlesex Archaeological Society.

Whiteside, J.A., S. Fleagle, and A. Kalenak (1981) Fracture and refractures in intercollegiate athletes, an eleven-year experience. *Am J Sports Med*, *9*: 369–377.

Wiggins, R., A. Bolyston, and C.A. Roberts (1993) Report on the human skeletal remains from Blackfriars, Gloucester (19/91). Cited in Roberts and Manchester 1995, q.v.

Wiley, J., J. Pesington, and J.P. Horwich (1974) Traumatic dislocation of the radius at the elbow. *J Bone Joint Surg*, *53B*: 501–507.

Willey, P. (1990) *Prehistoric Warfare on the Great Plains. Skeletal Analysis of the Crow Creek Massacre Victims*. New York: Garland Pub. Co.

Willey, P. and T.E. Emerson (1993) The osteology and archaeology of the Crow Creek massacre. *Plains Anthropol*, *383*(*Memoir 27*): 227–269.

Williams, J.G.P. (1979) Wear and tear in athletes—an overview. *Brit J Sports Med*, *12*: 211–214.

Willis, T.A. (1924) The age factor in hypertrophic arthritis. *J Bone Joint Surg*, *6A*: 316–325.

Wilkinson, R.G. (1997) Violence against women: Raiding and abduction in prehistoric Michigan. In: Frayer, D.W. and D.L. Martin (eds.), q.v. (in press).

Wilkinson, R.G. and K.M. Van Wagenen (1993) Violence against women: prehistoric skeletal evidence from Michigan. *Mid-Cont J Archaeol*, *18*: 190–216.

Wilson, F.D., J.R. Andrews, T.A. Blackburn et al. (1983) Valgus extension overload in the pitching elbow. *Am J Sports Med*, *11*: 83–88.

Winchell, F., J.C. Rose, and R.W. Moir (1995) Health and hard times: A case study from the Middle to Late Nineteenth Century in eastern Texas. In: Grauer, A. (ed.), *Bodies of Evidence*. New York: Wiley-Liss, pp. 161–172.

Wolff, J. (1892) *Das gesetz der transformation der Knochen*. Berlin: A. Hirshwald.

Woo, S.L.Y., S.C. Kuei, D. Amiel et al. (1981) The effect of prolonged physical training on the properties of long bones: a study of Wolff's law. *J. Bone Joint Surg*, 6A: 780–787.

Wood Jones, F. (1910) Fractured bones and dislocations. In: Elliot-Smith, G. and Wood Jones, F. (eds.), *The Archaeological Survey of Nubia*, Vol. II, *Report on the Human Remains*. Cairo: National Printing Department.

Woods, R.J. (1995) Biomechanics and osteoarthritis of the knee. Ph.D. thesis, Ohio State University, Columbus, OH.

Wright, V. (1990) Post-traumatic osteoarthritis—a medico-legal minefield. *Br J Rheumatol*, *29*: 474–478.

Wyman, J. (1874) Report on a collection of Peruvian crania. *Annual Report*, *Peabody Museum*, Salem, MA, pp. 1–37.

Index